Department of the Environment

LANDSLIDING IN GREAT BRITAIN

London: HMSO

First published 1994

ISBN 0 11 752556 1

"D.K.C. Jones is Professor of Geomorphology and Convenor of the Department of Geography at the London School of Economics and Political Science.

E.M. Lee is Managing Engineering Geomorphologist at Rendel Geotechnics, specialist geotechnical consultants."

Contents

List of Plates

List of Photos

List of Figures

List of Tables

Preface

Landsliding is a hazard that many people in Great Britain fail to take seriously. The reasons for this are fairly self-evident. Large tracts of the country appear too flat to pose any real threat of major slope instability, while those areas which do contain high, steep slopes are generally characterised by sparse populations, so the risks appear low. However, appearances can be deceptive. Not only are there numerous reported landslides each year which cause damage, destruction, disruption and sometimes death, but there are also many thousands of old landslides which can be made to move again if disturbed by human activity. Many of these are extremely well camouflaged, which makes their detection difficult unless special searches are undertaken using trained personnel. This statement is not made with a view to being sensationalist. There are enough highly exaggerated, pseudo-scientific, emotionally biased `Domesday' texts on the market to satisfy those desires. Nor is there any cause for panic reactions, for the existence of numerous old slides is not a completely new revelation but merely confirms a growing view amongst earth scientists who have variously studied different portions of the British landscape. But neither can the existence of old landslides be seen as a cause for complacency, merely because the British Isles are not subject to the catastrophic failures that have affected other regions of the world. Slow, progressive and relatively inconspicuous movements can produce very significant damage and disruption which can be extremely embarrassing and costly to those who happen to be affected.

There is clearly a need for a book that describes the nature of the threat presented by landsliding in Great Britain which is written for the non-specialist. This book attempts to fill that need. Its purpose is to provide an informed, readable and relatively non-technical appreciation of landsliding in Britain based on current knowledge. The objective is to provide an explanation of what the various types of landslide look like, where they occur, why they occur, what is known about the major phases of movement in the past, their impact on the British economy and the ways in which landslide-imposed costs can be limited in the future.

This book grew out of a major Department of the Environment commissioned project entitled *Review of Research on Landsliding in Great Britain*, undertaken by Geomorphological Services Limited in association with Rendel Palmer & Tritton under the overall direction of Dr J C Doornkamp and Professor D Brunsden, and mainly carried out over the years 1984-1987. One of the important products of this wide-ranging study was a census of all available published information on landsliding which now forms the computerised *National Landslide Databank* and can be accessed by any interested party. The statistics presented in this book have been drawn from this database.

As with any book, publication would not have been possible without the assistance of a great many people. First and foremost, sincere thanks have to be given to Dr David Brook and Dr Brian Marker of the Minerals and Land Reclamation Division, DOE, for their cheerful patience and helpful comments over the years. Grateful thanks must also be given to the various academic geomorphologists who helped collect the data as part of the original census: Dr Colin Ballantyne of the University of St Andrews; Dr Roger Cooper of Birkbeck College, London; Professor Eddie Derbyshire of Leicester University; Drs Colin Rouse and Mike Bridges of University College, Swansea; Dr John Gerrard of Birmingham University; Dr Gareth Hearn, now of Scott Wilson Kirkpatrick and Dr "Chuff" Johnson of Manchester University. But to create the final products required endless hours of labour from a great many people associated with Geomorphological Services Limited, and most especially: Dr Jim Griffiths and Dr Roger Moore, to whom we are especially grateful for detailed comments on drafts of this book; Trevor Dibb for preparing the statistical data from the landslide database; Mrs Susan Craig; Clive Smith and Steve Wiltshire for the original data handling; Ian Glenister, Jane Pugh and Samantha Pratt for the text diagrams; Jane Pugh, Allison Aspden and Gary Llewellyn for the original cartography and Judy Mynard, Caroline Driver and Claire Lee for typing endless drafts, reports and the text of the book. The errors are almost certainly ours, the efforts largely theirs, and it is to all of them that this book is dedicated.

David K.C. Jones
Mark Lee

1.2.94

CHAPTER 1

Introduction

Landsliding is not a hazard that most people readily associate with Great Britain. Conspicuous examples of active slope instability are relatively infrequent and largely confined to cliffed coastlines, especially in those areas where the rocks are weak and easily undermined by the erosive power of the sea. By contrast, many inland terrains are characterised by subdued topography with gentle slopes and have the appearance of being relatively free from landsliding. Even the uplands of the west and north for the most part retain the smoothed outlines imposed by the repeated passage of glaciers and ice-sheets in the past and rarely display the bare scars and chaotic jumbled ground considered typical of slope failure. In addition, the occurrence of landslides is seldom deemed to be of sufficient magnitude to attract anything more than local media attention, and in most instances is relegated to the role of by-product of more conspicuous hazard forces, such as storm-lashed seas or intense rainstorms. Those few events that do get reported at length mainly involve the short-term blockage of roads and railways, often at sites where it is obvious that human activity has steepened slopes thereby increasing the likelihood of instability.

During the past three decades, however, detailed studies have revealed that appearances can be very deceptive, and that old landslides are considerably more common and widespread than had previously been thought. For example, a survey of the South Wales Coalfield undertaken by the British Geological Survey in the late 1970s[1] identified nearly 600 separate landslides. An even more detailed investigation of part of central Devon, west of Exeter, completed in 1983[2] revealed over 300 sites with indications of localised shallow slope instability within an area of 25km^2. As a consequence of such studies it is now recognised that the distribution and concentration of inland landsliding have traditionally been seriously underestimated for many areas. Such a conclusion has serious implications for many sectors of the economy concerned with land development and construction, as will be discussed later.

The prevailing lack of awareness of landslide hazard is understandable. The common perception of landsliding centres on the notion of large volumes of rock and soil moving rapidly downslope en masse under the influence of gravity, accompanied by much noise and dust and creating a swathe of destruction. Indeed, this conforms with the only legal definition available from case law and is reinforced by the popular usage of the word 'landslide' to mean overwhelming, as in 'landslide victory'. That such phenomena do exist is undeniable, but they merely form the more conspicuous examples of a broad spectrum of gravity-dominated slope movements known collectively as **mass wasting** or **mass movement,** which also include very much more numerous extremely slow, cumulative displacements, the best known example of which is **creep***, an imperceptible process that produces the

* Soil creep may be defined as the very slow downslope movement of superficial soil and rock debris due to heaving and settling movements caused by fluctuations in moisture and temperature, freeze-thaw cycles and solution. The process is imperceptible except to observations of long duration and is usually confined to the uppermost metre of material immediately underlying a slope.

characteristic narrow steps on hill-sides known as terracettes *(Plate 1)*. The general view of landslides as predominantly rapid downslope movements automatically suggests that such features will be best developed in mountainous areas where recent geological uplift and intense erosion by rivers, glaciers or the sea, have created high, steep, unstable hillslopes with plenty of potential energy. The likely frequency and magnitude of failures will be increased further if the areas are subject to intense rainstorms or periodic ground shaking by earthquakes.

In truth, the largest and most dramatic examples of landslides are to be found in unstable mountainous environments and this inevitably places them overseas. Catalogues of such events are readily available[3] *(Table 1.1)* and usually include:

- the Vaiont Dam disaster in the Italian Alps (1963) when 250 million cubic metres (250 million m^3) of rock slid into the impounded lake causing waves of water up to 100m high to overtop the dam drowning 2600 people in the valley below;
- the Hope Valley slide (1965) in the Canadian Rockies *(Photo 1.1)*, where the 47 million m^3 of debris still entombs the bodies of several travellers on the highway below;
- the Huascaran rock avalanche in the Peruvian Andes (1970) when the ice-cap on the summit of Mt. Huascaran collapsed, causing 50–100 million m^3 of snow, rock and ice to descend 2700m at great speed (up to 400 kilometres per hour), obliterating two towns and killing 20–25,000 people;
- the Mayunmarca rockslide in the Peruvian Andes (1974), when 1,000 million m3 of debris killed 450;
- the Salashan failure in Gansu Province, China (1983), which obliterated three villages killing over 200 people;

1.1 The Hope Valley landslide in the Canadian Rockies, 1965 (D K C Jones).

1.2 The Catak landslide, Turkey, which killed at least 66 people in June 1988 (D K C Jones).

- the Catak landslide, Turkey (1988), when the sudden collapse of a 200m high hillside destroyed a roadside coffee-house, killing 66 *(Photo 1.2)*.

To these well reported recent catastrophes must be added the numerous large failures that undoubtedly have occurred in remote areas without impact on human societies or without reports of occurrence*. In addition, huge numbers of destructive failures have been caused by strong earthquakes, most notably the major earthquakes of 1556 and 1920 in Central China which caused death tolls of 855,000 and 200,000 respectively, the majority as a direct consequence of widespread slope failures in the exceptionally thick deposits of windblown silt (loess).

By contrast, the potential for major destructive landslides in Great Britain appears slight. The terrain is, for the most part, subdued with moderate relative relief mainly restricted to northern and western regions; the climate is temperate and the area is free from large magnitude earthquakes. Thus the likelihood of a catastrophe generated by slope failure seems remote. Indeed, few people have been killed by landsliding in Britain and the significance of those impacts which have occurred appears minor when compared with the relatively high-level of costs resulting from wind storms, intense rainfall events, floods and sea surges. The single, tragic, exception occurred in the Taff Valley, South Wales, in October 1966, when part of the 67m high tip 7 of the Merthyr Vale colliery, perched high on the valley-side above Aberfan, collapsed and 107,000m^3 of colliery spoil slid and flowed downslope into the village below[4] *(Photo 1.3)*. Twenty houses and a school were overwhelmed and 144 people died. The large number of fatalities in this one human-induced

* The best recent example of such an event was the collapse of a large portion of the summit of Mt. Cook, New Zealand, in December 1991.

Table 1.1 **Some major disasters caused by landslides in the Twentieth Century**

Place	Date	Type of landslide	Est.vol. (million m³)	Impact
Java	1919	Debris flow		5100 killed 140 villages destroyed
Kansu, China	Dec. 1920	Loess flows		c. 200,000 killed
California, USA	Dec. 1934	Debris flow		40 killed, 400 houses destroyed
Kure, Japan	1945			1154 killed
SW of Tokyo, Japan	1958			1000 killed
Ranrachirca, Peru	June 1962	Ice and rock avalanche	13	3500+ killed
Vaiont, Italy	1963	Rockslide into reservoir	250	about 2600 killed
Aberfan, UK	Oct. 1966	Flowslide	0.1	144 people killed
Rio de Janeiro, Brazil	1966			1000 killed
Rio de Janeiro, Brazil	1967			1700 killed
Virginia, USA	1969	Debris flow		150 killed
Japan	1969–72	Various		519 died, 7328 houses destroyed
Yungay, Peru	May 1970	Earthquake – triggered debris avalanche – flow		c. 25,000 killed
Chungar	1971			259 people killed
Hong Kong	June 1972	Various		138 killed
Manijima, Japan	1972			112 killed
S.Italy	1972–3	Various		c.100 villages destroyed 200,000 affected
Mayunmarca, Peru	April 1974	Debris flow	1000	town destroyed, 451 killed
Mantaro Valley, Peru	1974			450 killed
Mt. Semeru	1981			500 killed
Yacitan, Peru	1983			233+ killed
W. Nepal	1983			86 killed
Dongxing (Salashan), China	1983		3	4 villages destroyed, 227 killed
Armero, Colombia	Nov. 1985	Lahar		c.22,000 killed
Çatak, Turkey	June 1988		0.4	66 killed

incident is clearly atypical and according to the leading landslide expert Professor John Hutchinson[5] "exceeded the deaths caused by natural landslides in the U.K. during the previous few centuries".

However, fatalities are increasingly coming to be recognised as a poor measure of landslide hazard impact, for they imply sudden, rapid and unexpected movements of ground materials that by chance happen to involve humans. Better evaluations of the true cost of hazardous events can only be achieved if the scope of assessment is enlarged to include estimated values for destruction, damage, disruption and delay, in order to arrive at a realistic appreciation of significance. It is undeniable that

1.3 The Aberfan disaster, October 1966 (Copyright Holton Film Library).

1.4 The landslide at Pentre, 1916, which destroyed a billiard hall and several houses (Photo courtesy of Sir William Halcrow and Partners.

1.5 Road subsidence at St Lawrence, Isle of Wight caused by mudslide movements, 1926.

sudden, rapid movements of ground materials have the potential to impose great economic costs through destruction of property *(Plate 2 and Photo 1.4)* or disruption of transport networks. The Holbeck Hall landslide of June 1993 is probably the most dramatic example in recent years, involving the overnight loss of around 60m of Scarborough's coastal cliffs and the destruction of a 30 room cliff top hotel (*Plates 4 and 48, Photo 1.7*). But it has to be recognised that undramatic slow-moving failures can prove to be equally costly because of the way they deform roads, railways and structures *(Plate 3 and Photo 1.5)*, damage property *(Plate 4)*, disturb foundations *(Photos 1.6 and 1.7)* and disrupt underground infrastructure (e.g. water-supply networks, sewage pipes, gas lines, etc). Experience over recent decades has emphasised the cumulative significance of these undramatic and often relatively

1.6 Foundations disturbed as a result of landslide movement, Ventor, Isle of Wight (N H Noton & Associates).

1.7 Extensive damage to Holbeck Hall Hotel as a result of the landslide of June 1993 (E M Lee).

inconspicuous slope movements in imposing costs on the British economy, not least because of their more widespread distribution and much greater frequency.

More significantly, the occurrence of a number of unexpected slope failures during engineering programmes has drawn attention to the existence of a widely distributed legacy of old (ancient) landslides now largely concealed through the effects of subsequent slope resculpturing, sediment deposition and vegetation growth. These features pose very real threats to the construction industry, for having moved once they are more prone to further movements than similar looking adjacent slopes formed on undisturbed materials. Thus, if affected by engineering activity, they can be reactivated, resulting in movements that may be both embarrassing and costly to engineers and developers.

The most famous example of such an event was when work on the A21 Sevenoaks Bypass had to be halted in 1966 because bulldozers cut through innocent looking grass covered lobes of material which proved to be the remains of ancient landslides. The inadvertent removal of mass from the lower portion of these landslides led to their reactivation despite the fact that they appeared to have remained stable and stationary for most of the last 10,000 years[6]. The problems turned out to be so severe that the affected portion of the route had to be realigned, at considerable cost. Comparable difficulties were encountered with apparently innocent looking slopes during construction of the M4 Motorway near Swindon (1969) and the M6 at Walton's Wood, Staffordshire (1961)[7]. These and similar events gave considerable impetus to the development and refinement of ground mapping and subsurface investigation techniques designed specifically to identify the existence of ancient landslides prior to engineering works. However, unexpected impacts still occur as is illustrated by the problems encountered during the construction of the Daventry Bypass in 1979[8] *(Plates 21 & 22)*, and the reservoir scheme at Malvern which was abandoned after a landslide in 1981, at a cost of £250,000 to the ratepayers[9].

These examples of conspicuous engineering problems did, for the most part, involve the movement of ground material by sliding. However, it is essential to recognise at the outset that the terms **landslide** and **landslip** are often considered interchangeable and represent the most over-used and loosely defined terms employed in slope studies. They are, in fact, potentially misleading but convenient umbrella terms used to cover all gravity dominated downslope movements of relatively dry ground materials (including dumped waste and peat), with displacement achieved by one or more of three main mechanisms: **falling, flowing** (turbulent motion) and **sliding** (movement of materials as a relatively coherent body over a basal discontinuity or shear surface).

In reality, these three basic mechanisms combine together and with geological and topographical factors to produce a bewildering spectrum of slope failures that tend to be generally labelled 'landsliding', while at the same time displaying exceptional variability in terms of form and behaviour. For example, in the context of Great Britain, size varies enormously with volumes ranging from 1 to 1 million m^3 and displacements measured in millimetres to kilometres; some movements are rapid, others exceedingly slow; in some circumstances the majority of displacement is achieved in a single, short-lived event while in other cases movement is gradual, cyclical or pulsed; sometimes the displaced material moves in well defined masses to create the familiar irregular terrain of scars, ridges, humps and hollows, while in other circumstances the earth materials may appear to completely lose coherence and move like dry sand or wet cement. There have been many attempts at providing a generally accepted scientific classification of these diverse slope failures but the debate continues, for the variable combination of slope forming materials and agents responsible for movement "opens unlimited vistas for the classification enthusiast"[10]. Similarly, although dissatisfaction has been widely expressed regarding the general usage of the terms landslide and landslip, neither of the suggested alternatives (mass movements, slope movements) convey the same sense of drama and danger, and so the terms persist.

There are many available texts concerned with landsliding phenomena[11] but none to date have focussed exclusively on Great Britain. Indeed, the only recent scientific paper to address this subject from a national perspective was published in a Japanese journal[5]! Most published studies have focussed on individual conspicuous landslides or relatively localised groups of failures, although regional treatments do exist for Scotland[12] and for the South Wales Coalfield[13]. Landsliding as a potential hazard has also been generally neglected in the British context, as is clearly illustrated by A.H. Perry's book ***Environmental Hazards in the British Isles*** (1981) which devotes a mere 5 pages out of 189 to landsliding, a third of which are concerned with the tragic failure at Aberfan.

However, the last decade has witnessed a growing appreciation of the fact that landslide features are considerably more widespread and numerous than had been previously thought and that research needed to be directed towards evaluating their extent and significance. The Department of the Environment (DOE) therefore commissioned a desk study investigation in December 1984, entitled ***Review of Research into Landsliding in Great Britain***, which had a number of goals including the gathering together and synthesising of all freely available information on landsliding in Britain. The analysis of results obtained by this survey was completed in December 1987 and subsequently up-dated in 1991, and revealed that reported landslides were indeed surprisingly numerous: the initial estimate of 1600 eventually turning into a final one close to 9,000! It is the results of this survey that provide the basis for much of the statistical information presented in this book.

Landslides are a natural phenomena which only become hazards where they interact with human activity, i.e. when development encroaches onto unstable slopes. In other circumstances they can be valued resources that need to be safeguarded. Indeed, 27 mass movement sites have been designated as Sites of Special Scientific Interest (SSSIs) under the Wildlife and Countryside Act 1981, including the Alport Castle landslide (*Photo 2.11*) and Llyn-Y-Fan Fach in the Black Mountain of Dyfed which contains the best examples of debris flows in Wales (*Photo 1.8*). Landslides can also create unique landscapes and habitats, as in the Landslip Nature Reserve on the East Devon coast; repeated rockfall activity is necessary to maintain the dramatic appearance of cliffs such as the Seven Sisters or the White Cliffs of Dover; continued landslide activity is necessary for maintaining fresh exposure of internationally

1.8 Landslides can be valued for their conservation interest. Here debris flow lobes mantle the slopes of the Black Mountain Site of Special Scientific Interest, Wales (Cambridge University Collection: copyright reserved).

recognised rock outcrops such as the Bartonian Stratotype in the cliffs between Highcliffe in Dorset and Milford Cliff in Hampshire. In many coastal locations landsliding is essential for the continued supply of coarse sediments to nearby beaches.

In order to understand the significance of landsliding in Great Britain, it is necessary to have some appreciation of what is known concerning the character and distribution of landsliding. This is covered in Chapter 2 and includes discussion of the most important landslide sites in Great Britain, the character of landslipped ground and present knowledge regarding the distribution of landslides as revealed by the DOE survey. In Chapter 3 the various types of landsliding are described and this naturally leads on to Chapter 4 where the causes of landsliding are discussed. Present knowledge regarding the pattern of landsliding is examined more fully in Chapter 5, including evidence for particularly strong associations with specific geological and topographical situations. The influence of environmental change over the last 200,000 years or so on slope stability, together with the more recent effects of human activity, are discussed in Chapter 6. In Chapter 7, the consequences of landsliding are briefly discussed and this logically leads on the Chapter 8, where the focus is on available methods of identifying and dealing with unstable ground. In the final chapter, by way of conclusion, an assessment is presented regarding the state of current knowledge about landsliding in Great Britain and research needs for the future.

NOTES

1. Conway et al., 1980
2. Grainger, 1983
3. Brabb & Harrod, 1989
   Whittow, 1980
4. Miller, 1974
5. Hutchinson, 1984a
6. Skempton & Weeks, 1976
7. Early & Skempton, 1972
8. Biczysko, 1981
9. Malvern Gazette & Ledbury Reporter, 1981
10. Terzaghi, 1950
11. Anderson & Richards, 1987
    Bromhead, 1986
    Brunsden & Prior, 1984
    Crozier, 1986
12. Ballantyne, 1986
13. Morgan, 1986

CHAPTER 2

Background to Landsliding in Great Britain

In view of the prevailing lack of awareness of landsliding in Britain, the logical starting point for this review is an examination of the most important sites of recent activity. Once it is accepted that numerous sites of contemporary instability exist, some of which are both extensive and dramatic, then it should prove relatively easy to extrapolate from these crumbling slopes with their obvious scars of bare earth and irregular ground, to their older equivalents which are very much more numerous, but display a less pronounced ground form.

It is generally agreed that four broad categories of contemporary landsliding exist:

(i) Coastal landslides, largely the result of wave attack stimulated by the prevailing conditions of progressively rising sea-levels (currently varying between 1.1mm and 6.0mm per year depending on location) which have generated cliff retreat rates averaging 0.1–1.5m per year along most of the unprotected 'soft rock' coastline of eastern and Southern England, locally rising to 2m per year on Holderness (Humberside), in north Norfolk, along parts of the Suffolk coastline, near Herne Bay (Kent) and Middleton (Sussex), with even higher values on record for Warden Point, Sheppey (5m per year)[1] and Selsey Bill, Sussex (8m per year; pa)[2];

(ii) Inland landslides on natural slopes for the most part unaffected by human activity, which are in many cases concentrated at sites where the base of the slope is being undercut by rivers (e.g. the Ironbridge Gorge in Shropshire, Lower Teesdale);

(iii) Inland landslides on natural slopes, largely produced by the disturbance of pre-existing ancient landslides as a consequence of human intervention (e.g. Sevenoaks Bypass, 1966);

(iv) Landslides in cuttings, fills and waste dumps produced wholly as a consequence of human activity (e.g. Aberfan, 1966).

The actual causes of slope failure will be discussed at some length in Chapter 4, but it is important to note here that a fundamental distinction must be drawn between "first-time" failures, on slopes which show no signs of having been affected by landslides in the past, and "reactivated" failures. The majority of failures in groups (i), (ii) and (iv) above are essentially "first-time" failures, although in those situations where destabilising forces repeatedly become dominant, either due to seasonal changes or where erosion removes the support provided by previously slipped material (e.g. a coastal cliff), an intermediate category of "repeated failure" can be recognised. By contrast, the movements forming group (iii) are the consequences of the slope-forming materials having already failed at some time in the past, since when they have remained relatively stable or "dormant" until disturbed by natural erosion or human activity. Unfortunately, as will be explained in Chapter 4, once materials have been involved in slope failure they are much more prone to further movements than identical materials forming slopes that have not failed. Thus one of the urgent tasks of landslide hazard mitigation in Great Britain is to establish the distribution and extent of ground that is underlain by ancient landslides, so that precautions may be taken to avoid cost-inducing failures during future construction works.

Major Landslides in Great Britain

The preceding comments regarding the potential significance of ancient dormant landslides should not be taken to indicate that major active landslides do not exist in Great Britain. They do, especially along the coast, and frequently display the extensive tracts of chaotic, jumbled, irregular topography so characteristic of slipped ground. As such tracts of unstable terrain are easily identified, they have been demarcated on geological and topographic maps, described in books and articles, observed over time, photographed from the ground and from the air, and in certain cases, investigated scientifically in order to provide information about mechanisms of movement and rates of displacement. They therefore include what may be termed the "best known landslides", conspicuous landforms often of dramatic appearance which are frequently cited in literature. The short discussion that follows is intended to provide an introduction to the largest and best known examples of this group, and to some of the better known belts of landslipped ground; 21 are coastal and to varying degrees influenced by contemporary marine erosion, 2 were originally coastal but have been abandoned by the sea for some thousands of years, and 8 are from inland situations *(Figure 2.1)*.

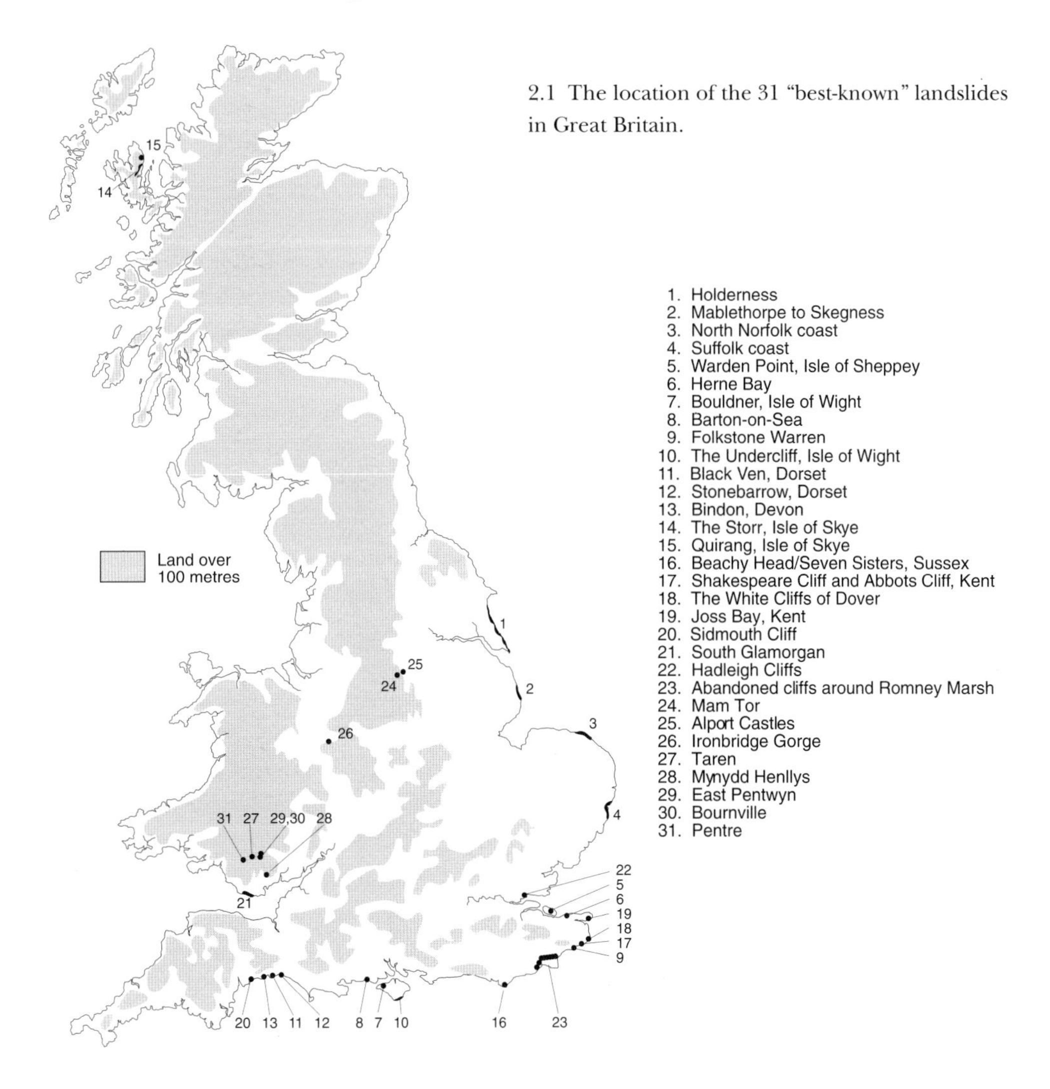

2.1 The location of the 31 "best-known" landslides in Great Britain.

2.2 The geological fabric of Great Britain

(a) The distribution of igneous, metamorphic and sedimentary rocks.

(b) The distribution of rocks older than 280 million years which together form Highland Britain as traditionally defined by MacKinder in 1902. The remaining areas in England and Wales form Lowland Britain.

(c) Proposed fourfold division of the geological fabric.

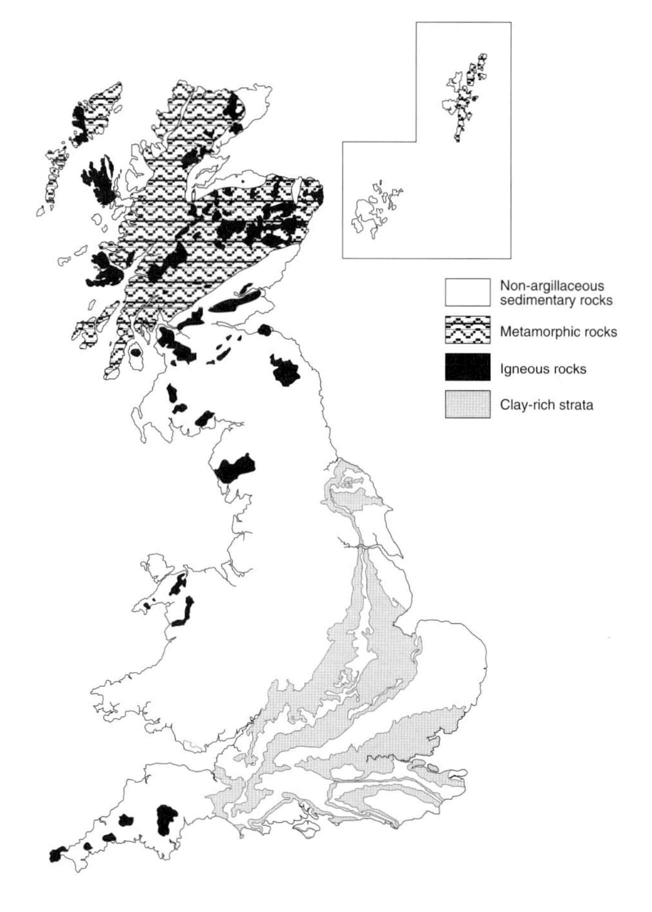

The distribution of these 31 important sites of landsliding is of interest, for they are neither uniformally spread nor situated in the most hilly terrains, nor located in the wettest areas. Their clustering in the southern half of Britain owes something to the fact that research and site investigations have been strongly focussed on this area, but in the main serves to illustrate that ground materials are undoubtedly the dominant control on landsliding, profoundly influencing the type of movement and therefore the character of slipped ground. Indeed, geological considerations are of such fundamental importance in determining the distribution and character of both contemporary and ancient landsliding, that it is necessary to briefly examine the broadscale patterns of ground materials that are exposed to erosion in Britain, prior to discussing the major sites of landsliding. This discussion is merely intended as introductory, for a more detailed consideration of geological and topographical factors is to be found in Chapter 5.

The geological framework of Great Britain

The geological fabric of Great Britain contains an amazing diversity of rock types *(Figure 2.2a)*: crystalline igneous rocks and volcanic lavas produced by the cooling and solidification of molten magma; sedimentary rocks of variable age and character – argillaceous (clayey), arenaceous (sandy), carbonaceous (organic), calcareous (lime-rich) and varying combinations of constituents; metamorphic rocks created through the alteration of pre-existing rocks under conditions of elevated temperatures and/or pressures, but without melting. The age of geological strata *(Table 2.1)* is also of relevance to landsliding in that the older the rocks the more likely they are to have been well compacted (consolidated), altered, cemented and contorted by geological events, with the result that they tend to be more resistant to

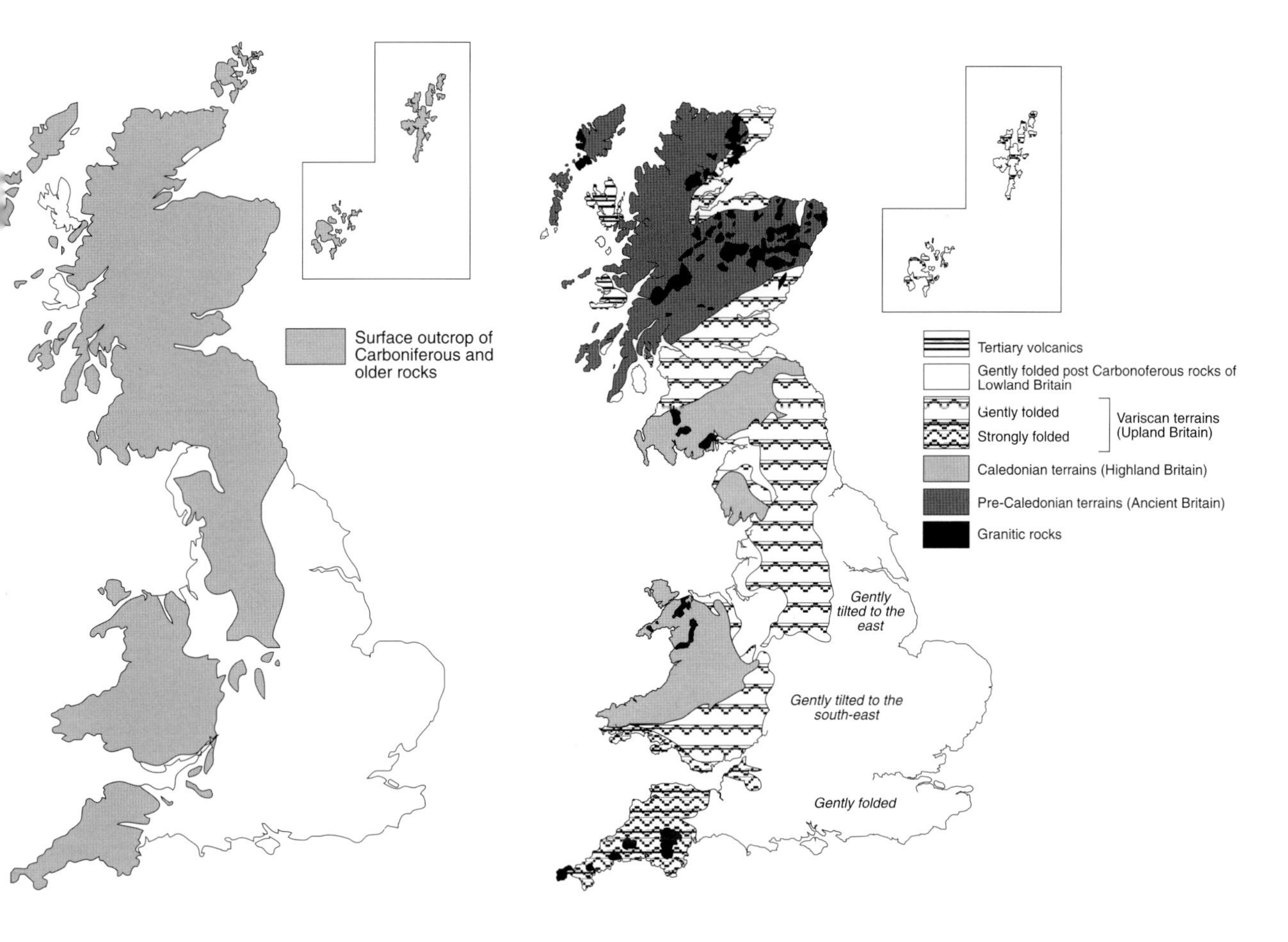

denudation and therefore form upstanding areas. Studies of Great Britain have traditionally focussed on a fundamental two-fold division of the geological foundations into whether the outcropping rocks are older or younger than the Permian (286 million years; *Figure 2.2b and Table 2.1*). The older rocks underlie the craggy outlines of Highland Britain in the north and west where uplands predominate. The remaining lower and more subdued landscapes of Lowland Britain are developed on younger strata, exclusively sedimentary rocks and mainly clays. Beds of sandstone and limestone often form belts of more elevated land that are terminated by steeper escarpments that overlook vales developed in clays. It is the juxtaposition of sandstones/limestones with underlying clay strata that frequently present conditions susceptible to slope instability, as will be discussed in Chapter 5.

This basic division of Great Britain usefully identifies the weaker sedimentary strata of Lowland Britain, but is misleading because it groups together many very different types and ages of rock under the general heading Highland Britain. A preferable alternative is the division of Highland Britain into three parts: Upland Britain, Highland Britain and Ancient Britain *(Figure 2.2c)*. Upland Britain consists of rocks formed over the period 408 – 286 my (Upper Palaeozoic-Devonian and Carboniferous periods) which were folded and faulted towards the end of the Carboniferous during the Variscan episode of earth movements (orogeny). They are overwhelmingly toughened sedimentary rocks – shales, sandstones and limestones – except in the South-West Peninsula where intrusive igneous rocks (granites) are found and in the Cheviots and Central Scotland where extrusive volcanic rocks (e.g.

Table 2.1 **A summary geological column for Great Britain, with ages shown in millions of years (my)**

Era	Period	Epoch	Age (myears)	Mountain Building Phase	Two-fold division of Makinder	Four-fold division used in this book
Cenozoic	Quaternary	Holocene	0.01		Lowland Britain	Lowland Britain
		Pleistocene	2.4			
	Tertiary	Pliocene	5			
		Miocene	25			
		Oligocene	38	ALPINE		
		Eocene	54			
		Palaeocene	65			
Mesozoic	Cretaceous		144			
	Jurassic		213			
	Triassic		248			
Palaeozoic	Peremian		286	VARISCAN	Highland Britain	Upland Britain
	Carboniferous		360			
	Devonian		408			
	Silurian		438	CALEDONIAN		Highland Britain
	Ordovician		505			
	Cambrian		590			
	Pre-Cambrian					Ancient Britain

not to scale

basalts) occur. The remainder of Northern England, Wales and Southern Scotland is underlain by rocks of Lower Palaeozoic age (590 – 408 my: Cambrian, Ordovician, Silurian), predominantly sedimentary rocks with some volcanics, much toughened and altered by geological disturbances during the Caledonian mountain-building episode. Finally, over much of Northern Scotland to the north of the Highland Boundary Fault, outcrop the very old (> 590 my), much altered and highly fractured rocks of Ancient Britain. Here metamorphic and igneous rocks predominate although some much hardened sedimentary rocks also occur.

Thus the geological foundations of Great Britain displays some interesting spatial variations, most especially the general increase in antiquity and durability in a north-westerly direction. However, this varied geological fabric is frequently not exposed at the surface but lies concealed beneath a discontinuous sheet of generally weak, lithologically variable, young sediments known as superficial deposits *(Figure 2.3)*. These surface sediments represent the debris produced by the operation of geomorphological processes (wind, water, ice, gravity, the sea), mainly during the Quaternary (the last 2.4 million years). It is estimated that thick (i.e. greater than 2m deep) spreads of superficial deposits cover 41% of Great Britain (33% of England, 59% of Scotland and 30% of Wales) and in certain areas, such as East Anglia and North-east England, wholly conceal the underlying bedrock for hundreds of square kilometres with extensive blankets of material up to 70m thick. The nature and composition of these materials is extremely varied. They include floodplain alluvium and alluvial lowlands fringing the coast, coastal spits of sand and gravel, river terrace deposits, windblown sand and silt (loess) and even the spreads of debris produced by landslides, especially the widespread downslope sludging of materials that occurred during past cold climate episodes (solifluction). But the overwhelming bulk of superficial deposits are of glacial origin, produced by the ice-sheets that repeatedly expanded over the northern half of Britain during the Pleistocene *(Table 2.1 and Figure 2.4)*. These materials were either derived directly from the ice as beds of "boulder clay" (till) with subordinate lenses of sand and gravel, or laid down as spreads of sand and gravel by meltwaters issuing from decaying ice-sheets (outwash or fluvio-glacial deposits).

The character of landsliding is greatly influenced by the nature of the bedrock, the degree to which it is obscured by superficial deposits and the nature of the superficial deposits themselves. But the potential for differing kinds of landsliding can also vary for individual bedrocks, depending on the extent to which they have been eroded by past ice-sheets and glaciers *(Figure 2.5)*. In Southern England, beyond the furthest known advance of the ice-sheets (the glacial limit), the surface layers of exposed rocks tend to be physically and chemically altered (weathered) due to lengthy contact with the atmosphere, percolating water and vegetation, and this alteration often renders the rocks more prone to landsliding. By contrast, exposed rocks in glaciated Britain tend to have been eroded to varying degrees by the passage of glaciers and ice-sheets in the past, and their loosened and weakened surface layers (weathered mantles and weathered zones) removed to form

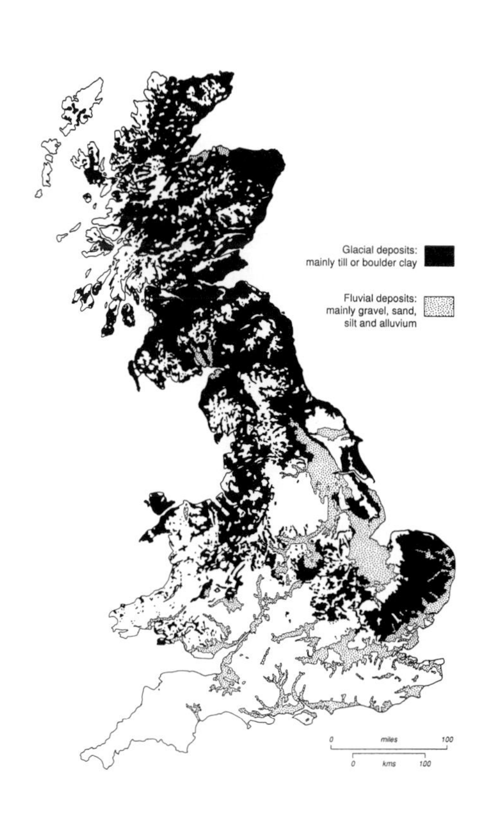

.3 Generalised distribution of thick superficial deposits in reat Britain (after Dearman & Eyles, 1982). For a more letailed portrayal see the 1:625,000 scale Quaternary Map of he United Kingdom (1977).

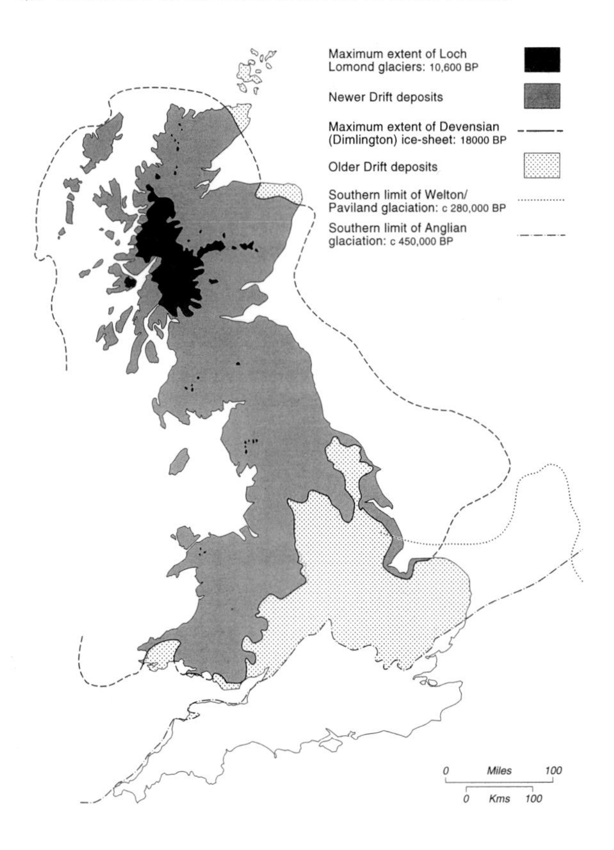

2.4 The limits of ice sheet advances in Great Britain

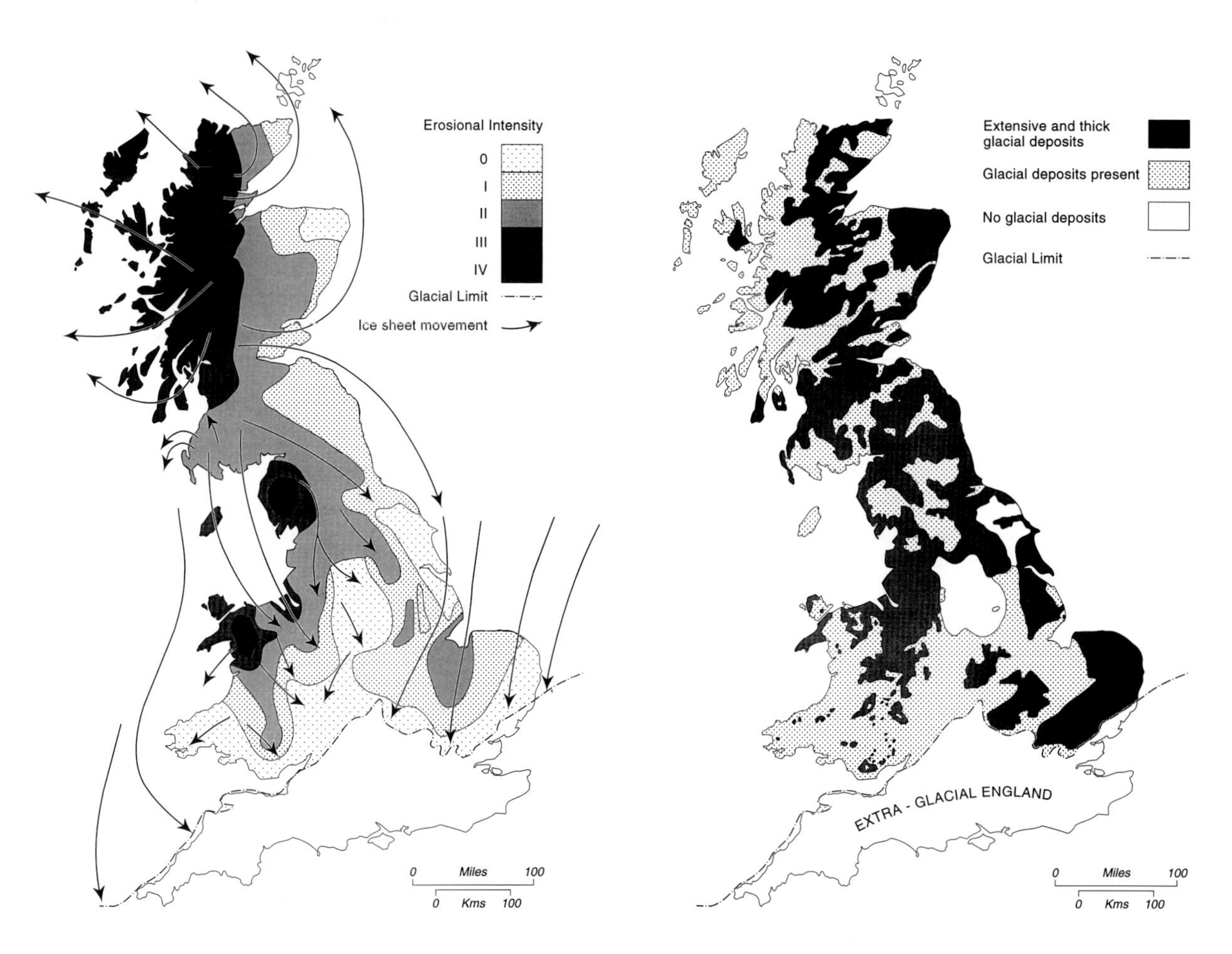

2.5 The effects of the ice sheets on the landscape of Great Britain.

(a) Patterns of ice sheet movement, the ice sheets and glacial erosion in Great Britain (after Boulton et al, 1977).

Zone 0	No erosion
Zone I	Ice erosion confined to detailed or sub-ordinate modifications. Suitable valley slopes in uplands ice steepened.
Zone II	Extensive excavation along main flow lines. In Uplands, many preglacial valleys converted to troughs where coincident with flow lines. Some ice smoothing of obstacles but ridge crests largely unaffected.
Zone III	Preglacial forms no longer recognisable due to extensive erosion. In Upland areas, transformation of valleys to troughs comprehensive giving compartmentalised relief with isolated plateau or mountain blocks.
Zone IV	Intensive erosion with complete domination of streamlined "glacially smoothed" forms even over structural influences. Ice moulding extends to high summits.

(b) The extent of glacial deposits in Great Britain.

superficial deposits elsewhere. The frequency and intensity of these glacial erosion episodes increases north-westwards *(Figure 2.5)* so that relatively sound, fresh rock becomes increasingly widespread in this direction.

Thus the distribution of the 31 "most important landslides" *(Figure 2.1)* must be seen to reflect the interrelationship of a number of important factors:

(i) local topography, especially the existence of long, steep slopes;

(ii) the character and durability of exposed bedrocks;

(iii) the character and extent of superficial deposits;

(iv) the intensity of erosion in the relatively recent past (i.e. last 250,000 years); and

(v) location with respect to the main areas of human occupancy, which influences perception of impact potential to the extent that the most 'significant' or most accessible major slides tend to be studied in the greatest detail.

The relevance of these factors will be made clear in the following site descriptions.

Major coastal landslides

The coastline in Britain is very long and well known for its varied character, rapidly changing rock type and local intensity of marine erosion, so it should come as no surprise that the major areas of coastal slope failure are correspondingly diverse. Four broad categories of coastal landslides can be recognised on the basis of ground-forming materials[3]:

(i) Landslides developed in weak superficial deposits *(Figure 2.3)*;

(ii) Landslides developed in stiff clay *(Figure 2.2a)*;

(iii) Landslides developed in stiff clay with an overlying layer (cap-rock) of hard rock;

(iv) Landslides developed in hard rock.

This classification will be used to organise the following discussion, but before proceeding further it is essential to recognise that all coastal erosion involving the retreat of cliffed coastlines is achieved by landsliding stimulated by wave attack. Indeed, all cliffed coastlines are testimony to the cumulative efficacy of landslides, for they are in reality, the coalescent scars of innumerable individual failures. However, not all clifflines are undergoing rapid rates of retreat. Those formed of the tougher rocks underlying Upland Britain, Highland Britain and Ancient Britain *(Figure 2.2c)* are retreating very slowly so that landsliding is of limited frequency and often relatively small scale. By contrast, where the rate of coastline retreat is high, failures are obviously more frequent and sometimes of larger scale. Thus it is of interest to note that although there are numerous stretches of eroding coast *(Figure 2.6)* the vast majority of the 22 major coastal landslides that will be described, together with both examples of abandoned coastline slides, occur on the soft sedimentary rocks and superficial deposits of Lowland Britain *(Figures 2.2c and 2.3)*. Indeed many of the most dramatic examples occur in areas untouched by the ravages of Pleistocene ice-sheets.

(i) Major coastal landslides in weak superficial deposits

The east coast of England from Flamborough Head to Essex is largely developed in thick sequences of glacial till interbedded with sands and gravels *(Figure 2.3)*. These deposits can be rapidly eroded by the sea, so that coastal landslides are ubiquitous except where the cliffs are protected by sea defences. For example, the entire 60km length of the undefended Holderness coastline (Humberside) *(Figure 2.1)* has retreated at rates of 1–6m pa since 1852[4] and it is known that 200km^2 of land has been consumed by the sea over the last thousand years, including at least 26 villages listed in the Domesday survey of 1086. Similar, albeit less dramatic, retreat rates are recorded for Lincolnshire between Mablethorpe and Skegness and the Norfolk coast from Appsburgh to Weybourne. At those locations on the Norfolk coast where the ground is sufficiently elevated to form high cliffs, dramatic failures have developed,[5] with particularly impressive examples recorded near Trimingham and Cromer. Further south, on the Suffolk coastline between Lowestoft and Thorpness, severe erosion has been a major long-term problem and has particularly affected Pakefield *(Photo 2.1)*, the area around Southwold and, most famously, the cliffs at Dunwich where all that remains is a fragment of the cemetery.

(ii) *Major coastal landslides developed in stiff clay*

Stiff clays are particularly prone to landsliding with classic examples occurring along the southern shore of the Thames estuary in Kent. Here cliffs up to 40m high developed in London Clay have repeatedly failed in response to marine erosion, which results in average retreat rates of up to 2m pa. At Warden Point, Isle of Sheppey, huge failures have repeatedly occurred in response to rapid basal erosion[6] *(Plate 5)*, while at nearby Herne Bay, urban developments have been threatened by major failures involving the whole height of the cliff[7]. In both instances, active marine erosion at the cliff base resulted in extensive, deep-seated failures, bench-shaped in plan, which usually extend along the coast for distances of between four and eight times the cliff height. The displaced material moves downward and seaward on a curved failure surface (**rotational landslide**) producing a characteristic, highly irregular terrain of back-tilted ridges and depressions which is subsequently broken up by later secondary movements.

A quite different form of landsliding occurs on nearby cliffs formed in the same London Clay but where the rate of marine erosion at the base of the cliff is reduced

2.6 The distribution of eroding, advancing and stable coastlines.

2.1 Severe coastal erosion led to houses being destroyed at Pakefield (Reproduced by permission of the Director, British Geological Survey: NERC copyright reserved).

to 0.3 – 0.8m pa[8]. Under these circumstances, movements take place in the form of seasonally active (winter) **mudslides** which act as debris-transporting "conveyor belts", each operating within a distinct embayment, thereby giving the cliffs a distinctive, scalloped plan form *(Photo 2.2)*.

Although minor failures occur elsewhere along the coast where clay strata outcrop, the largest failures are associated with interbedded clays and sands. There are

2.2 Mudslide embayments developed in the London Clay cliffs, Herne Bay, Kent (Acknowledgements to Photoflight Ltd).

conspicuous failures of this type on the north coast of the Isle of Wight, especially at Bouldnor where the cliffs are up to 50m high. On the opposing coast of Christchurch Bay, at Barton-on-Sea, landslides extend for 5km on cliffs up to 30m high developed in Barton Clay and Barton Sand overlain by Plateau Gravel. The repeated failure of these cliffs has persistently threatened property, highlighting the problems of coastal zone management in unstable situations[9] *(Plate 6)*. Expensive stabilising measures were undertaken in the 1960s but largely destroyed in 1974–5 when further slips occurred, and although new stabilisation measures have been emplaced the situation remains sensitive.

(iii) *Landslides developed in stiff clay with a hard cap-rock*

The largest coastal landslides and certainly the most visually dramatic, occur in situations where a thick consolidated clay (shale) stratum is overlain by a rigid cap-rock of sandstone or limestone, or sandwiched between two such layers. Seven examples are particularly noteworthy, five on the south coast (extra-glacial Lowland Britain) and two lone sites on the Isle of Skye.

Immediately to the north of Folkestone on the Channel coast, the 130–150m high cliffs of soft white limestone (Chalk) have foundered on the underlying Gault Clay for a distance of 3 kilometres to create Folkestone Warren. Although originally a purely natural feature, the creation of Folkestone Harbour between 1810 and 1905, seriously disrupted the north-eastward movement of beach shingle along the shore, thereby leaving the base of the cliffs exposed to erosion and exacerbating the frequency and magnitude of landslide movements[10]. Twelve major slips have been recorded since 1765, culminating in huge movements throughout the Warren in 1915, including one slipped block which moved the Folkestone-Dover railway seaward by 50m and derailed a train *(Photo 2.3)*. Indeed, the railway has been closed 30 times due to landsliding since it was constructed in 1844 and the continued operation of the route is only achieved due to the presence of large and expensive engineered structures.

A similar situation exists, albeit with much more minor contemporary movements, along the southern coast of the Isle of Wight. Here a belt of slipped ground up to a 700m wide, known as the "Undercliff", extends for 12km and is inhabited by more than 6,000 people, mainly in the small town of Ventnor[11] *(Photo 2.4)*. As damage continues to be recorded in this area *(see Plate 4 and Photo 1.5)*, it has proved an ideal location for one of the recently completed DOE sponsored studies into techniques by which land use planning and development can be best accomplished in areas of unstable ground[12].

The remaining three southern examples occur on the coast that straddles the Devon-Dorset border, an area considered by many as preeminent in its display of varied landslide phenomena. Between Lyme Regis and Charmouth lies Black Ven *(Plate 7)*, probably the most active and photogenic large landslide in Great Britain and one that has been well studied in recent decades[13]. Within the huge double amphitheatre defined by a backscar of orange coloured Upper Greensand which rises to 145m O.D., occur a number of bluffs and benches formed in grey Liassic shales and limestone, across which great masses of saturated debris slide seawards in

2.3 The dramatic landslide of 1915 at Folkestone Warren, Kent (British Railways, Southern Region).

The Ventnor Undercliff, where the town has been built on an ancient landslide complex (Cambridge University Collection: copyright reserved).

a form of landsliding known as mudslides. Continuing surveys have revealed that the backscar has retreated at an average rate of 0.71m pa this century, although cliff collapse can cause local retreat of up to 30m in a single year[14]. It is estimated that about 267,000 tons of material is fed into the sea each year from these cliffs, the majority by mudslides. Since the 1950s, movements of material within Black Ven have been dominated by two huge mudslides that fed enormous lobes of debris which sometimes extended nearly 100m into the sea, with especially marked activity in the winters of 1957–8, 1969–73 and 1985–7. However, the winter of 1986–7 saw the reactivation of old degraded slides in the western part of the complex which resulted in loss to fields and the golf course, damage to property and the creation of a third major mudslide lobe *(Plate 8)*.

Immediately to the east of Black Ven, on the seaward termination of the next ridge, lies the spectacular bowl of Stonebarrow, sometimes known as Fairy Dell *(Photo 2.5)*. Here the slightly different geology results in the development of a prominent arcuate backscar in orange Upper Greensand above an irregular area of benches, bluffs and ridges of landslide debris which is abruptly terminated by prominent 40 to 50m high sea-cliffs in grey shales and limestones of Liassic age. Large failures in the Upper Greensand create landslide blocks on the benched area which are gradually broken up by further sliding and slumping until the material reaches the cliff edge where it falls onto the beach to be redistributed by waves and tides[15]. Eventually, marine erosion causes the sea-cliffs to retreat sufficiently far for the

2.5 The Stonebarrow (Fairy Dell) landslide complex, Dorset (Cambridge University Collection: copyright reserved).

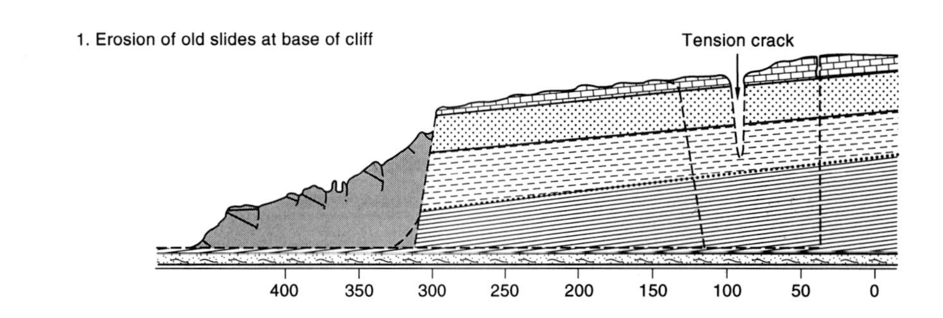

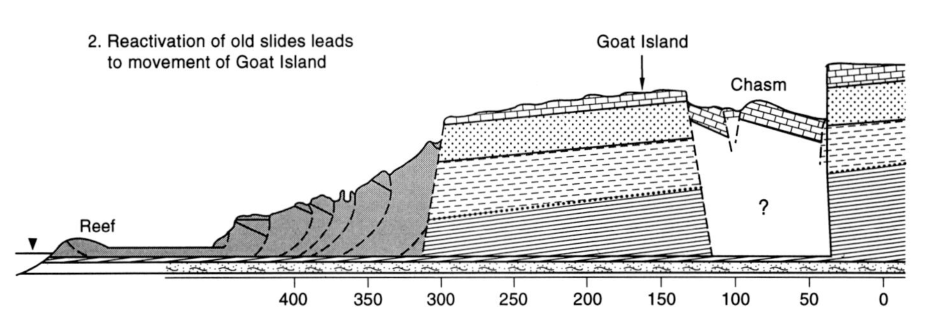

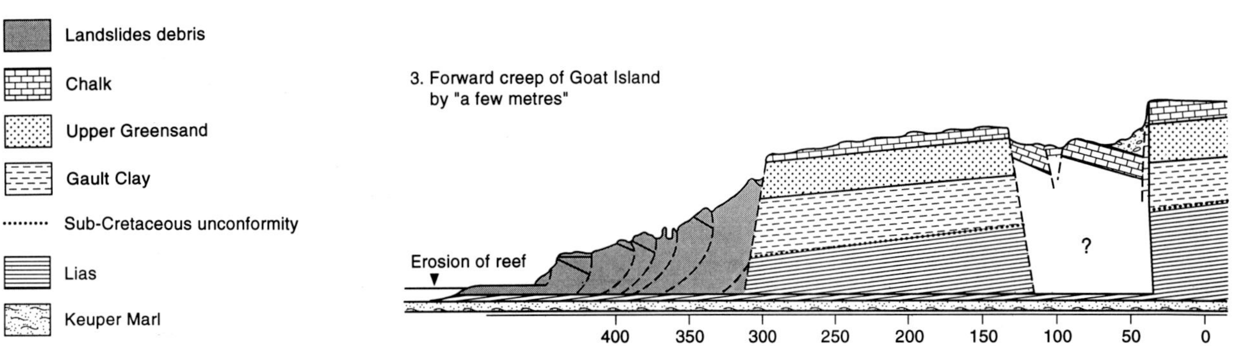

2.7 The development of the great Bindon landslide of Christmas Day, 1839 (after Pitts & Brunsden, 1987).

backscar to become destabilised and the landsliding cycle begins again[16]. The average backscar retreat rate this century has been estimated at 0.4m pa giving rise to about 160,000 tons of beach material each year, over half through cliff falls[17].

To the west of Lyme Regis occurs the third important site, the Bindon landslide[18]. This represents the location of arguably the most dramatic landslide event in recorded history. Here, in the early hours of Christmas morning 1839, a sound "like the rending of cloth", accompanied by "deafening crashes of falling rock" and "flashes of fire and a strong smell of sulphur" heralded a "most extraordinary and terrific explosion of nature"[19], when a huge raft of rock (Goat Island) moved seaward to become separated from the remainder of the cliff-top by a chasm 1,200m long, 92m wide and 46m deep *(Figure 2.7 and Photos 2.6 and 2.7)*. The movement of such a huge coherent mass of Chalk and Upper Greensand caused the underlying Liassic shales to be upheaved along the foreshore as a bulge or reef, which took many years to erode, during which time it was used as a temporary refuge by shipping.

2.6 View of the great Bindon landslide of Christmas day 1839.

2.7 The Bindon landslide, east Devon (Cambridge University Collection: copyright reserved).

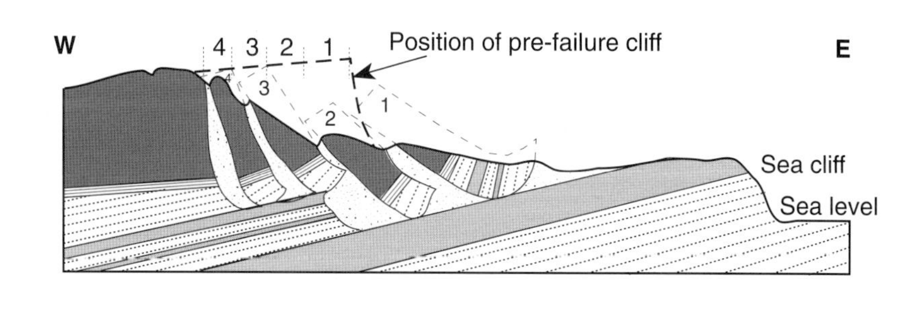

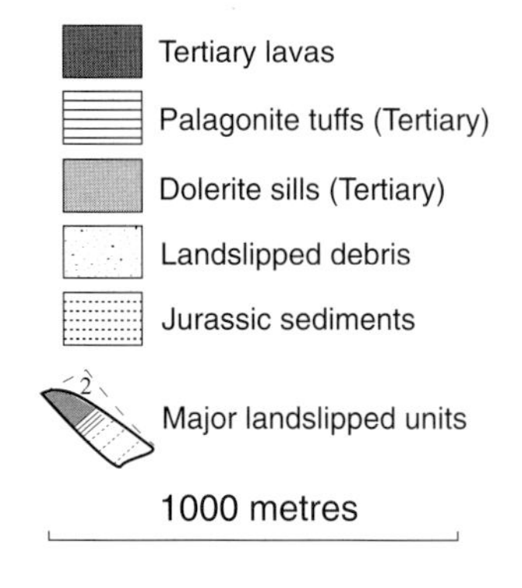

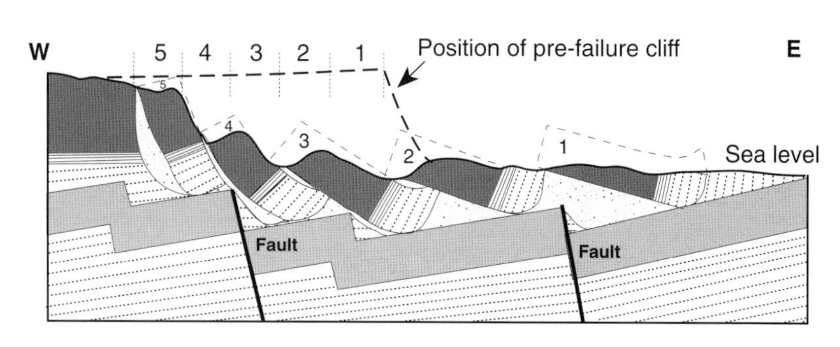

2.8 Diagrammatic section through the major landslides of the Trotternish Escarpment, North-east Skye: The Storr (top) and Quiraing (bottom).

Impressive though these features are, they fail to rival the Trotternish slides on the Isle of Skye which are widely acknowledged to be the most extensive area of slipped ground in Great Britain. Here deposits of Jurassic clay have locally been preserved beneath a thick and resistant sheet of basaltic lava composed of innumerable individual flows. Erosion by glaciers and the sea in the Pleistocene cut deeply into these deposits and led to the development of such steep slopes as to eventually cause masses of the lava sheet to break away and slide over the underlying clay (Figure 2.8). The resultant Storr landslide has created a huge area of jagged buttresses, jumbled towers, hollows, pinnacles and hidden dells *(Plate 9)* which many claim to attain an almost mystical quality in certain weather conditions. The nearby Quirang landslide is similar *(Figure 2.8)* but even more extensive, measuring almost 2km from headscarp cliff to toe.

The huge Trotternish slides are problematic, not merely because of their great size but also because of uncertainty as regards their origin and classification. This area of Scotland was heavily depressed by the weight of thick ice-sheets at several times in the Pleistocene, with the result that former marine levels (raised beaches) in the vicinity can be traced to elevations of 50m O.D. Thus it seems certain that both slides were actively stimulated by marine erosion in the past. However, the zones of instability extend upslope to elevations of over 200m, which suggests that glacial erosion may have been the initial cause of instability. Recent detailed mapping has revealed that both landslides consist of a combination of angular features and much more subdued, "rounded" forms[20]. The latter are considered to be ice-moulded landslides which clearly pre-date the last glaciation, while the more photogenic angular features *(Plate 9)* are post-glacial in age, and may well have developed as a direct consequence of the removal of support provided by the last ice-sheet.

Today, the Quirang landslip is considered "stable for the most part, though ... where the toe is being actively eroded (by the sea) there is continuous though not extensive movement"[21]. Thus Quirang appears to be undergoing reactivation by marine erosion and the same may well be true of the lower portions of the Storr complex, in which case these two slides are intermediate between the active coastal slides of Black Ven and Stonebarrow discussed above and the abandoned shoreline features described later.

(iv) *Landslides developed in hard rock*

Coastal cliffs developed in rocks are continually suffering minor collapse due to basal undermining by the sea. These events are most frequent in the soft rock clifflines of south-eastern England such as the famous Seven Sisters and Beachy Head Chalk cliffs of East Sussex, which are currently retreating at an average rate of 0.97m a year. Every cliff-line has its own magnitude-frequency distribution of failures dependent upon intensity of wave attack, strength of cliff-forming rocks, cliff height, orientation, etc., with relatively large numbers of very small falls at one end of the spectrum and very rare huge failures at the other extreme. According to Hutchinson[22] such catastrophic falls can reach volumes of up to 1,000,000m^3, although all evidence of their former existence is often removed in a few years by the sea.

Large falls have been recorded in a number of locations including the Seven Sisters coastline of Sussex, the Chalk cliffs to the north and south of Dover, Joss Bay in Kent[23], the Triassic sandstone cliffs of Sidmouth, Devon[22] and the Liassic limestone cliffs of South Glamorgan *(Photo 2.8)*. Unusually, the collapse of between 50,000 and 100,000m^3 of material from Gore Cliff in the "Undercliff" of the Isle of Wight in July 1928 was captured in a photograph to provide a remarkable record of such an event *(Photo 2.9)*.

Landslides on abandoned cliffs

Over the southern half of Great Britain sea-level has varied by less than 5m over the past 5,000 years, following a period of remarkably rapid rise from –121m at 18,000 B.P. (Before Present). As a consequence, some stretches of rapidly evolving coastal cliffline became abandoned by the sea because of the deposition of huge quantities of sediment in former bays, estuaries and other inlets by natural processes of sedimentation which eventually resulted in the creation of the alluvial lowlands of today (e.g. the Fens, Somerset Levels, Romney Marsh). In these situations the former cliffs have declined in steepness over time through the process of landsliding. Two instances of this phenomenon are particularly well known. At Hadleigh, in Essex, the former cliff-top is considered to have retreated 40m over the last 6,500 years[24] *(Figure 2.9)*, so that it now threatens the remains of Hadleigh Castle built in the 13th century *(Plate 10)*. A similar situation exists along the northern margin of Romney Marsh, Kent, to the west of Hythe, where the former cliff-line has suffered extensive failure since it was abandoned about 6,000 years ago[25] *(Plate 11)*, with periodic reactivation of movements following periods of high rainfall, as occurred inland of Burmarsh in early 1988, or human disturbance.

Inland landslides

Inland landsliding is not a prominent or conspicuous feature of the British landscape. There are few major sites of contemporary active landsliding, although numerous localised failures exist, usually associated with actively undercut river bluffs and eroding gullies.

The most dramatic area of inland landsliding is without doubt Mam Tor in Northern Derbyshire, the aptly named "shivering mountain", which is claimed to be

2.8 Major cliff fall in thinly bedded Lower Lias shales and limestones near Wick, Mid Glamorgan (Cambridge University Collection).

2.9 The Great Cliff Fall of July 1928, Gore Cliff, Isle of Wight.

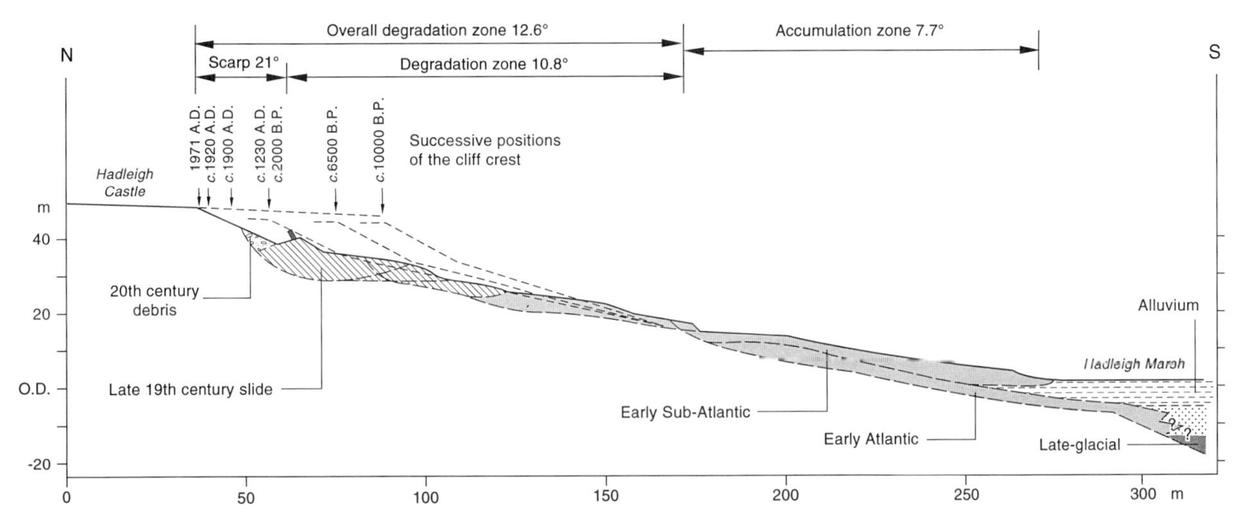

2.9 Evolution of the abandoned sea cliff at Hadleigh Castle (after Hutchinson & Gostelow, 1976).

.10 Mam Tor; the ιivering mountain, erbyshire (Cambridge niversity Collection: opyright reserved).

2.11 Major landslide at Alport Castles, Derbyshire, (Crown copyright/MOD).

the largest active inland landslide in Britain *(Photo 2.10)*. Ever since the turnpike road was constructed over the landslide debris in 1802 there has been a record of intermittent movements and highway dislocation, which during the 1920s were costing the equivalent of £100,000 a year, at present prices, to put right. Further displacements in the 1950s, 1960s and 1970s resulted in the eventual abandonment of the route in January 1979. The familiar irregular topography of nearby Alport Castles *(Photo 2.11)* is also indicative of large-scale landsliding, although only minimal movements are detectable at the moment.

2.12 The flanks of Ironbridge Gorge are mantled with landslide deposits (Cambridge University Collection: copyright reserved).

A second area of well-known inland movements is the Ironbridge Gorge of Shropshire *(Photo 2.12)*. Here several landslides are present on both the northern and southern valley sides. The original Iron Bridge, built in 1779, was reported as showing signs of distress as soon as 1784 and was squeezed by about a metre over the period 1795–1905; indeed the present reconstructed version continues to show movement of about 0.6mm per year. Instability has also been reported at three other locations in the near vicinity within the last century, most particularly at Jackfield, on the southern bank, where major movements between 1951 and 1953 caused a portion of the main road to be displaced 25m towards the River Severn and the river width to be reduced by 15m[26].

The third area with dramatic evidence of landsliding is the South Wales Coalfield. The events at Aberfan in October 1966 was by far the most tragic example, but landsliding involving both colliery waste and natural ground had long been recognised as a major problem in the area[27]. The main cause of instability is considered to be the impact of the most recent glacial episodes, when southward moving glaciers both deepened and steepened the river valleys that dissect the upland plateau. Melting of this last series of glaciers occurred about 18,000 years ago and left the over-steepened valley-sides unsupported, with the consequence that the outcropping sequence of sandstones and shales frequently failed as huge landslides. Taren landslide, in the Taff Vale, is one of the largest examples involving between 7 and 8 million m3 of material. The clearly defined steps of the landslide cover an area of steep hillside measuring 600m x 650m with a vertical amplitude of 220m from backscar to toe *(Figure 2.10)*. This high, steep landslide had to be investigated in considerable detail before the new A470 Taff Vale Trunk Road could be constructed across its toe with the greatest of care[28]. These investigations revealed that landslide debris at the foot of the slope had moved over and buried glacial deposits *(Figure 2.10)*, thereby confirming the failure as postglacial in age.

Taren may be an extremely impressive feature but it is not the largest landslide in the South Wales Coalfield. This occurs at Mynydd Henllys, in the extreme south-east and immediately to the east of Abercarn. Here, on the north-west facing slopes of the Nant Carn Valley, there exists an area of movement covering 2,500m x 800m and involving a mass of rock with a vertical thickness of 100–120m. Although large, the feature is not particularly visible, with minor back tilted blocks at the toe and fissures on the ridge demarcating the head.

The more active landslides in South Wales tend to be smaller and shallower features, although three deep examples are worthy of note. At East Pentwyn, on the eastern side of the Ebbw Fach valley at Blaina, there exists a prominent head scarp below

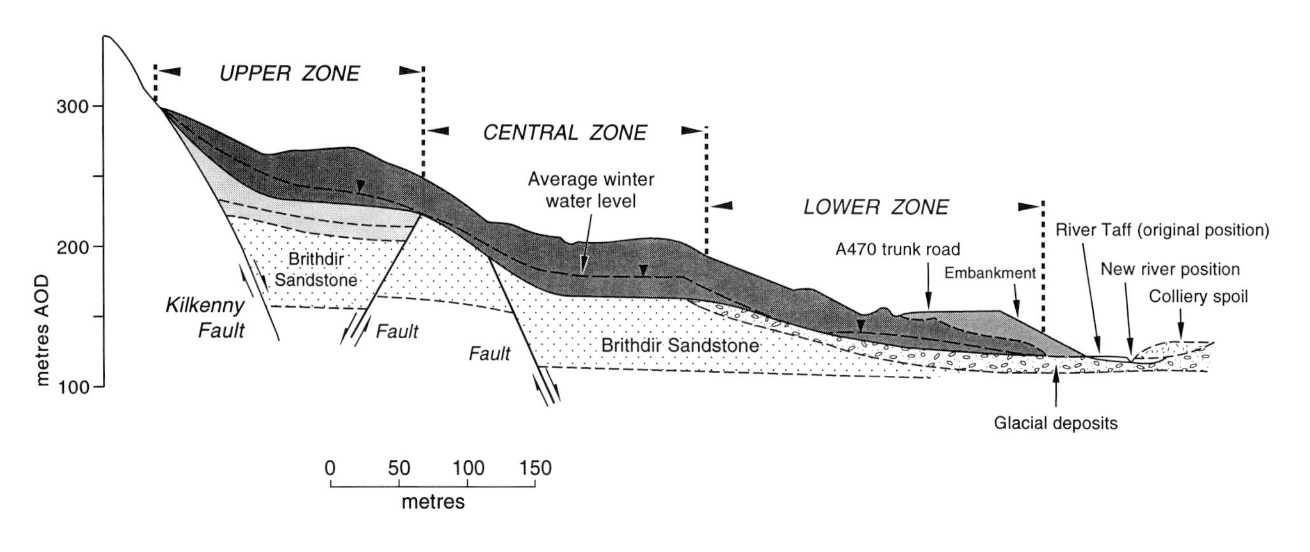

2.10 Section through the Taren landslide, South Wales, looking north (after Kelly & Martin, 1986).

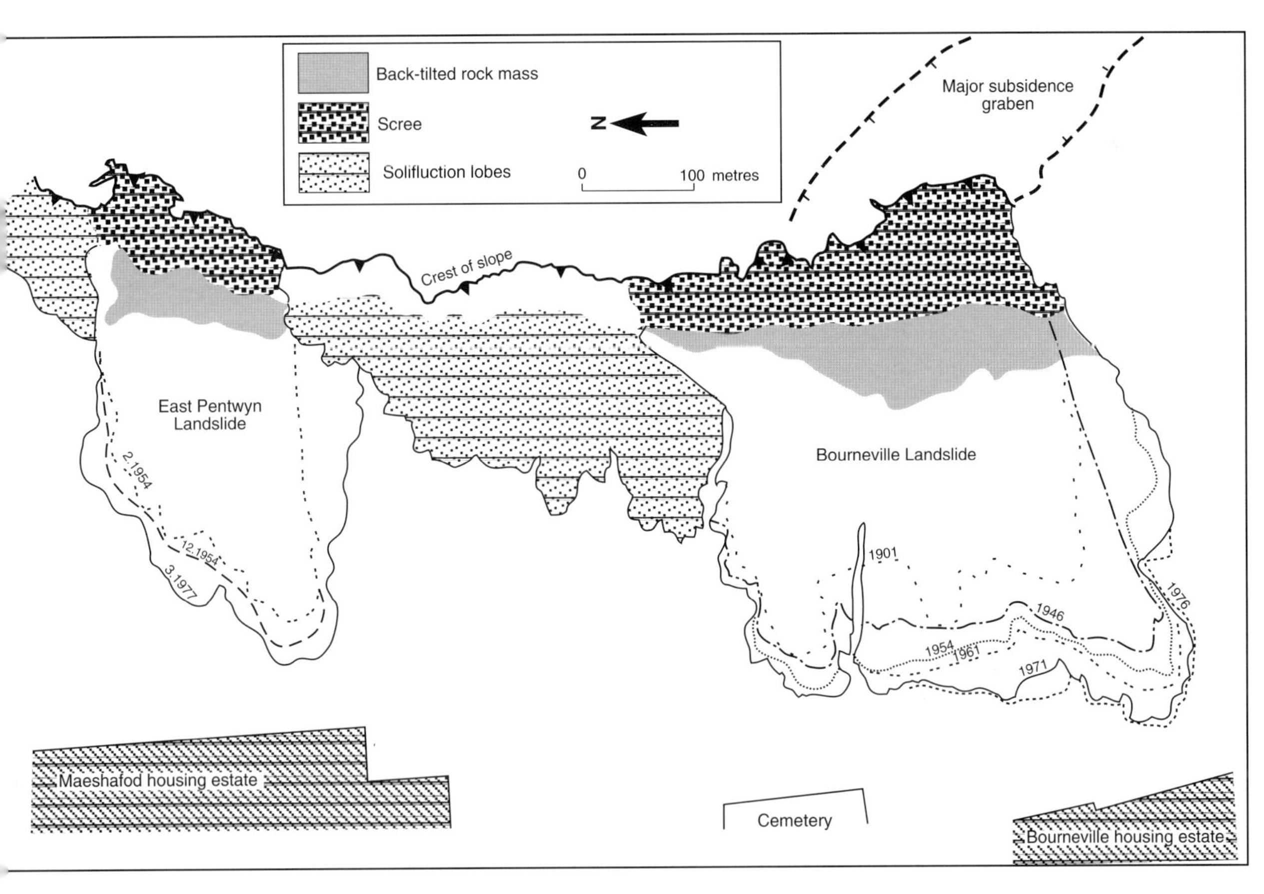

2.11 The East Pentwyn and Bournville landslides, Bliana, Gwent (after Gostelow, 1977).

which lies what is termed "a deep-seated multi-stage landslide of largely rotational form with a history of recent movements endangering residential properties"[29]. Initiation of major landsliding at this site began on the night of 21/22 January 1954 following three days of heavy rainfall, although analysis of aerial photographs has revealed the prior existence of minor landslides. Bulging and fissuring was followed by downslope movement of the main landslide mass to reveal a backscar from which sandstone blocks began to fall. From the toe a lobe of material, 3-4m high, extended downslope, from which issued copious quantities of water *(Figure 2.11)*. This feature advanced 9m in three weeks and threatened a row of houses. Stabilisation through drainage was attempted, but major reactivation occurred on 21 October 1980 with large falls of rock from the backscar and further advance of the lobes. Houses at risk were identified in late 1981 and 28 houses evacuated in December 1981[30]. The exact cause of these movements remained a matter of contention for many years. Initial investigation concluded 'the basic cause of rotational failure is considered to be mining subsidence which has weakened and tilted the ground'[31]. This view was not popular, but received subsequent support on the basis of the strong correlation that exists between cumulative strain, cumulative tilt and the initiation of the East Pentwyn slip[32]. However, it was not until 1989 that a DOE/Welsh Office study[33] confirmed the influence of shallow mining beneath the sandstone on the initiation of the East Pentwyn landslide.

Immediately to the south of East Pentwyn and on the same side of the valley, occurs a very similar, albeit larger area of failure, known as the Bourneville slide. Failure at this site was initiated at some time between 1870 and 1901, with 1893 the most likely date. There have been several phases of movement towards a row of houses, 55 of which were evacuated in December 1981 *(Figure 2.11)*. Once again, subsidence of shallow mine workings has been claimed to be the cause of movements.

Significant, but shallower, movements are located at Pentre, on the eastern flank of Rhondda Fawr, immediately to the south of Rhondda itself. There is historical evidence that Pentre has been the site of instability for over a century, but the contemporary, larger-scale activity appears to have been initiated in August 1916, since when a lobe of material up to 75m wide has extended down the valley side threatening and damaging property in the process[34]. For example, a skating rink was destroyed when the floor was raised so that it eventually met the roof, and several houses were destroyed along two roads at the base of the slope. A number of factors appear to have contributed to the exacerbation of instability at this site. Excavation on the lower slopes during the later nineteenth century, the operation of sandstone quarrying on the upper slopes, deposition of colliery spoil on the floor of the abandoned quarry, and mining subsidence may all have had some influence. Detailed investigations have been carried out at this site and stabilisation is now completed.

Discussion

The list of 31 "significant" or "important" landslides *(Figure 2.1)* is by no means exhaustive, and there are many other examples that could (many may argue should) have been included. A switch in emphasis from size and dramatic appearance, to impact and cost would undoubtedly have resulted in inclusion of some or all of the following:

- the rapidly eroding coastal cliffs of glacial till and fluvio-glacial gravels at Whitby, recently stabilised at a cost of over £2 million[35];
- the three large rockslope failures in the Highlands of Scotland that have (i) persistently damaged the West Highland railway line in Glen Douglas, (ii) closed the branch line in Glen Ogle (1965) and (iii) contributed to the abandonment of the proposed extension to the Loch Sloy Hydro-electric dam in the 1970s[36];
- the landslides that so affected the Sevenoaks Bypass in 1966[37]; and
- the marked concentration of minor failures that persistently affect construction activity in Bath and it's environs[38].

Leaving aside these questions of selection, it is of interest to note that all the chosen examples of inland landsliding are located in Upland Britain *(Figure 2.2c)*, and for the most part in areas where glacial erosion was either negligible *(Figure 2.5)* or confined to valley accentuation. Even an expansion of the list as suggested above only results in two out of thirteen inland examples in Lowland Britain. Thus the distribution of major active coastal and inland landslides are very different.

The restriction of prominent inland landslides to upland regions must not be taken to indicate that landsliding does not occur in other inland areas, especially on the hills, valley flanks and escarpments of Lowland Britain. In reality, it does, sometimes affecting very extensive tracts of land, but it is usually much less obvious visually and characterised by more subtle movements, as will be discussed in the following sections.

The Extent of Landsliding

The examples of landsliding briefly described above are not isolated phenomena. They are merely the most conspicuous, largest or most investigated examples of a phenomenon that range in size from, at the one extreme, huge catastrophic failures which occur with great rarity (otherwise known as large magnitude-low frequency events), down through the size spectrum to the other extreme represented by countless, tiny, wholly unremarkable movements involving the displacement of less than $1m^3$ of earth material. The magnitude-frequency characteristics are most obvious in terms of coastal landsliding, where large conspicuous landslides such as Black Ven *(Plate 7)* are merely dramatic manifestations of much more widespread activity which includes substantial numbers of medium-sized failures, huge numbers of small failures and innumerable tiny displacements.

The same is true of inland landsliding where minor collapse of river banks and bluffs is fairly widespread. But inland landsliding differs from coastal landsliding in one important regard: whereas the material displaced in coastal landslides tends to be quickly removed by the sea, landslides on inland slopes can survive for tens of thousands of years in areas where there is limited resculpturing by gravity, water or ice. This means that contemporary examples of inland landsliding are far outnumbered by the remains of ancient landslides, many of which are difficult to identify because their outlines have been smoothed by the passage of time, or their extent obscured by woodland and scrub.

Indeed, it has to be recognised that presently prevailing conditions in inland areas are not as conducive to landsliding as they were on many occasions in the Pleistocene *(see Table 2.1)*, as will be discussed more fully in Chapter 6. The characteristic landslide topography of irregular or "rippled" ground beneath a prominent arcuate head scarp, so well displayed by Mam Tor *(Photo 2.10)* and Alport Castle *(Photo 2.11)* in the Southern Pennines, can also be widely recognised in more subdued form on other slopes in the vicinity *(Plate 12)*. The same is true of the South Wales Coalfield where the five "well known" landslides described in the last section are merely the most conspicuous of 596 landslides identified by the British Geological Survey and, more recently, by Sir William Halcrow and Partners[39].

It is also clear that the small number of inland landslide sites displaying evidence of major contemporary instability, such as Mam Tor, are not so much the isolated products of unusual local circumstances, but rather the survivors of much more widespread instability in the past. This is especially true of many relatively insignificant landslides visible today which, on closer examination, turn out to be the last active elements of much more extensive areas of failure, the vast proportion of which are now relatively stable. The same is true of the widely reported stretches of deformation on roads traversing slopes, a great many of which turn out to be the result of localised reactivation of small landslide units set within a much more extensive belt of ancient landsliding[40].

This situation can be graphically illustrated by reference to the large, isolated hill-mass of Bredon Hill in Hereford and Worcester. The northwestern flank of Bredon Hill displays a complex form with a prominent crestal scarp overlooking a belt of irregular ground above a discontinuous bench or second escarpment, midway down the flank slopes, which is divided into a number of spurs by a series of deep steep-sided valley re-entrants. At the head of one of these re-entrant valleys there occurs a seasonally active mudslide which feeds material downslope to a series of overlapping lobes of debris on the valley floor *(Plate 13)*. Although mudslides are a common form of mass movement, especially in coastal locations involving the failure of clay strata (e.g. Black Ven), the Woollashill mudslide is one of relatively few active examples in the South Midlands, thus warranting scientific description in 1951[41] and monitoring since 1968[42]. In the early 1980s, problems with the deformation of a trunk water pipeline in another re-entrant valley nearby led to detailed ground surveys which revealed that similar mudslides had existed in all the re-entrant valleys but were now largely stable. These mudslides drew material from the extensive belt of irregular ground between the mid-slope bench and the main scarp, an area which is now recognised to be a mass of landslides lying below a landslide head-scarp *(Figure 2.12)*: indeed all the flanks of Bredon Hill are now considered to be mantled by an extensive, continuous belt of ancient landslides of varying size and character[43], of which only the very small Woollashill mudslide element displays readily visible signs of contemporary activity.

The irregular ground (morphology) on the upper parts of Bredon Hill is very similar to the forms extensively preserved along virtually the full length of the nearby Cotswold escarpment and on many other escarpments in Lowland Britain. It is a morphology that can be recognised from the ground and on aerial photography *(Photo 2.13)*, thereby making it possible to identify and delimit areas of former

2.12 The pattern of landsliding on the western flank of Bredon Hill (after Morris, 1974).

Hill summit developed on Inferior Oolite

Slip troughs

Indistinct camber features

Zone of rotational landslide blocks

Main backscar to extensive zone of multiple rotational landslides

Old mudslides developed in Middle Lias clays

Middle Lias (Marlstone) bench

Scarp slope developed in Middle Lias clays capped by Marlstone

0.5 kilometre

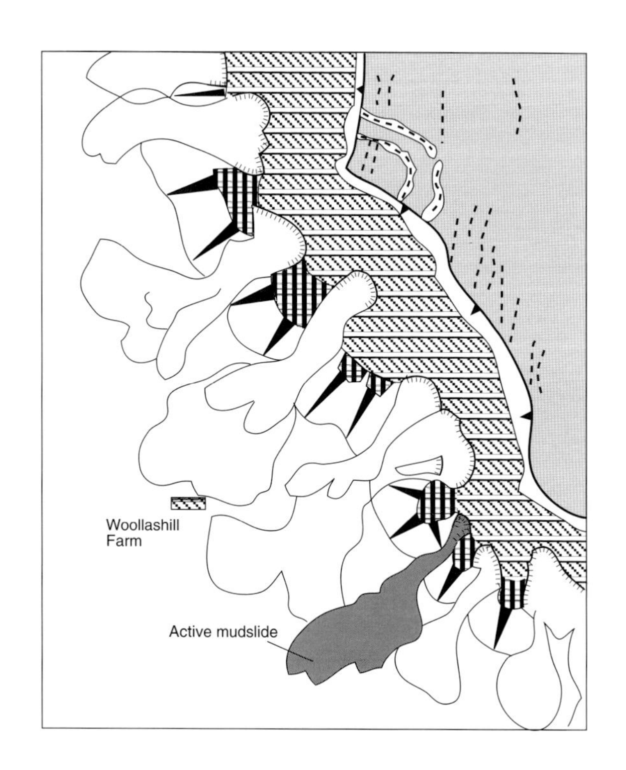

2.13 Characteristic landslide morphology can be easily recognised on aerial photography: Newbiggen Pasture, North Yorkshire (Cambridge University Collection: copyright reserved).

instability. The exact nature of the physical features varies with the type of failure and the nature of the materials involved, and their clarity with the age of movement, vegetation cover and degree of human re-sculpturing.

Landslide Morphology

In Great Britain the most widely developed landslide morphology is similar to that on Bredon Hill and consists of a bluff or head scarp, often arcuate in plan although it may be scalloped, beneath which lie a series of sub-parallel ridges which may be separated by enclosed hollows or marshy depressions, as at Stonebarrow Hill, Dorset *(Figure 2.13a)*. These ridges represent masses of coherent material that have become displaced from the backscar (landslide blocks) and slid to their present position, frequently undergoing some rotation in the process (rotational landslides). The ridges may be knife-edged and distinct in the case of recent movements or if a very resistant cap-rock layer is involved, but tend to become more subdued and rounded with age *(see Figure 2.13b for comparison)*. Surface slopes on these ridges

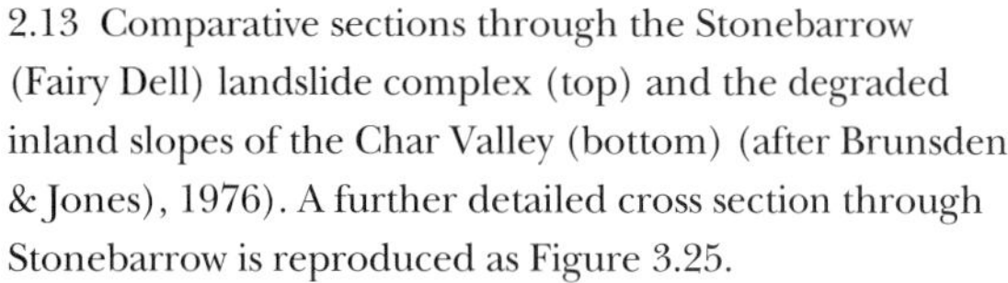

2.13 Comparative sections through the Stonebarrow (Fairy Dell) landslide complex (top) and the degraded inland slopes of the Char Valley (bottom) (after Brunsden & Jones), 1976). A further detailed cross section through Stonebarrow is reproduced as Figure 3.25.

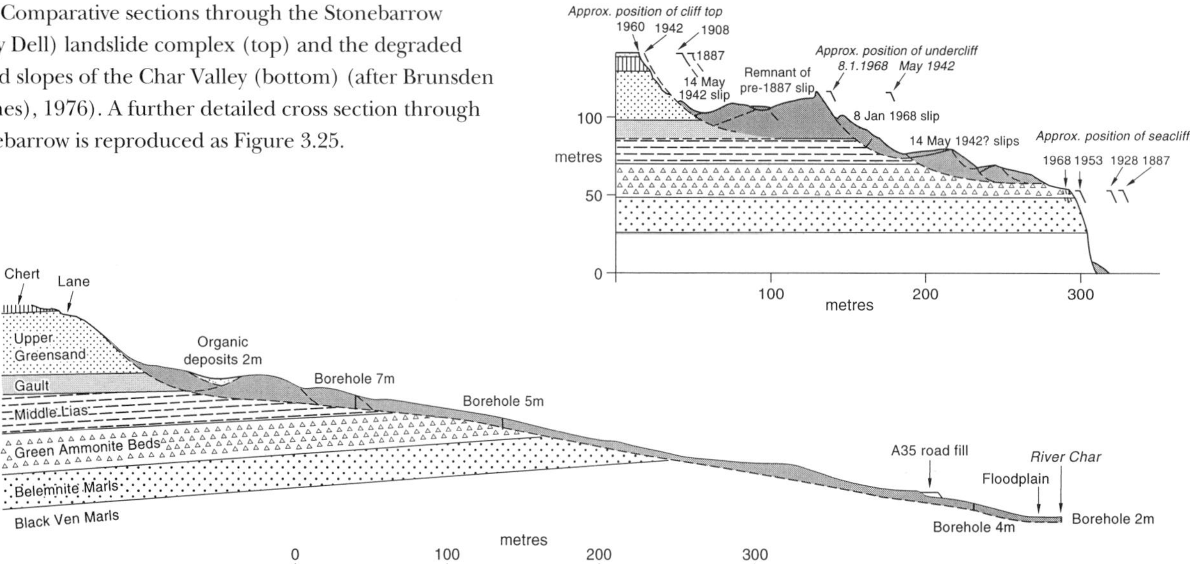

may decline away from the valley (reverse slopes) and towards linear depressions or enclosed basins, both of which are diagnostic features of landslides. Downslope, the ridges usually become smaller and less visible, largely because of the further disintegration of the original landslide blocks by shallower and more localised movements (block disruption), and they often are progressively replaced by low elongate bulges (lobes) oriented downslope towards the valley floor. The latter are generally considered to be old mudslides: very small subdued versions of the huge examples visible on Black Ven *(Plate 7)*.

The morphology associated with other forms of mass movement will be described in Chapter 3. For the moment it is merely necessary to note that evidence for contemporary inland landsliding is extremely restricted compared with the huge tracts of "rippled ground" which indicates the presence of ancient landslides.

The DOE Landslide Survey

Unexpected encounters with landslides during a number of major engineering projects in the 1960s and 1970s inevitably raised questions as to the extent of ancient landsliding in Britain and, in turn, the need for greater consideration of the potential problems posed by unstable ground to land use planning and development. The DOE commissioned review of 1984–87, undertaken by Geomorphological Services Ltd, in association with Rendel Palmer & Tritton, was designed to provide some of the answers by collecting, collating and synthesising the many and varied sources of information on landslides available in the public domain – books, maps, theses, articles, reports and the files of those engineering companies and public utilities that would provide access to data. It was, therefore, a census of **reported** landslides. At the outset it was envisaged that the census would reveal the existence of about 1,600 separate reported landslides, a gross underestimation that clearly testified to the poor prevailing perception of landsliding in Great Britain or, to be more precise, inadequate appreciation of the extent of landslipped ground. In reality, the data collection revealed a grand total of 8,365. Subsequent updating of information to 1991 has raised the total to 8835 as shown on the distribution map *(Figure 2.14)*, with 6120 in England, 1200 in Scotland and 1515 in Wales *(Table 2.2)*.

The vast majority of these 8835 recorded landslides are inactive or stable at the present time. Excluding the 1302 coastal landslides, the 7533 reported inland slides are distributed as follows: 1025 in Scotland, 1401 in Wales and 5107 in England. The breakdown of reported inland landslides by county/Scottish region is shown in Table 2.2 and the resultant densities *(Table 2.2; Figure 2.15)* reveal that the five highest values are Mid-Glamorgan (50.75/100km^2), West Glamorgan (28.04/100km^2), Derbyshire (26.95/100km^2), Avon (20.95/100km^2) and Gloucestershire (17.82/100km^2).

Both the distribution map *(Figure 2.14)* and the density map *(Figure 2.15)*, which represent the most comprehensive available national overview of landslides, reveal a spatially variable distribution with marked concentrations of reported inland landslides in certain areas *(see Appendix A)*:

14 Computer drawn map showing the distribution of the 8,835 recorded landslides in Great Britain. It should be noted that in this portrayal each dot represents a reported landslide, irrespective of size.

2.15 The density of inland landsliding in Great Britain by county and Scottish administrative region.

Table 2.2 The pattern of recorded landsliding in Great Britain, by County and Administrative Region.

County/Region	National Grid Series 1:10,000/ 1:10,560 scale BGS geology map coverage(%)	Total coverage of 1:10,000/ 1:10,560 scale BGS geology map (%)	Number of recorded landslides	Number of recorded coastal landslides	Density of inland landsliding (per 100km^2)
Bedfordshire	11.5	11.5	16	0	1.30
Berkshire	2.1	2.1	9	0	0.71
Buckinghamshire	5.6	5.6	10	0	0.53
Cambridgeshire	31.3	31.3	21	0	0.62
East Sussex	46.5	46.5	110	28	4.57
Essex	19.9	19.9	202	37	4.49
Greater London	25.3	25.3	36	0	2.28
Hampshire	12.7	12.7	68	9	1.56
Hertfordshire	0.0	0.0	5	0	0.31
Isle of Wight	0.0	0.0	90	76	3.68
Kent	27.3	27.3	252	93	4.26
Norfolk	12.1	12.1	38	29	0.17
Oxfordshire	38.5	38.5	41	0	1.57
Suffolk	2.5	2.5	54	40	0.37
Surrey	25.5	25.5	86	0	5.12
West Sussex	46.5	46.5	108	1	5.38
SOUTH-WEST ENGLAND	**19.9**	**19.9**	**1146**	**313**	**2.04**
Avon	70.8	70.8	286	2	21.10
Cornwall	9.7	12.8	47	47	0.00
Devon	19.2	19.2	446	359	1.30
Dorset	1.5	1.5	287	144	5.39
Gloucestershire	50.3	51.7	477	6	17.82
Somerset	16.9	16.9	104	19	2.46
Wiltshire	17.4	17.4	85	0	2.37
SOUTH-WEST ENGLAND	**21.6**	**22.2**	**1732**	**577**	**4.78**
Derbyshire	83.4	88.7	709	0	26.95
Hereford & Worcester	40.5	41.1	84	0	2.14
Leicestershire	33.9	41.6	161	0	6.31
Lincolnshire	13.9	13.9	32	0	0.54
Northamptonshire	89.3	89.3	148	0	6.25
Nottinghamshire	36.8	37.5	7	0	0.32
Shropshire	11.4	13.7	86	0	2.46
Staffordshire	44.7	57.5	192	0	7.07
Warwickshire	78.3	93.8	41	0	2.07
West Midlands	42.3	58.5	3	0	0.33
MIDLANDS	**41.6**	**46.0**	**1463**	**0**	**5.09**

ble 2.2 (cont'd)

ounty/Region	National Grid Series 1:10,000/ 1:10,560 scale BGS geology map coverage(%)	Total coverage of 1:10,000/ 1:10,560 scale BGS geology map (%)	Number of recorded landslides	Number of recorded coastal landslides	Density of inland landsliding (per 100km²)
ıeshire	76.8	84.8	61	0	2.62
eveland	8.9	75.3	25	3	3.77
ımbria	21.2	54.7	253	5	3.64
ırham	74.4	88.8	188	14	7.14
eater Manchester	35.2	100.0	47	0	3.65
ımberside	30.4	90.4	54	25	0.83
ıncashire	15.8	70.8	69	14	1.80
erseyside	0.0	69.2	8	7	0.15
orth Yorkshire	11.5	87.6	533	42	5.91
orthumberland	19.1	68.0	301	3	5.92
outh Yorkshire	47.3	93.4	96	0	6.15
ne & Wear	68.7	100.0	20	10	1.85
est Yorkshire	0.0	94.5	124	0	6.08
ORTHERN NGLAND	**26.5**	**78.7**	**1779**	**123**	**4.32**
wyd	32.7	53.9	247	3	10.05
yfed	7.4	13.1	96	57	0.68
went	42.9	53.0	164	3	11.70
wynedd	6.2	6.2	140	37	2.66
id Glamorgan	62.7	88.0	521	4	50.75
owys	2.8	3.7	103	0	2.03
outh Glamorgan	81.7	85.5	14	9	1.20
est Glamorgan	58.8	82.3	230	1	28.04
ALES	**17.6**	**24.8**	**1515**	**114**	**6.72**
orders	11.5	75.6	46	10	0.77
entral	30.9	52.1	102	0	3.88
umfries & Galloway	6.2	61.9	29	4	0.39
fe	67.5	92.6	34	8	1.99
rampian	6.0	24.8	23	7	0.18
ighland	0.0	80.0	502	83	1.60
othian	86.2	91.4	94	2	5.24
rkney	0.0	100.0	1	1	0.00
hetland	8.6	100.0	43	27	1.12
trathclyde	8.5	59.9	268	32	1.70
ayside	19.1	37.8	55	0	0.73
estern Isles	0.0	0.0	3	1	0.07
COTLAND	**9.5**	**61.6**	**1200**	**175**	**1.31**
OTALS	**20.1**	**47.8**	**8835**	**1302**	**3.28**

(i) The Scottish Highlands;

(ii) The North Yorkshire Moors;

(iii) The Pennines, especially the Northumbrian Fells, the Northern Pennines, the Forest of Bowland, the Central Pennines and the High Peak;

(iv) The East Midlands Plateau;

(v) The Ironbridge Gorge;

(vi) The Vale of Clwyd and Denbighshire Moors;

(vii) The South Wales Coalfield;

(viii) The Cotswolds, including Bredon Hill and other outliers;

(ix) The area around Bath;

(x) Exmoor;

(xi) The slopes bounding the East Devon Plateau;

(xii) The Weald, especially the Upper and Lower Greensand escarpments, the Central Weald and the Lower Chalk escarpment at Folkestone;

(xiii) The London Basin.

However, there are also some areas with unexpectedly low reported concentrations including:

(i) the majority of the Southern Uplands of Scotland;

(ii) the Lake District;

(iii) most of Wales;

(iv) the majority of Devon;

(v) Cornwall, the only county for which no record of an inland landslide was found.

The patchiness of the distribution therefore raises questions as to the extent to which the concentration displayed in the map *(Figure 2.14)* reflects the true pattern of landslides on the ground as against spatially variable reporting. It now seems certain that the pattern merely highlights those landslides which happen to have been investigated, mapped and reported, and the extent to which the total available corporate knowledge of landsliding was tapped by the survey. It is undoubtedly true that many reports of landslides published in obscure journals and old newspapers were not accessed by the survey , and the same is true of the data held in the files of numerous individual professionals, companies and even some national organisations. It must also be stressed that there must be numerous other landslides that have not yet been recorded because they exist in remote areas, are concealed by woodland, are relatively insignificant or have yet to be actually recognised as landslides. This is clearly illustrated by the results of the recently completed (1988) Applied Earth Science Mapping of the Torbay area[44] which raised the total of known and reported landslides from 4 to 304, although it has to be recognised that the vast majority of these were coastal failures (290). Even in the South Wales Coalfield, which has been the subject of a major landslide inventory exercise[45], by the British Geological Survey, a detailed mapping programme in the Rhondda valleys[46] resulted in an increase in the number of recorded landslides from 102 to 346. Clearly, in some areas, the harder you look the more examples you find. Indeed, extrapolation leads to the inevitable conclusion that the actual number of landslides in Great Britain is many times in excess of the 8835 recorded so far by this survey.

However, the example from Torbay quoted above clearly shows that where detailed investigations have been carried out, knowledge of landslides may be good and large numbers of occurrences noted. Thus the survey of the South Wales Coalfield by the British Geological Survey[45], the recording of landslides in the Southern Pennines by R H Johnson[47] and the investigations by Gregory[48] in the North Yorkshire Moors, Butler[49] in the Cotswolds and Mitchell[50] in the northern Pennines, to name but a selection, will have provided considerable local detail in a sea of uncertainty. The pattern of landslides displayed on the map *(Figure 2.14)* must therefore be recognised as an artifact of investigation, reflecting varying degrees of ignorance. How else is it possible to explain that 70% of the recorded landslides in Wales are located in the South Wales Coalfield.

Perhaps the best illustration of this phenomenon is the spatial distribution of landslides as revealed by the search of 1:10,000/10,560 scale geology maps *(Table 2.2)* carried out as part of the DOE survey. A total of 3160 landslides in England and Wales was obtained from these maps. However, inspection of Table 2.2 reveals an extremely variable extent of map coverage; large areas have no map coverage at this scale and therefore no landslides can be shown *(Figure 2.16)*. Other parts, especially in Northern England, are still only covered by old County Series maps which were prepared at a time when the portrayal of landslides was not seen as a priority requirement in the preparation of geology maps. Finally, there are the modern National Grid Series maps which do portray landslides, although the proportion of actual landslides that may be identified will vary depending on surveyors' skills in recognising landslides and the density of other geological data that has to be shown. This inevitably raises questions as to the actual densities of landslides that may exist in areas between the known concentrations but for which presently available data is either sparse or non-existent. The concentrations of landslides revealed by mapping in Clwyd, Powys and the South Wales Coalfield *(Figure 2.16)* clearly suggest that present estimations of landsliding in Wales are significantly less than would be revealed if the principality were wholly surveyed to the same standard and seriously underestimate the actual total. These arguments apply equally to the reported distributions in England and Scotland. Put simply, landslides have only been recorded in those areas where they have been encountered or searched for.

In addition to the uncertainty regarding the extent to which existing landslides have been reported in any specific area, there is also the problem of what a 'reported landslide' actually means. It has to be noted that there is no uniformity of interpretation with respect to the term **'a landslide'**. Both the Storr on Skye and the collapse of 10m^3 of material in a road cutting could be reported as **'a landslide'**. The difficulties are further complicated by the fact that the same or similar landslides can be described in very different ways by different authors/surveyors. This problem of 'operator variance' is particularly apparent in the case of larger landslides and

2.16 The distribution of recorded landslides in Wales compared with the available British Geological Survey detailed geological map coverage.

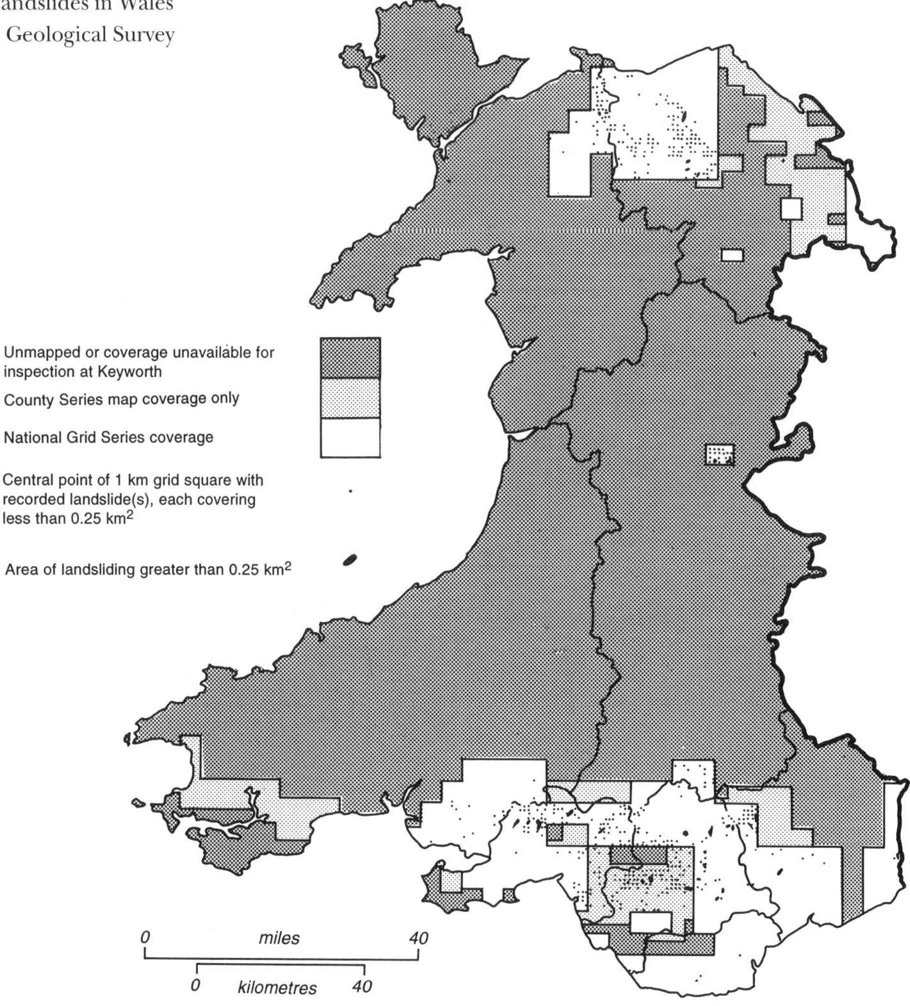

makes comparison of the work of different authors/sources extremely difficult. This problem can be illustrated with reference to the example of Bredon Hill cited earlier *(Figure 2.12)*, which is known to be ringed by landslides. At one extreme, the widespread and varied instability features could be interpreted as many hundreds of individual small slides. But other authors could group the features into a much smaller number of larger slides, or amalgamate further to create a yet smaller number of 'areas of landsliding' or even, at the other extreme, subsume all the features into a single, extensive 'zone of landsliding'. Under such circumstances, both the distribution map *(Figure 2.14)* and the database can become biased due to the dominance of information provided by highly detailed investigations of what may be relatively minor features*. This also means that any statistical analysis has to be treated with caution.

---

* A very good example of this phenomena was the 309 minor failures recorded in 25km^2 of Central Devon by Grainger (1983) which were included in the original database (1987) but subsequently recalibrated to 4 areas of landsliding in the 1991 up-date.

These difficulties can only be overcome if the data is adjusted to take account of size and extent of landslides. In reality, such an adjustment has proved difficult to accomplish because much of the information on landslides currently available in the public domain is exceedingly vague as to the size of features. Nevertheless, some initial steps have been taken to overcome these problems, as are described below using an area centred on the Midlands as an example *(Figure 2.17 a–c)*.

In Figure 2.17b, the landslide distribution map for the Midlands *(Figure 2.17a)* has been recalibrated by superimposing a 2km x 2km grid over the entire area and calculating the number of 'landslide units' falling within each 2km x 2km square.

Definition of a 'landslide unit' has been dependent upon constraints imposed by the methodology employed in the data collection exercise, where landslides extending over an area greater than 0.25km^2 were drawn on 1:250,000 scale maps, while smaller features were merely marked by a dot. A 'landslide unit' was defined as 0.1km^2 and every dot assigned the value of one unit. The area of larger landslides was measured to the nearest 0.1km^2 and given that value in landslide units. The maximum value is therefore 40/4km^2. The actual distribution of values clearly emphasises the high concentrations in the Cotswolds, Northamptonshire and the Southern Pennines of Derbyshire and North Staffordshire.

The conversion into a contour map *(Figure 2.17c)* has merely involved the recalculation of the landslide unit values in terms of landslide units per 100km^2.

2.17 The density of landsliding in the Midlands

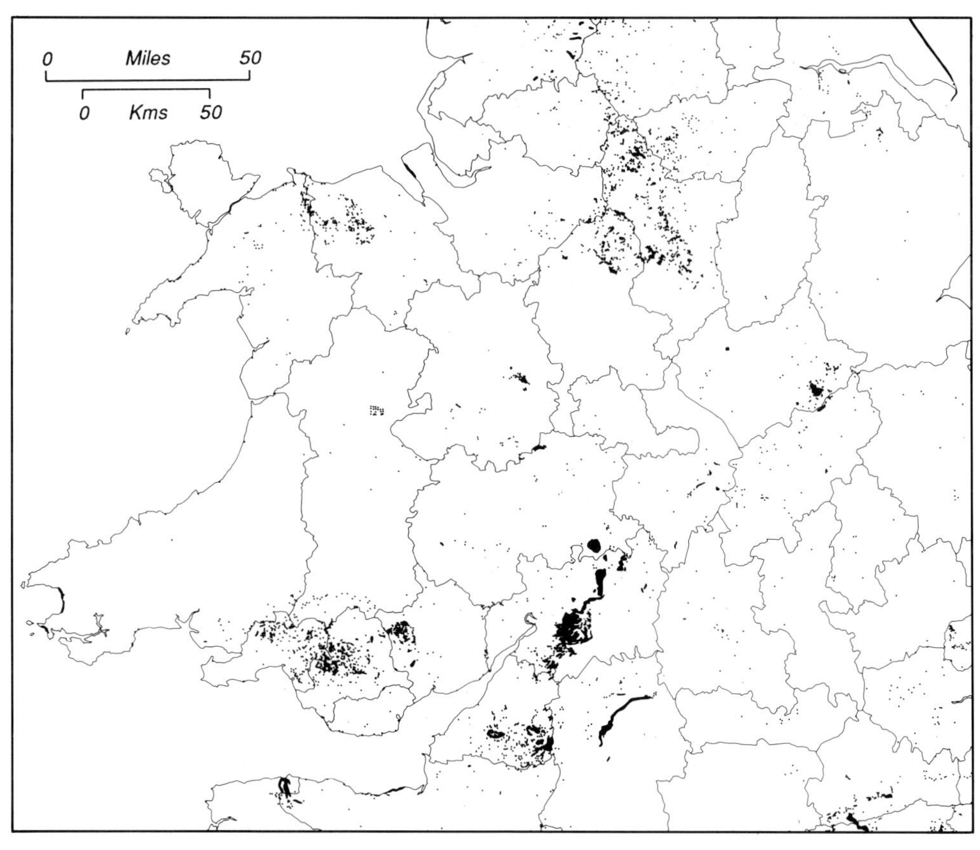

) The distribution of
corded landslides

(b) In 4km^2 grid cells

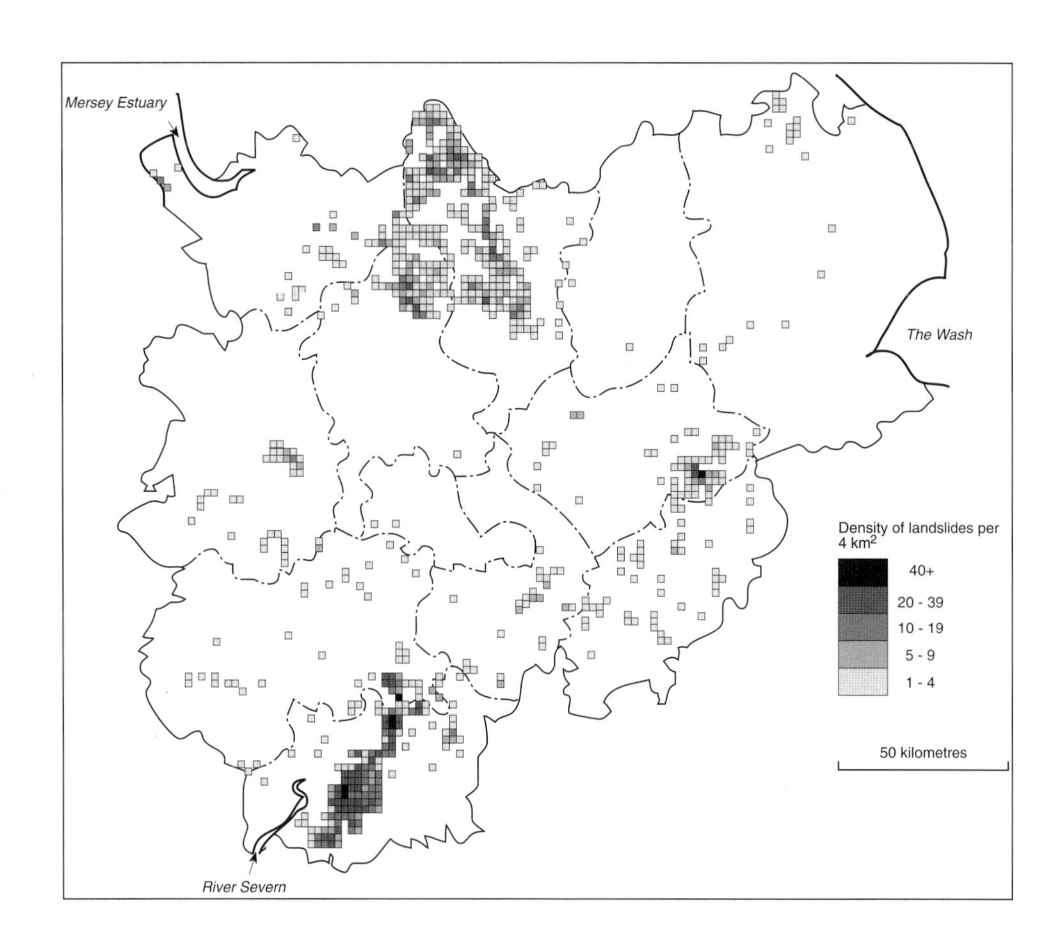

(c) Smoothed over 100km^2 grid cells

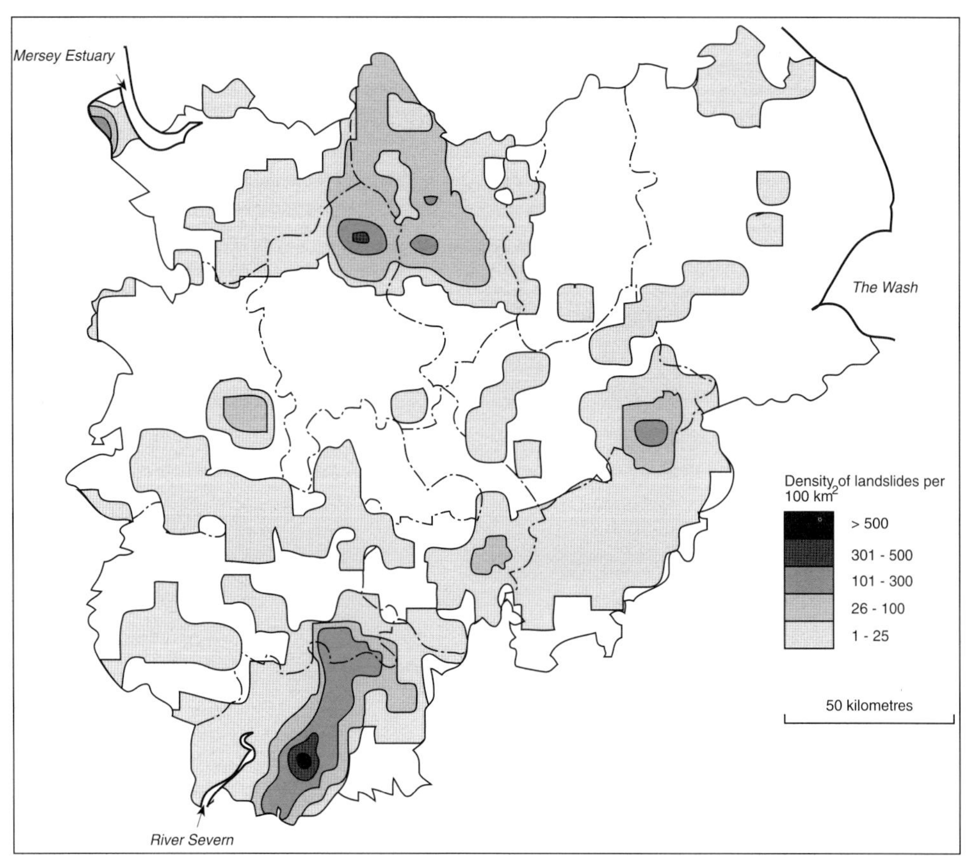

This process smoothes away much of the local variation and reinforces the high values in Northamptonshire, the Southern Pennines and most especially, in the Cotswolds. Indeed, the Cotswolds come out as one of the main concentrations of known landslides in England and Wales, together with South Wales and parts of the Pennines.

While such maps are useful in indicating the areas of known occurrence, questions remain as to the number and distribution of landslides that are still to be identified. Nevertheless, the large number of recorded landslides and their obvious concentration in those areas that have been well investigated, has confirmed to the DOE the potential significance of landsliding to land use planning and development. Thus the large volume of information produced by the review of research on landsliding in Great Britain[51] has contributed directly to the recently produced Planning Policy Guidance Note (PPG) 14, *Development on Unstable Land*[52], which included, for the first time, sections on landsliding. This subject will be considered further in Chapter 5 and succeeding chapters, but first it is necessary to examine exactly what the known distribution represents in terms of types of failure.

NOTES

1. So, 1966
2. Anon, 1951
3. Hutchinson, 1984a
4. Valentin, 1971
5. Hutchinson, 1976
6. Hutchinson, 1968
   Hutchinson, 1984a
7. Hutchinson, 1973
   Bromhead, 1978a
   Bromhead, 1978b
8. Hutchinson, 1973
9. Clarke, M.J. et al, 1976
10. Hutchinson, et al., 1980
11. Chandler, M.P., 1984
    Chandler, M.P. & Hutchinson, 1984
    Hutchinson et al., 1981a
    Hutchinson et al., 1985a
12. Geomorphological Services Ltd, 1991
    Lee et al, 1991a
13. Conway, 1974
    Brunsden, 1984
14. Bray, 1990
15. Brunsden and Jones, 1976
16. Brunsden and Jones, 1980
17. Bray, 1990
18. Pitts, 1974
    Pitts, 1981
    Pitts and Brunsden, 1987
19. Conybeare et al., 1840
20. Ballantyne, 1990
21. Anderson and Dunham, 1966
22. Hutchinson, 1984a
23. Hutchinson, 1972
    Hutchinson, 1980
24. Hutchinson and Gostelow, 1976
25. Hutchinson et al, 1985
26. Henkel and Skempton, 1954.
27. Knox, 1927
28. Kelly and Martin, 1986
29. Morgan, 1986
30. Pullen, 1986
31. Gostelow, 1977
32. Franks, C.A.M. et al., 1986
33. Halcrow, 1989
34. Franks, C.A.M. et al., 1986
35. Clark, A.R. and Guest, 1991
36. Smith, D.I., 1984
37. Skempton & Weeks, 1976
38. Kellaway and Taylor, 1968
39. Conway et al., 1980
    Halcrow, 1988
40. Brunsden and Jones, 1972.
41. Grove, 1953
42. Gerrard and Morris, 1980
43. Whittaker, 1972
44. Geomorphological Services Ltd, 1988
    Doornkamp, 1988
45. Conway et al., 1980
46. Halcrow, 1986
    Halcrow, 1988

47. Johnson, R H, 1965a

Johnson, R H, 1965b
Johnson, R H, 1980
Johnson, R H, 1987
Johnson, R H and Vaughan, 1979
Johnson, R H and Walthall, 1983

48. Gregory, 1963.
49. Butler, 1983.
50. Mitchell, 1991
51. Geomorphological Services Ltd., 1986-1987
52. DOE, 1990

Chapter 3

Landslide Types

Introduction

All slopes are under stress, due to the force of gravity. Should the forces acting on a slope exceed the resisting strength of the materials that form the slope, then the slope will fail and a landslide will occur. As has already been discussed in Chapter 1, the term landsliding is merely convenient short-hand to cover the great range of gravity-dominated processes that transport relatively dry earth materials downslope. In reality, these processes produce such a bewildering variety of features that the general terms **mass wasting, mass movement** or **slope movements** are much to be preferred, leaving the term **landslide** to cover those situations where coherent masses of material actually move downslope by sliding*. However, the term has come into such widespread usage that any proposal for change is almost certain to fail, especially with respect to public perception of the hazard.

That there should be such a range of mass movement processes and forms is not surprising considering the variables involved: slope height and shape (morphology); vegetation type and cover; slope-forming materials, their character, disposition and inter-relationships; climate; groundwater characteristics and their variation in time and space; and the nature of the triggering events that stimulate instability. The resulting spectrum of mass movement features is understandably extremely diverse and although there have been many attempts to provide a generally acceptable classification[1], there still remain points of dispute, with the result that certain features may be classified differently by different authors. The discussion that follows is therefore directed essentially to the types of failure that occur in Great Britain, and written from a British perspective. It adopts the classification scheme utilised by the DOE commissioned Landslide Study[2], modified slightly to incorporate aspects of the most recent proposals[3].

Most classifications are based on the combination of the **mechanism(s) of movement** and the **types of material involved in movement.** Three principal mechanisms are widely recognised: **falling, flowing** and **sliding.** Although these can and do operate separately and in isolation, they more commonly occur in association with one mechanism dominating. More confusing is the fact that in reality the boundaries between the different mechanisms are blurred so that individual failures display characteristics that are common to two, or even all three, of the principal mechanisms. This point will be returned to following consideration of the three principal mechanisms.

---

* Mass wasting is defined as "the downslope movement of soil or rock material under the influence of gravity without the direct aid of other media such as water, air or ice". The term mass movement has been in use for over a century. It was originally proposed by Penck (1894) to distinguish movement under the influence of gravity from mass transport, where material is carried by a moving medium such as water, air or ice. The use of the term mass movement has been strongly advocated by Hutchinson (1968) to include all gravity-dominated movements of ground materials including creep, subsidence, shrinkage and heave. Slope movements (Varnes, 1978) includes all forms of gravity-dominated movements of ground material except subsidence, shrinkage and heave.

Falling

Falling occurs when material becomes detached from high-angle bluffs, scars and cliffs (free faces) and descends through the air because of gravity. Falls are therefore sudden, rapid events. The material involved is usually specified as a prefix: thus **rock** fall *(Figure 3.1)*, debris fall (*Figure 3.2;* usually weathered rock or superficial deposits composed of an admixture of coarse and fine debris), soil fall (fine-grained material), **loess** fall (fall from face of windblown silt), etc. Falling usually results in the disintegration of the displaced mass, with debris accumulating at the base of the free face in a series of steep (up to 35°) debris cones, or talus cones, which may coalesce to form scree slopes. Sometimes, however, initial falling may lead to material tumbling over lower slopes in a process known as **avalanching.**

Rockfalls have also been classified on the size of failure with a progression from **pebble** fall, through **boulder** fall to **block** fall. The most complete volumetric nomenclature put forward to date[4] involves a progression from debris fall ($< 10m^3$), through boulder fall ($10–100m^3$, but including some large boulders), blockfall ($100–10,000m^3$), cliff fall ($10,000–1,000,000m^3$) to bergsturz, huge failures involving more than $1,000,000m^3$ of material which may move considerable distances. While such a descriptive scheme is useful, despite the criticisms that have been levelled at the terminology employed, it should only be used as an adjunct to the more important consideration of processes involved.

It is easily appreciated that true falls can only occur in those situations where free faces are near vertical or overhanging. But there are several other situations where rock and superficial deposits can become detached from a free face and undergo free fall. In certain instances where rocks contain well defined discontinuities (joints) inclined outwards from a free face at a steep angle, beds of rock may slide out of the face and fall **(plane failure)**. In those cases where the rocks have two sets of well defined intersecting joints, the result is the detachment of a mass of rock with triangular cross-section known as **wedge failure**[5] *(Figure 3.3)*. The debate continues as to whether these failures should be classified as falls or slides, depending on whether the terminology should be based on initial movement mechanism, final movement mechanism or the dominant movement mechanism. For simplicity, they will be included under fall in this discussion, although in later sections they will be considered under both headings. In many instances, steep rock faces are underlain by expansion or dilation joints that almost parallel the surface and in such cases, individual surface layers may be spalled off by a process known as **exfoliation.** But the most common mechanism is **toppling** *(Figure 3.4)*, where blocks become dislodged through a forward rotation about a pivot point low down on the block due to forces exerted by the movement of adjacent rock units or external factors such as tree root growth, freezing of water trapped in joints, wind, mass movement upslope or even humans[6].

Progressive weakening of the rock, superficial deposit or soil mass, by weathering or undercutting are the most important factors in generating falls. This was highlighted by work on the Chalk cliffs of Kent[7], where 40 falls were reported between 1810 and 1970. It was demonstrated that the incidence of falls was closely related to the average monthly effective rainfall and the average number of days with air frost per

3.1 Rock fall.

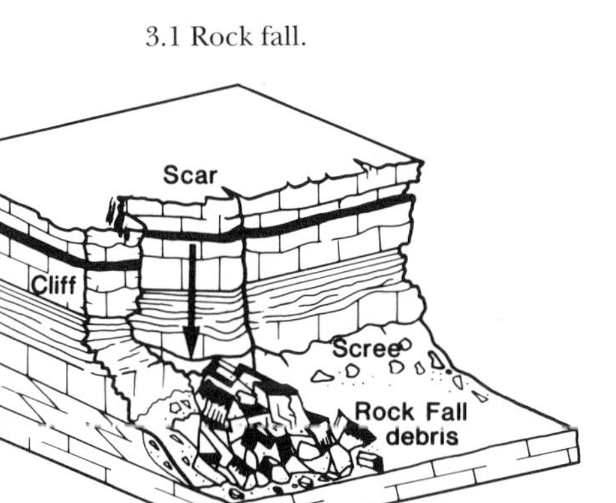

3.2 Debris fall.

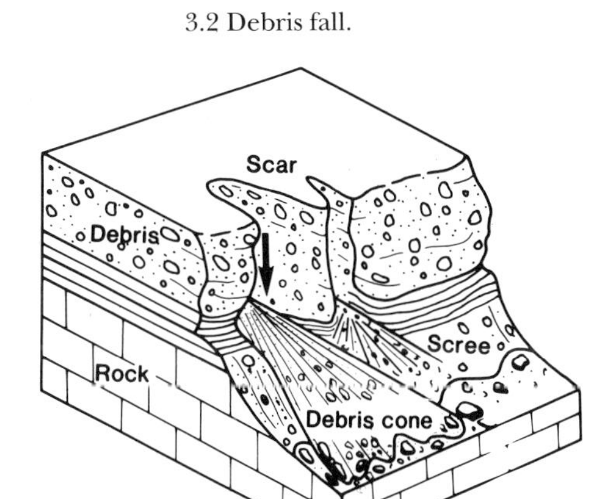

3.3 Wedge failure.

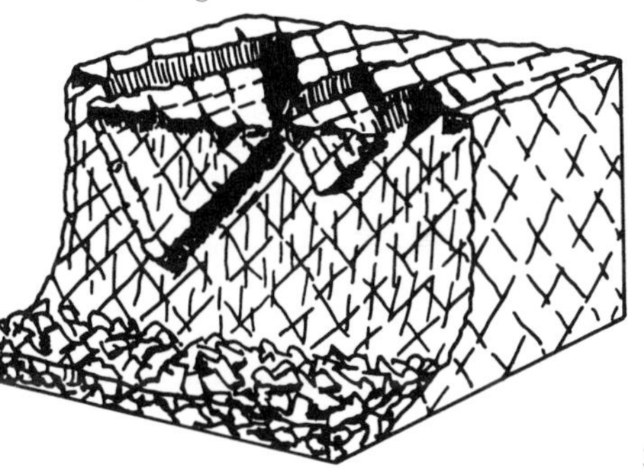

3.4 Examples of toppling failures (after Hoek & Bray, 1977).

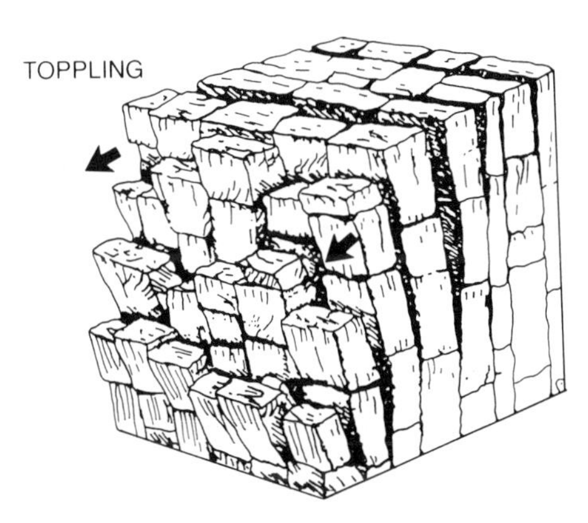

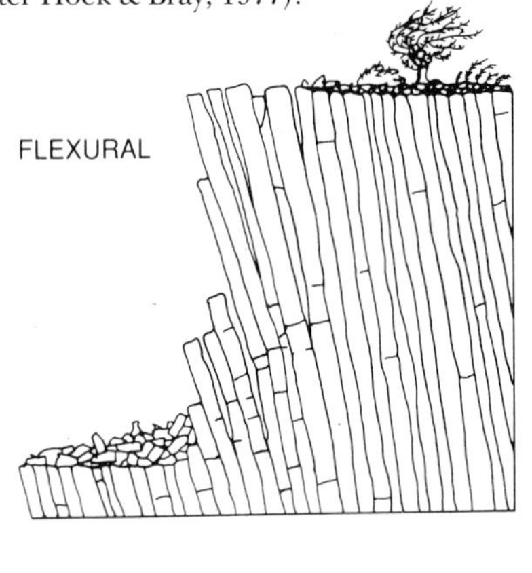

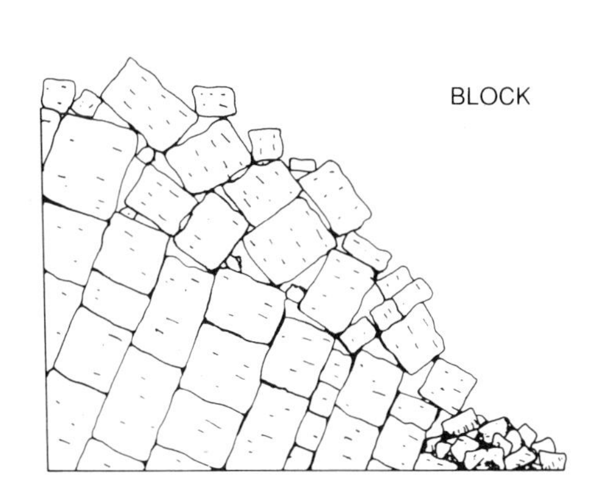

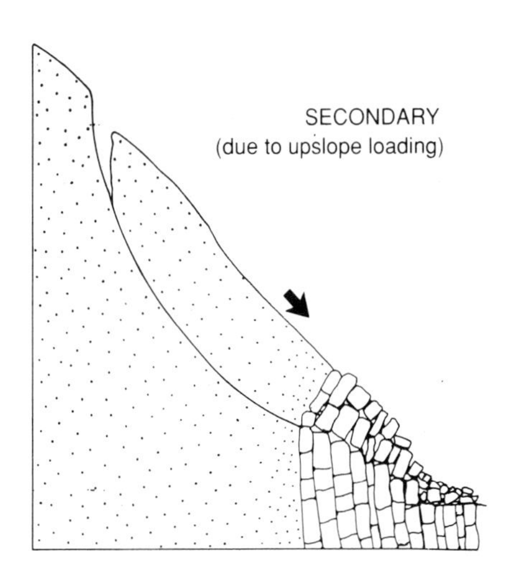

month. The immediate cause of failure was found to be the development of high water pressures within the pores and discontinuities, which were exacerbated when the water became frozen.

Flowing

Many will undoubtedly find it conceptually difficult to accept flowing as a mechanism for landsliding in Great Britain. Although obvious instances where transport of material by flowing can be identified – the failure of sediments on the floors of lakes and the sea (sub-marine flows), the movement of newly deposited volcanic ash (lahars), the mass movement of snow (avalanches), the collapse of sand faces (sand runs) and dry fine-grained sediments (powder flows), and the downslope distribution of debris produced by catastrophic falls and slides (rock avalanches, debris avalanches) – few appear to have particular relevance in a British context. However, flows can also occur on land without catastrophic triggering events and are usually divided into two groups: valley confined flows, where the movement of debris is down a pre-existing valley, and hill-slope flows. It is the latter that occur in Great Britain.

Both hill-slope and valley-confined flows are the product of extremely wet conditions, when relatively loose materials become so charged with water that moisture contents exceed the liquid limit values. The resultant movement of material is as a viscous fluid with turbulent motion, and may be classified as **debris** flow, **sand** flow, **mud** flow, or **earth** flow depending on the material involved (debris flows have high proportions of coarse material whereas mudflows consist of at least 50% particles sand-sized and smaller). Flows are the most elongate of all mass movement features *(Figure 3.5)* and consist of three main elements: a source area, a long track and a deposition zone consisting of a low angled fan or fans. The source area is normally one or more small hollows or landslide scars, for flows are usually initiated by other forms of slope movement. The track is usually long and sinuous, often a pre-existing streamline or gully, which appears relatively free of sediment after the passage of a flow, except for two bordering ridges (levées) of mixed debris. The depositional fans are normally on relatively flat ground at the base of the slope and are unusual because the coarsest debris, including large stones and boulders, frequently comes to rest on the surface.

Flows represent the wetter end of the spectrum of mass movement features close to the point where water becomes the mode of transport. The oft quoted reference to

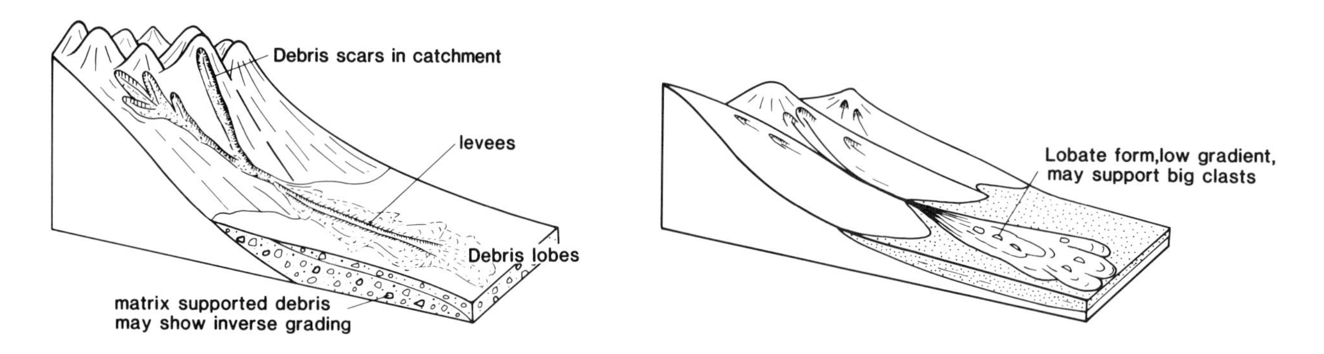

3.5 Flows (a) debris flow (b) mudflow.

the River Mississippi as "too thick to navigate and too thin to cultivate" indicates that there must come a point where fluvial (water) transport is replaced by mass movement. Surprisingly, it is when the water content value is about 21%. Water contents of less than 21% result in mudflows, 21–45% leads to what are known as hyperconcentrated-flows or mud-floods and more than 45% results in normal streamflow. Thus, although mudflows look exceptionally wet features, are usually produced during periods of intense rainfall, and can travel downslope at reasonable speeds, often as pulses of material, their behaviour is in many respects similar to that of wet concrete. The pulses or slugs of material are generated by slope instability, often into an active stream channel, and their passage down the track results in the creation of parallel levees due to frictional retardation at the margins. It is the turbulence and buoyancy effects that result in the largest sized material apparently floating on the surface of the moving feature and coming to rest on the surface of the depositional fan. However, it is the internal structure of the fan that is considered diagnostic, for whereas stream fans contain layers of pebbles in contact with each other, in debris flow fans the pebbles and stones are scattered through the mass and separated from each other by fine-grained material (matrix supported).

Sliding

In the case of sliding, instability results in a moving mass which behaves as a coherent body and always remains in contact with the ground. Movement is progressive and tends to propagate outwards from a point, and displacement is achieved by the development of a subsurface basal rupture known as a **slip surface, shear surface** or **shear plane.** As slides display considerable variety in terms of size and form, it is usual to employ a basic division into two broad categories dependent on the shape of the basal shear surface, for it is this that determines the nature of movement. If the shear surface has the general form of an inclined plane sub-parallel with the ground surface, the material slides downslope as a sheet or slab, known as a **translational slide.** Small translational slides may be sub-divided on the basis of the moving materials (**rock** slide *(Figure 3.6)*, **debris** slide, **soil** slide), but it is more usual to classify on the basis of the morphological characteristics of the moving mass. Thus a sheet of moving material will be termed a **planar slide** or **slab slide** *(Figure 3.7)*, the displacement of a particularly large rigid mass (block) which results in the formation of a chasm (graben) to the rear, is often known as a **block**

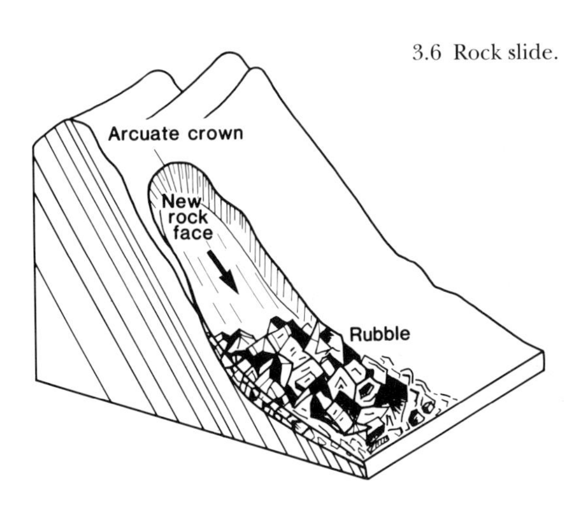

3.6 Rock slide.

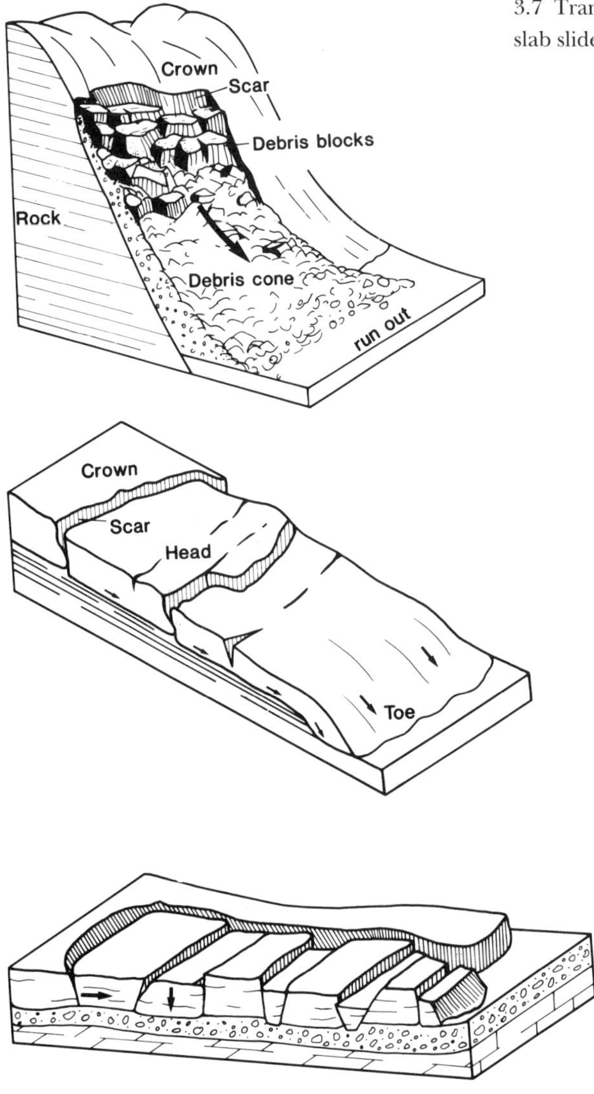

3.7 Transitional slides (a) debris slide (b) slab slide (c) spreading failure

glide (*Figure 3.8*, e.g. the Bindon landslide, East Devon[8]; *Photos 2.6 and 2.7*), and the even more widespread displacement of relatively weak materials that result in a series of upstanding 'blocks', separated by collapsed 'grabens', are known as **spreading failures**. Finally, there is the **mudslide** *(Figure 3.9)*, an elongate feature widely developed on argillaceous (clayey) outcrops and consisting of a mass of moving debris with moderately high water content (e.g. Black Ven, West Dorset; *Plate 7*). Mudslides have a distinctive form consisting of three main elements. At the upslope head of the feature is a bowl, backed by a head scarp formed in stable ground, from which material collapses to feed the mudslide. The central part of the mudslide is called the track and consists of a long, relatively narrow belt of moving material defined by a well marked basal shear surface and two lateral vertical shears which clearly demarcate the boundary between moving material and relatively stable ground. As the material moves from the source area into the track it becomes increasingly broken up and softened and often takes on the appearance of a porridge-like mass. The third morphological element is the basal accumulation

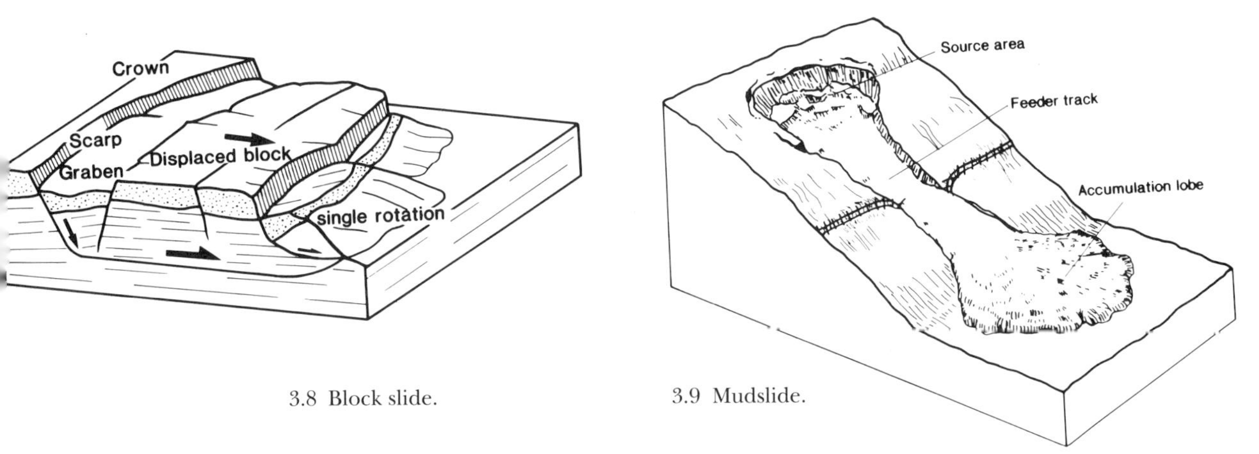

3.8 Block slide.

3.9 Mudslide.

zone, or toe, where material is dumped in one or more gently inclined lobes with relatively flat upper surfaces (1–5°) but steep margins (15–25°).

In many cases, however, ground displacement takes place by sliding over a basal shear surface which is curved (concave), with the result that the moving mass becomes tilted, or rotates, as it moves. These are **rotational landslides** *(Figure 3.10)* and are extremely widely developed. Initially, the ground may appear to subside downwards along well defined tension cracks, but as movement progresses rotation becomes more apparent. The result is the landform assemblage widely considered to be characteristic of landsliding, consisting of an embayment, defined by a curved back scar, within which lies very irregular ground composed of ridges, benches, minor scarps and hollows. Rotational failures may be further classified according to whether there is one failed block (**single**) or several which reflect repeated failure with a sequence of rotated blocks sharing a common basal shear surface (**multiple**) *(Figure 3.11)*. Rotational failures on a slope which are triggered in succession by the destabilising effect of an initial failure are known as **successive.** Rotational landslides may be further classified according to whether the initial movement was at the base of the slope and the destabilising effect riffled back up to slope (**retrogressive**) or high on the slope with destabilising effects working downslope (**progressive**). The term slumping was considered synonymous with any form of rotational landsliding in some earlier classification schemes[9], but has subsequently decreased in popularity so that use of **slump** is now restricted to individual, small rotational failures.

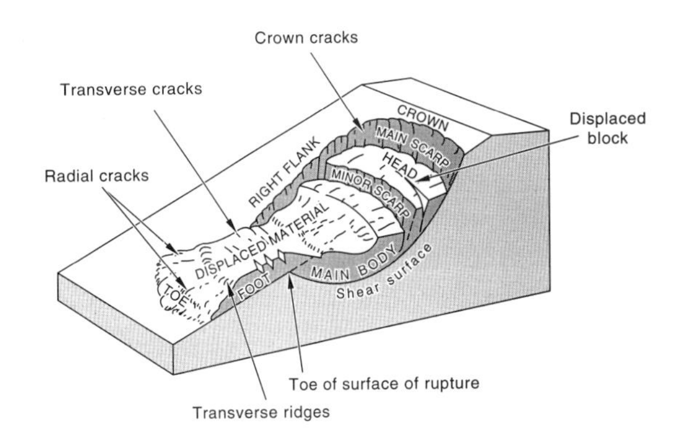

3.10 Single rotational slide.

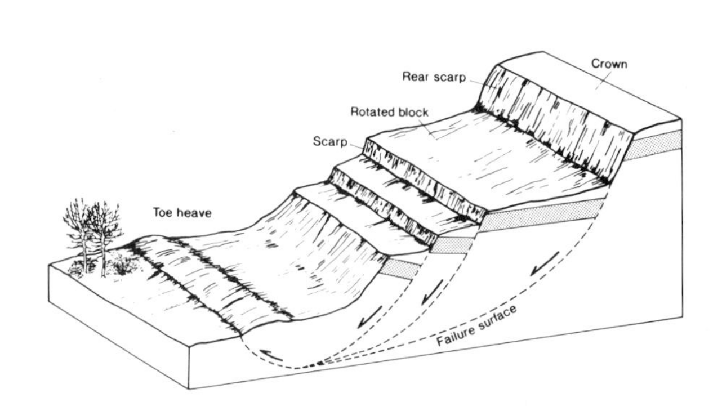

3.11 Multiple rotational slide.

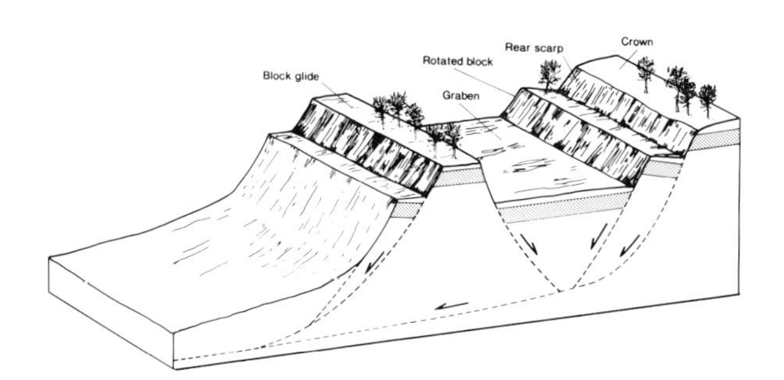

3.12
Compound landslide

The classic definition of a rotational landslide involved movement over an arcuate shear surface of nearly circular form. However, subsurface investigations have revealed that many basal shear surfaces are not circular but consist of a steeply inclined curved section above a gently inclined plane *(Figure 3.12)*. This has given rise to the term **non-circular rotational landslides.** But in addition, many failures that are predominantly translational in form have been found to have rotational elements at their heads. As a consequence, a third group of slides is now recognised which were originally given the confusing title of **translational slides (non-rotational)** but are now increasingly known as **compound slides**. The dramatic Alport Castles slide, Derbyshire *(Photo 2.11)* is a classic example of this type of failure.

Discussion

In reality there is a continuum of mass movement features from falls through slides to flows. In many instances it is difficult to determine whether masses of material fell or slid, and similarly there are a number of instances where material both slid and flowed, thus necessitating the term **flowslide** (e.g. the Aberfan failure of 1966). It is also true that individual failure movements may involve more than one mechanism. Rotational landsliding on a slope may be accompanied by translational sliding or flowing at lower elevations (e.g. below the Lower Greensand escarpment at Sevenoaks, *Figure 3.13*). Similarly, very large falls can result in various types of flow involving fluidisation with either water or air. Thus there is the need to recognise the existence of **complex landslides** where ground displacement is achieved by more than one type of mass movement. This should not be confused with **landslide complex,** which is an area of instability within which occur many different types of mass movement (e.g. Black Ven and Stonebarrow on the west Dorset coast and the Isle of Wight Undercliff).

Landslide Types in Great Britain

The frequency of occurrence in Great Britain of the various types of mass movement, as recorded in the recent DOE survey, is shown in Table 3.1. The fact that for almost 60% of the 8835 recorded landslides there was no freely available information on type is a sad reflection of the quality of data currently available in the public domain. However, as many of these landslides would have been identified from geological map sources, it is undoubtedly true that more information on these features could exist and may be available to further detailed enquiries. Concentrating on the 3621 examples for which type information is available (specified landslides) reveals that slides predominate (52.0%), with rotational slides the largest group (27.5%). But it must be noted that over 21% of these specified landslides are included under the heading "complex landslides" (777) and that further information on the occurrence of failure mechanisms exists in 683 cases. If the individual component mechanisms identified within these various sites are considered separately and added to the original data set *(Table 3.1)*, then the final totals are 9988 identified mass movement features of which 4680 (46.8%) have information on type. These are predominantly slides (63%), followed by flows (18%) and falls (16%).

The nature and history of types of failure in Great Britain was analysed in detail in the DOE study[2], and will not be repeated here. Instead, it is proposed to discuss merely the general characteristics of landsliding, focussing on particularly illuminating examples.

Falls

The statistics presented in Table 3.1 give a totally false impression of the frequency of falls in Great Britain. In reality, they are extremely widely distributed and become ubiquitous along eroding cliffed coastlines, although, fortunately, they rarely achieve great size.

Any very steep, vertical or over-hanging face developed in ground materials (soil, superficial deposits, rock, spoil) is prone to suffer falls. Material may become detached through the combination of loss of support (basal erosion or undercutting), reduction in internal strength (weathering) or externally imposed stresses (freezing of water in joints, tree root growth, etc.). Once detached, the material descends to accumulate in debris cones mantling the lower slopes, which may coalesce to form inclined belts of debris, generally known as **talus slopes**, with the term **scree slopes** reserved for coarser accumulations. Thus the spectacular screes of Wastwater in the Lake District *(Plate 14)*, Eglwysed Mountain, near

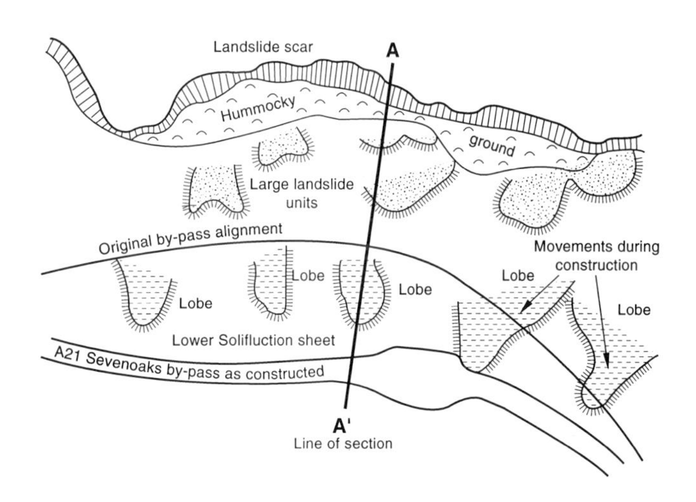

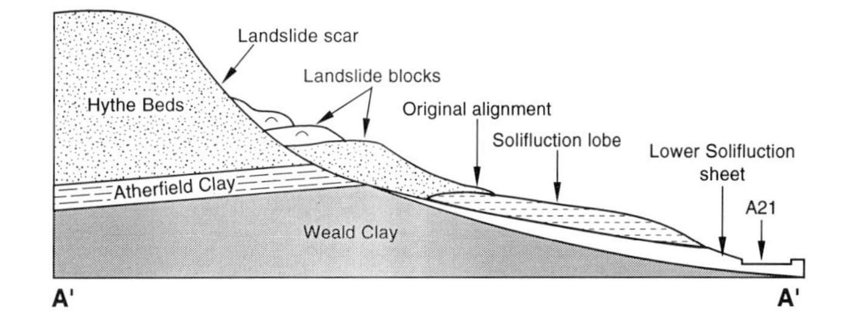

13 Diagrammatic section through the landslides ong the Lower Greensand escarpment at Sevenoaks, ent, which caused considerable problems during ad construction (after Brunsden et al., 1988).

Table 3.1 **Numbers and relative significance of different types of landslide as revealed by the DOE sponsored survey**

Failure Types	Number	Percentage	Percentage of Specified Landslides	Composition of 777 Complex Landslides	Total Landslide Attributes	Percentage	Percentage Specified (4680)
Unspecified	5214	59.0	–	94	5308	53.1	–
FALLS							
Rockfall	363	4.1	10.0	203	566	5.7	12.1
Topple	56	0.7	1.5	82	138	1.4	3.0
Sagging	30	0.3	0.8	6	36	0.4	0.8
	449	**5.1**	**12.3**	**291**	**740**	**7.5**	**15.9**
ROTATIONAL SLIDES							
Single	644	7.4	17.8	200	844	8.5	18.1
Multiple	197	2.2	5.5	175	372	3.7	8.0
Successive	153	1.7	4.2	76	229	2.3	4.9
	994	**11.3**	**27.5**	**451**	**1445**	**14.5**	**31.0**
TRANSLATIONAL SLIDES							
Compound	153	1.7	4.2	384	537	5.4	11.5
Translational	734	8.3	20.3	222	956	9.5	20.4
	887	**10.0**	**24.5**	**606**	**1503**	**14.9**	**31.9**
FLOWS	370	4.2	10.2	464	834	8.4	17.8
COMPLEX SLIDES	777	8.8	21.5	–	–	–	–
CAMBERED/FOUNDERED STRATA	144	1.6	4.0	14	158	1.6	3.4
TOTAL	**8835**	**100.0**	**100.0**	**1920**	**9988**	**100.0**	**100.0**

Llangollen (Clwyd) *(Photo 3.1)* and the Cuillin Hills of Skye, are just the most conspicuous examples of tens of thousands of smaller talus deposits which represent the accumulated debris of countless individual falls over many thousand years. Such falls will display a well developed magnitude-frequency distribution, with huge numbers of very small events at one extreme and rare very large collapses (i.e. > $100,000m^3$) at the other.

Before proceeding further, it is necessary to draw a distinction between falls from bluffs, scars, crags and cliffs produced by erosion in the past (e.g. the walls of glacially overdeepened valleys, the headwalls of corries, abandoned marine clifflines, former river meander scars) and those from faces actively undergoing contemporary erosion. In the case of the former, falling assists in slope evolution so as to eventually produce more gently inclined stable slopes, the lower parts of which will be developed on talus. However, in those situations where there is active or continuing basal erosion, falls yield materials which are continuously removed by erosion (rivers, the sea) so that no protective accumulation can develop at the slope base to limit further undercutting. Thus at these locations, falls are transient features, for both scars and debris *(see Figure 3.1)* are quickly removed as the slopes continue to evolve. This adds considerably to the potential for greatly underestimating the

3.1 Scree slopes at Eglwysed Mountain, near Llangollen (Crown copyright/MOD).

number of falls that actually occur. In inland areas, this situation is restricted to currently active river scars, banks of meandering rivers and the slopes of evolving gullies. However, along the present coastline this condition prevails throughout the entire length of eroding clifflines.

Thus, everywhere the coast is known to be retreating *(Figure 2.6)*, falling occurs but often leaves no lasting signs of its occurrence. For example, the sea cliffs at Stonebarrow, Dorset *(Photo 2.5)* are retreating at an average rate of 0.4m pa[10], mainly through falls which accumulate on the foreshore until removed by the sea to expose the cliffs to further attack. The same is true of the famous Seven Sisters coastline, Sussex, where the vertical white Chalk cliffs are retreating at an average of 0.91m pa[11]. Debris is rarely visible, however, indicating the extremely rapid removal of about 100,000m^3 of fallen Chalk each year along this short stretch of exposed shoreline *(Plate 15)*.

It is now thought probable that most of the currently eroding clifflines have been evolving through instability since sea-level rose to near its present value 5,000–6,000 years ago. Over this period, these clifflines have retreated between 0.5km and 3km, depending on materials and exposure to wave attack, with even larger values recorded for East Anglia and Holderness where the low cliffs are composed of weak glacial deposits *(see Chapter 2)*. The Straits of Dover was widened considerably and the former ridge of Chalk linking Purbeck to the Needles breached to separate the Isle of Wight[12]. In the shorter term, however, many of these coastal cliffs appear to experience cycles of instability, when phases of rockfall activity are followed by periods of quiescence, during which time the sea removes the accumulated debris prior to recommencing erosion at the cliff base. Such cycles have been suggested for the coasts of Dorset[13] and Sussex[14] with periodicities that vary depending on geological conditions, cliff height and available wave energy.

At a more detailed level, the evolutionary cycle of Chalk cliff failure has been established by studies at Joss Bay, Kent[15]. This revealed that cliff foot erosion leads to the gradual opening of a tension crack down pre-existing joints to a point about mid-height in the cliff, at which time a high-angled shear surface develops through the root of the mass, which collapses to create a talus cone that is quickly removed by the sea *(Figure 3.14)*.

The higher clifflines of Chalk, Upper Greensand and Lower Greensand occasionally experience huge falls with volumes of about 1,000,000m^3 (e.g. Abbott's Cliff, Kent in December 1911), and similar size failures affect high cliffs formed of other materials, albeit more rarely (e.g. the Triassic sandstone cliffs near Sidmouth, June 1893[16] or the Lias cliffs of South Glamorgan, *Photo 2.8*). In the case of the huge Chalk falls, the material can be carried seaward across the shore platform for distances of up to four times the cliff height (i.e. 300–600m), suggesting that the very high pore pressures generated upon impact are sufficient to create flowslides *(see below)*.

A variety of failure mechanisms have been recorded on the Lias cliffs of South Glamorgan, including rockfalls, topples and rockslides. Monitored cliff recession rates are in the order of 3.2–6.8cm a year, achieved mainly by failures ranging in size from 200–5000m^3, although extreme events can occur, such as a rockfall near Nash lighthouses in July 1983 which involved some 20,000m^3 of material.

Even on those stretches of cliffed coast considered 'stable' *(Figure 2.6)*, the precipitous cliffs actually undergo very gradual retreat, largely by falls.

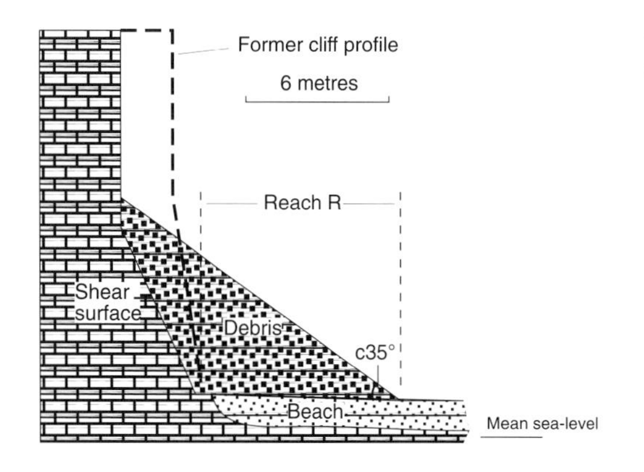

3.14 Cross section of a fall in Upper Chalk at Joss Bay, Kent (after Hutchinson, 1983a).

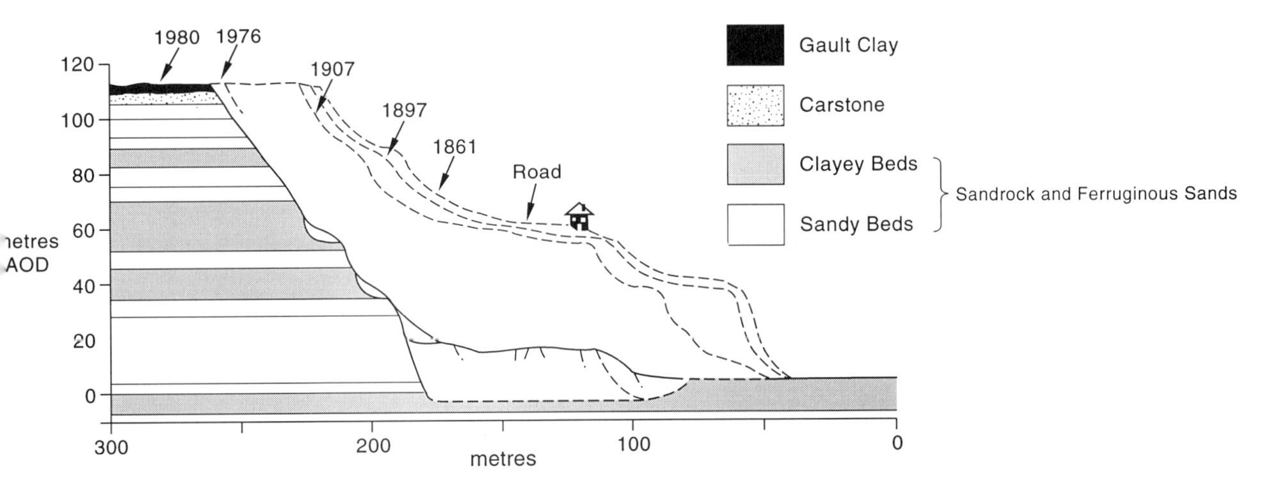

3.15 The retreat of the Lower Greensand sea cliffs at Chale, Isle of Wight (after Hutchinson et al., 1981a).

Unfortunately, published accounts of failures on these often remote clifflines are rare but the same magnitude-frequency distribution almost certainly applies, with the displacement of numerous, individual joint-defined blocks at one extreme and occasional large failures at the other.

Having established the association between falls and evolving coastal cliffs, it is important to note that many of the falling failures are actually topples, especially in heavily jointed rocks, and that not all the falls are associated with undercut, near vertical cliffs. Other forms of failure, such as rotational sliding, create steep slopes which suffer from falls and topples, as can be seen along the backscar of Black Ven *(Plate 7)*, Stonebarrow *(Photo 2.5)* and was captured on film in the classic photograph of a large toppling failure at Gore Cliff, Isle of Wight *(Photo 2.9)*.

Where strata with differing water-bearing characteristics crop out at the coast, issues of water from perched water tables held up on impermeable layers (aquicludes) can result in the removal of fine materials, thereby weakening the slope enough to cause collapse. This seepage erosion is an important contributor to cliff instability in many parts of southern England, most especially at Chale Bay, Isle of Wight, where the alternating sands and clays of the Lower Greensand give rise to seepage at a number of levels in the cliffs *(Plate 16)*. These cliffs represent one of the most active areas of coastal cliff instability in Great Britain, with the seepage erosion causing numerous falls and debris slides. Of interest, documentation of the pattern of cliff retreat at this site has revealed an increase in instability over the last 119 years[18] *(Figure 3.15)* with an average rate of cliff retreat between 1861–1980 of 0.4m pa.

While falls are common features on coastal cliffs, the limited availability of precipitous slopes inland restrict their occurrence. Apart from sites of active river erosion, falls are for the most part confined to bluffs, scars, crags and over-steepened rockslopes produced by the Pleistocene glaciations, gorges, and the steep scars produced by other forms of slope failure. Falls and toppling failures are also fairly common in working and abandoned quarries and have been known to cause deaths. Other sites of human-induced failures include road cuttings[19] and excavations for foundations, both of which can result in toppling failures *(Figure 3.4)* and wedge failures *(Figure 3.3)*.

Although inland falls are not especially numerous and usually fairly small, some large examples of varied character do exist. These are mainly the product of very different stability conditions in the past. The most obvious difference was the greater incidence of freeze-thaw activity during the colder phases of the Pleistocene which accelerated the rate of sculpturing of exposed rockfaces, thereby resulting in relatively frequent falls and rapid rates of scree production[20]. This topic will be discussed at greater length in Chapters 5 and 6. But it is also true that the glaciations profoundly affected patterns of slope instability in upland areas by the creation of over-steepened valley sides *(Photo 3.2)*, corrie backwalls *(Photo 3.3)* and scars which often collapsed when the supporting mass of ice was removed, together with the exposure of previously buried rock masses which were quickly attacked by freeze-thaw processes. Such features are best displayed in the higher areas of Wales and the Highlands of Scotland *(Photo 3.4)*.

3.2 Debris flow modified talus in the Lairig Ghru, Cairngorms (C Ballantyne).

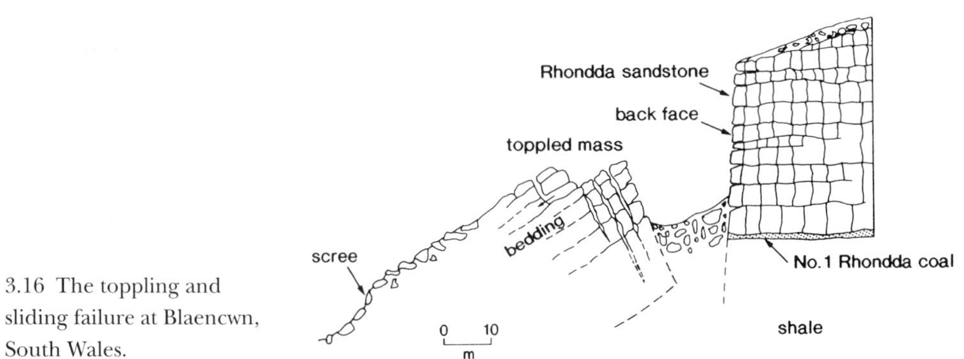

3.16 The toppling and sliding failure at Blaencwn, South Wales.

3.3 Rockfall talus in Glas Tholl, An Teallach, Wester Ross (C Ballantyne).

There are also several examples of old topples[21], some of which attain great size. One at Nant Gareg-Iwyd in South Wales involves an area of about 4,000m^2 *(Figure 3.16)* where blocks of Rhondda sandstone have suffered forward rotation through

3.4 Rock slope failure on Beinn Alligin, Wester Ross (C Ballantyne).

20-90° due to loss of support following sliding in the underlying shales. The Blaencwm landslide *(Plate 17)* involves an unusual combination of toppling and sliding that developed during the 1980s. This is the largest slide to have occurred in

17 The toppling and
iding failure at Blaencwm,
outh Wales.

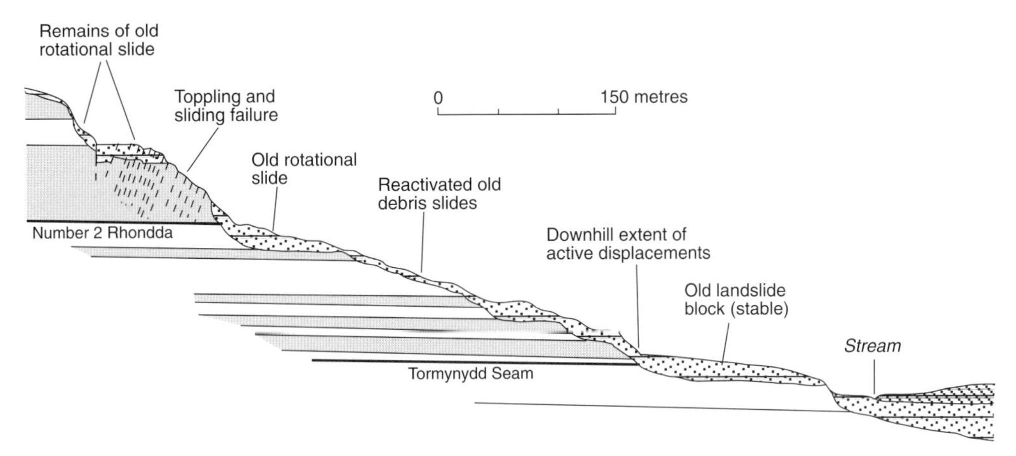

South Wales for 35 years and the movements have reactivated part of the series of overlapping landslides on the slopes below *(Figure 3.17)*. A good example of a coastal toppling failure has been described at Great Southwell, on the Isle of Portland *(Figure 3.18)*, involving the forward rotation of blocks of Purbeck Beds[22].

Finally, the statistics *(Table 3.1)* record the presence of 36 sagging failures. This unusual category of failure involves two main groups of features in Britain: those where loss of support at the base of a slope results in lateral spreading of the lower portion of the displaced block, accompanied by vertical displacement of the upper parts; and those where valleyward expansion of slope materials results in a form of multiple toppling, probably combined with sliding but with no element of mass creep[23]. The former appear relatively rare in Britain and is likely to be best developed in coastal situations, the latter is a feature of 'hard rock' terrains, with no fewer than 34 examples reported from the Highlands of Scotland, two of which are worthy of further discussion.

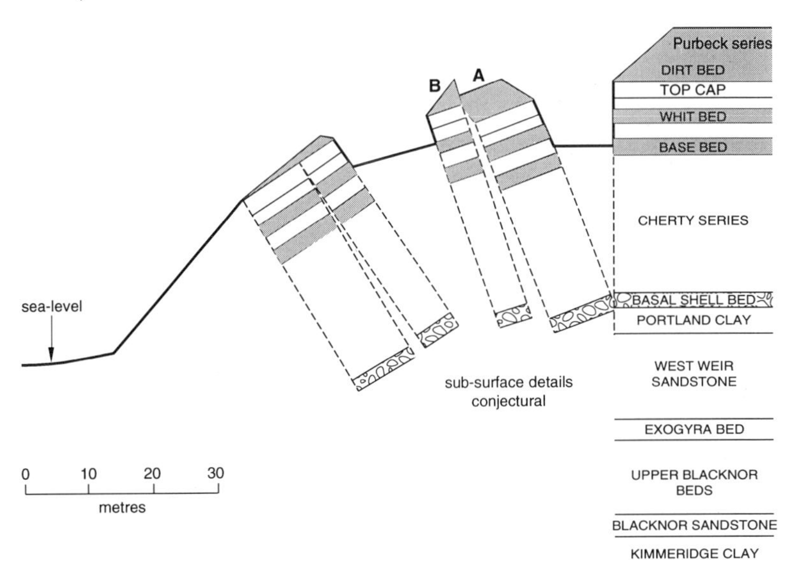

3.18 The Great Southwell Landslip, Isle of Portland: an example of toppling failure (after Bromhead, 1986).

At Carn Mor, very steep slopes (up to 40°) are formed in schist* which has been broken up into blocks by well-developed foliation** surfaces and joints parallel to the hillside and dipping at a steep angle into it. These blocks have suffered valleyward movement through rotation *(Figure 3.19)* to yield an irregular slope crossed by numerous, low (10–20m) reversed scarps or counterscarps. The whole surface layer of the slope, up to a depth of 50–60m has suffered movement due to sliding at the base of the hillside[21]. The original cause of these movements is considered to be the removal of ice-support about 10,500 years ago when high groundwater pressures in the over-steepened lower slopes resulted in sliding, thereby leading to retrogressive failure upslope. A similar situation occurs at Ben Attow[24] where the outward movement of the slope-forming rocks over an area of 2.2km^2 has created numerous well developed counterscarps up to 10m high running across the slope in a fine example of a sackung failure *(Photo 3.5)*.

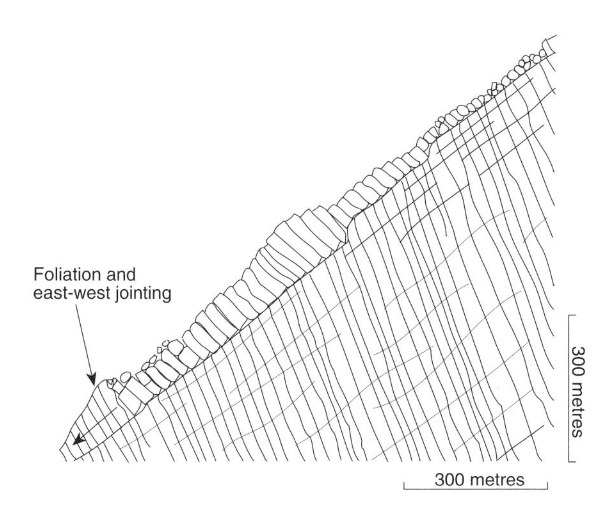

3.19 The toppling failure at Carn Mor in Glen Pean, Scotland (after Defreitas & Watters, 1973).

To conclude, therefore, the diverse forms of falls bring sharply into focus the problems inherent in undertaking surveys of instability features. Falls are so numerous in certain locations that to record every event would be extremely time consuming and costly. But what is the minimum size of a fall for recording purposes? At what locations are the presence of falls to be taken for granted and what should be recorded when occasional large falls occur off cliffs characterised by persistent falling? The data collected in the DOE commissioned survey *(Table 3.1)* relates to conspicuous or noteworthy occurrences which happened to be recorded. The actual distributions of the features is very much more extensive and their frequency far greater than these data suggest.

Flows

It was pointed out earlier in this chapter that flows are a relatively unusual form of mass movement in that the material involved possesses a high degree of fluidity. As a consequence the material is very mobile, suffers turbulence while in motion, and

---

* Schist is a strongly foliated crystalline rock, formed by metamorphism.

** Foliation is a general term for the planar arrangement of textural or structural features in a rock, and especially applies to the structures that result from the flattening of the constituent grains of a metamorphic rock.

3.5 Upslope-facing scarplets at Ben Attow (G Holmes and J J Jarvis).

can travel considerable distances at relatively high velocities, often over very low angled slopes[25]. Early attempts to classify these features on the basis of constituent materials have proved largely unsuccessful and the term **debris flow** is now generally applied to these unusual phenomena, irrespective of materials, although the terms 'mudflow', 'boulder mudflow', etc, may be employed for descriptive purposes. As this group lies intermediate between mass transport by water and mass movement by sliding there are bound to be features intermediate between debris flow (mudflow) and mudflood on the one hand and debris flow and debris slide on the other. This latter boundary has proved of some concern to mass movement investigations, with the result that a separate, intermediate category of **flowslides** has been established[26]. The significance of both debris flows and flowslides in Great Britain will be examined in this section.

(i) *Debris flows*

Debris flows are surprisingly common in mountainous areas of Great Britain, especially where there is an abundant supply of loose, coarse-grained surface debris. However, British examples tend to be small and undramatic in comparison with the huge devastating features developed in the Rocky Mountains, Andes and Himalayas. The three component elements of a debris flow are shown in Figure 3.5: the single or multiple source areas where shallow failures feed material into the channel; the track itself, which is often not eroded by the passage of the mass debris (often referred to as a 'plug', 'slug' or 'pulse'), thereby highlighting the fact that basal shearing is not a significant mechanism, and the single or multiple depositional lobes. The diagnostic feature of debris flows is, without doubt, the parallel ridges of debris (levées) that flank the track *(Photo 3.6)*.

3.6 Recent debris flows in the Lairig Ghru, Cairngorms (C Ballantyne).

Debris flows are initiated following either the saturation of coarse surface debris or shallow sliding of debris into a stream channel. The resulting mobile mass has shear stresses distributed throughout rather than concentrated along a single shear surface. The mobility of flows is related to the relatively high water content of the sediments, and motion will only continue whilst the slope is steep enough to maintain a velocity which prevents water draining from the debris. Once water is able to drain out, pore pressures decrease and the frictional strength of the material is restored. Thus, when debris flows stop, the strength of the material allows the formation of lobes of debris with steep margins at the front and sides.

Before proceeding further, it is necessary to note that the definition of debris flows employed here reflects the more restrictive use of the term following the publication of detailed research by Hutchinson and Bhandari[27] which was concerned with establishing the true characteristics of mudslides. Prior to this (1971), many shallow, wet-looking slope failures with arcuate headwalls and bulbous, lobate toes were classified as 'mudflows' *(e.g. Photo 2.2)*, and the term is still sometimes used incorrectly to describe the depositional products of mudslides. Thus the literature, and particularly the older literature, may incorrectly refer to slides as flows, thereby exaggerating the reported numbers and distribution of debris flows *(Table 3.1)*.

Confirmed debris flows have been reported from many mountainous areas in Wales and Scotland, including the Black Mountains of Southern Wales[28], Snowdonia[29], the Cairngorms[30] *(Photos 3.2 and 3.6)* and the Western Highlands[31] *(Photo 3.3)*. The most extensive investigation to date[32] has resulted in a map of Scotland showing the presence or absence of debris flow activity within 100km^2 grid squares *(Figure 3.20)* based on a search of aerial photographs. This map shows clearly how widespread debris flow activity is in the remoter highland areas where such activity had previously been unreported because of the remoteness and lack of impact on local communities and transport networks. However, there have been a number of debris flow events in Scotland that have caused damage. A severe thunderstorm over the Scottish Highlands in May 1953 generated numerous debris flows which caused

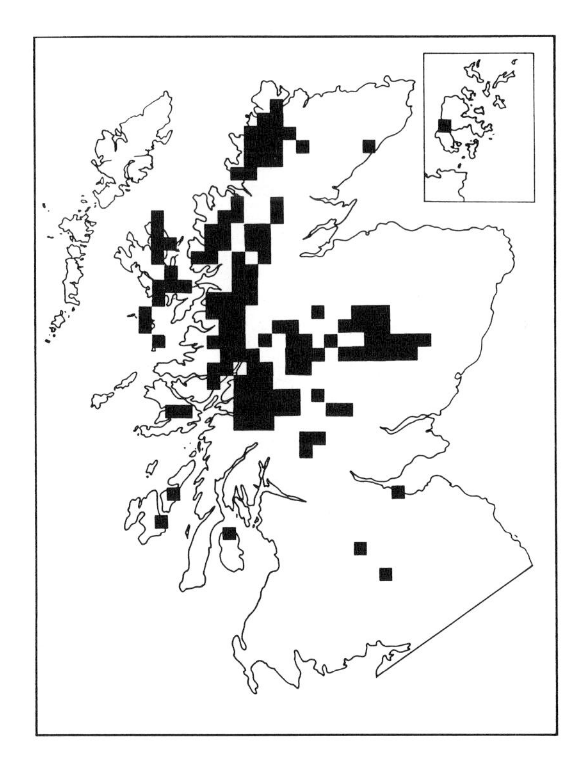

3.20 The extent of recorded debris flow activity in Scotland (after Innes, 1983).

extensive damage to forestry plantations and roads[33], and similar damage was reported in the Cairngorms following the heavy storms of July 1956[34].

The potential for debris flow activity is a function of the properties of the materials involved, particularly the **brittleness**, i.e. the decrease in strength between peak and residual strength of the soil[28]. Work in Scotland[32] has recognised a clear concentration of debris flow activity on rocks which break down into sandy soils, such as the Torridon Sandstone, as opposed to others, such as schists, which weather to cohesive soils*. This susceptibility of sandy soils to debris flow may reflect the high infiltration rates associated with such soils, which allow the rapid build up of excess pore pressures during heavy rainstorms. Indeed, observations in the Highlands of Scotland, the Ochil Hills[35] and Snowdonia[29] indicate that debris flows in Great Britain are, almost without exception, associated with intense storms.

Geomorphological considerations indicate that debris flow activity is probably widely developed in many upland regions, especially Wales, the Pennines, the Lake District and the Highlands of Scotland. As no detailed census of these relatively small features has yet been undertaken, it is impossible to estimate the quantities that may exist or even the number that may occur each year. Any report of debris flows in lowland areas needs to be treated with caution, for although debris flows undoubtedly exist in Lowland Britain, especially along the coast, there is a history of misclassifying 'slides' as 'flows' in these areas which may produce errors.

(ii) Flowslides

Flowslides are large, fast-moving, potentially extremely destructive and rare. The term is normally reserved for a disintegrating subaerial slide in loose coarse material with a highly porous structure which suffered collapse as a consequence of some disturbance[36]. When the structure collapses, the overburden load is transferred onto the pore fluids (either air or water) and excess pore pressures are generated, giving the material a temporary semi-fluid nature[23].

The Aberfan disaster of October 1966 was a flowslide. The failure commenced at around 9.00 a.m. at the toe of Tip No. 7 and rapidly developed into a flowslide which accelerated downslope into the village, destroying property and killing 144 people, including 116 school children[37]. The slide involved 106,000m^3 of material, of which 38,000m^3 crossed the railway embankment and caused the destruction in the town *(Figure 3.21)*. The maximum distance travelled was just under 600m at speeds of 16–32km/hour. The main, destructive, slide was initiated by a small failure at the base of the tip moving over pre-existing shear surfaces which had been triggered by high pore-water pressures. Movement of this small slide caused the tip to collapse thereby generating the flowslide because of the presence of large volumes of loose, saturated material at the base of the tip and loose, wet material above it.

---

* The technical terms introduced in this paragraph are discussed further in the section on sliding and in Chapter 4.

21 Two examples of ›wslides developed in colliery ›oil: Abercynon and Aberfan, ›uth Wales (after Bromhead, ›86).

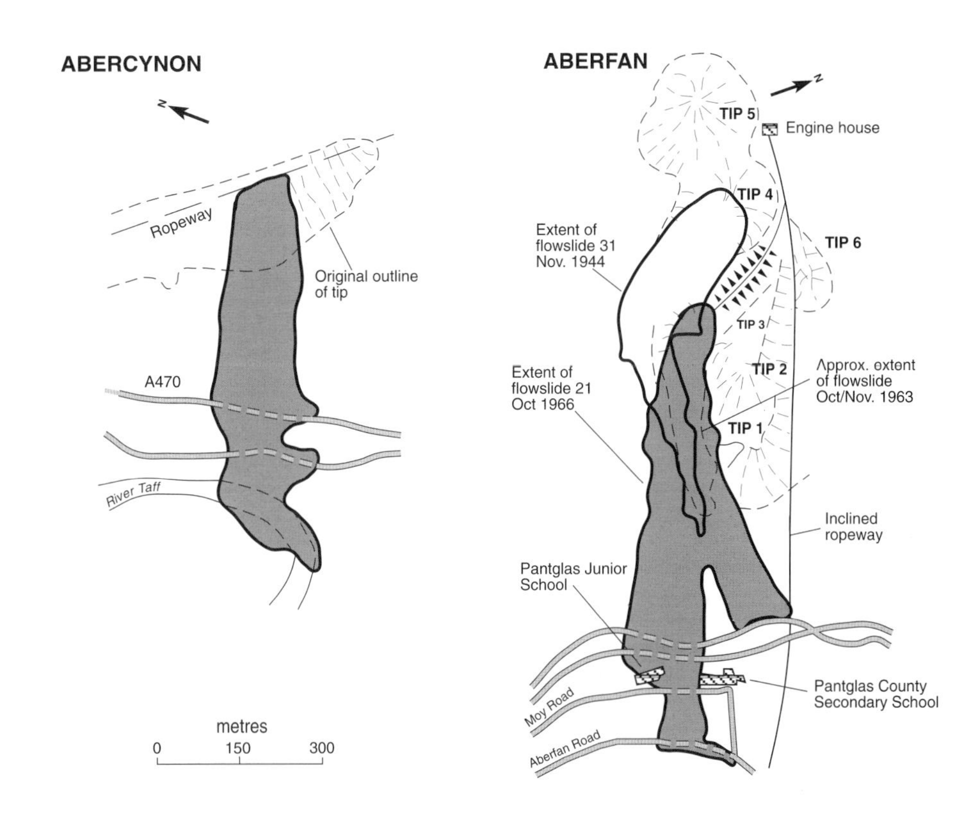

Regrettably, flowslides in colliery waste were not unknown in South Wales prior to 1966. In December 1939 a large flowslide developed a few kilometres down the Taff Valley at Cilfynydd Common, Abercynon. This flowslide involved an estimated 150,000m^3 of colliery waste, which travelled a maximum of 435m at about 16km/hour and in the process cut a power line, buried a 180m length of the main road and blocked and diverted the River Taff[37] *(Figure 3.21).*

Flowslides have also been reported as occurring below Chalk cliffs in southern England as a consequence of large rockfalls. One such example occurred in January 1905 from a 75m high cliff near Dover and ran-out seawards across the shore platform for nearly 400m. A similar feature was reported near Folkestone Warren in December 1915, with debris travelling over 500m from the cliff base[38]. These features move at great speed and it has been suggested[23] that the necessary excess pore fluid pressures are produced in the pore water by impact collapse of weak blocks of highly porous, saturated Chalk.

Impact-collapse flowslides have also occurred in kaolinised granite. An example of this type of failure occurred in November 1972 in a 45m high 'china clay' quarry in Cornwall, where a fall generated flowslide moved rapidly across the nearly flat quarry floor killing a man[39].

Slides

Mass movement by sliding involves the displacement of relatively coherent bodies of material whose margins are defined by well developed rupture surfaces, or rupture zones, known as slip surfaces or shear surfaces. Thus the material moves en masse, although displacement inevitably leads to internal stresses which result in the break-up of the moving mass, sometimes resulting in complete disintegration to jumbled debris if the materials involved lack coherence.

Sliding is particularly common in the British Isles, as is indicated by the high proportion of slides recorded in the recent DOE commissioned research *(Table 3.1)*. Simply stated, this is because of the availability of ground materials which facilitate basal shearing and the development of the all important basal shear surface or surfaces. The resulting slides are generally sub-divided into **shallow** or **deep-seated** categories according to the depth of the basal shear surface (slide depth). There is no accepted definition as to the maximum depth of a shallow failure, but the slide depth to length ratio (i.e. the maximum depth of the shear surface compared with the maximum downslope length of the shear surface) is normally between 0.01 and 0.15. In deep-seated failures the ratio is usually in the range 0.15 to 0.35 but can be as high as 0.50. Although there are exceptions, shallow sliding tends to be planar in form (**translational slides;** *Figure 3.7*) with curved shear surfaces (**rotational slides;** *Figures 3.10 and 3.11*) predominantly deep-seated failures.

To appreciate why slopes experience such a variety of sliding mechanisms, it is necessary to consider briefly the nature of slope-forming materials. This subject is examined in much greater depth in Chapter 4, so the purpose of this section is merely to illustrate why some slopes fail in translational slides and other by rotational movements.

The stability of slopes is determined by the relationship between the forces attempting to flatten the slopes (destabilising forces) and the resistance of slope-forming materials to such changes (resisting forces). This resistance, or strength, is determined by two measures, friction and cohesion. The frictional resistance to movement is the product of mineral particle contacts resulting from downward pressure (weight) and is usually referred to as the angle of internal friction (ø). It varies from material to material depending on sedimentological characteristics. Cohesion, on the other hand, measures the internal bonding of rock and engineering soil particles* produced by cements and chemical forces. It exists irrespective of loading, in direct contrast to frictional strength which increases with increased loading.

Earth material properties are very diverse, ranging from **cohesionless** or **non-cohesive** deposits (e.g. loose sands) to **cohesive** materials (e.g. stiff clays, hard sedimentary rocks, igneous rocks and metamorphic rocks: *see Table 4.1*). Shallow, planar slides develop in cohesionless materials but both types of failure can affect cohesive materials. The explanation for this variation is relatively straight forward.

In cohesionless soils the frictional strength of the material increases with depth in proportion to the effective overburden pressure, so that the only control on slope stability is slope angle. Should slope angle equal or exceed a critical value, then instability will result with failure occurring through the weakest horizon near the ground surface. Cohesive materials, by contrast, possess some strength at the ground surface. Their stability is determined by the relationship between shear stress and

---

* The term "engineering soils" is used to cover all ground materials excluding sound rock i.e. pedological soils, superficial deposits and weathered rock. Occasionally the term may be extended to include low strength clay-rich sedimentary rocks and unlithified sands. Geologists usually prefer to use the term "regolith" to cover the whole profile from ground surface to relatively sound rock.

.22 The development of deep-seated ailure in (a) the critical depth for ohesive materials (b) the variation of rincipal stress direction close to a slope.

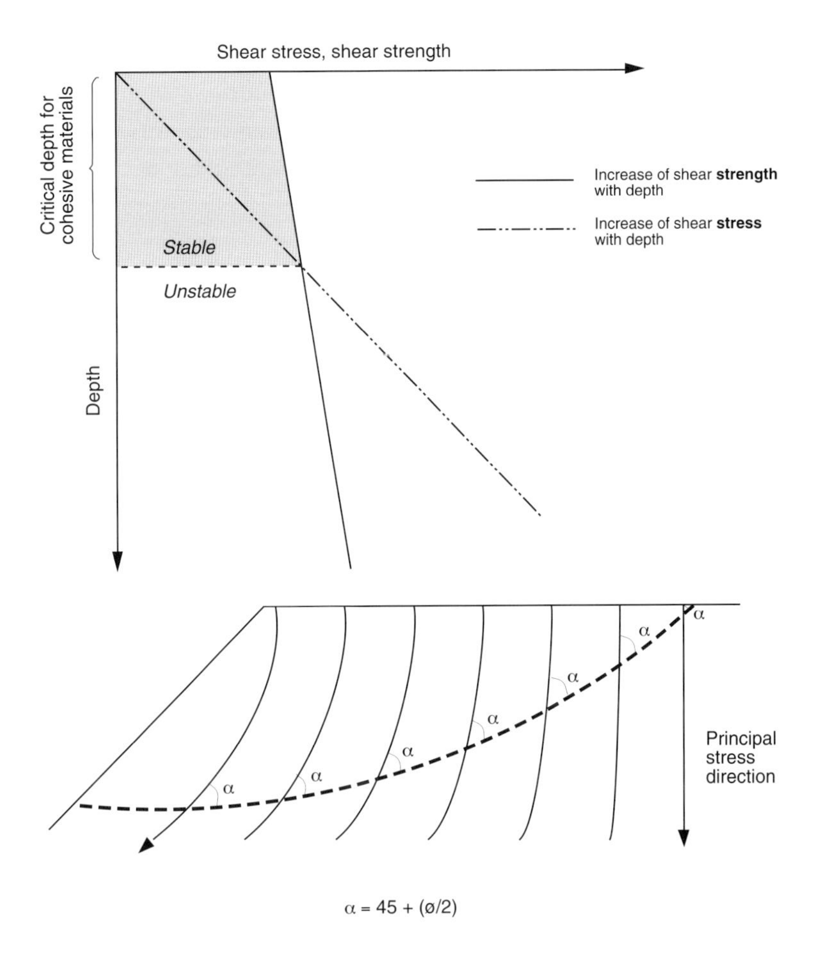

shear strength with depth. Should shear stress increase with depth at a greater rate than the increase in shear strength, then there will be a critical depth below which the slope is unstable *(Figure 3.22a; see Chapter 4)*. Thus slope height as well as slope angle are major controlling factors and deep-seated failures a more likely product.

Explanation of the curvature displayed by many deep-seated shear surfaces requires consideration of stress patterns within slopes. Beneath flat land the principal stress direction is vertical. However, along the edge of a valley, escarpment, bluff, cliff or excavation, the principal stress directions are curved towards the slope with depth because of the removal of passive support *(Figure 3.22b)*. It has been shown that shear planes tend to develop at a set angle ($45 + (\phi/2)$) to the direction of principal stress, thereby determining that they will be curved[40]. The simplest form is bowl-shaped with a cross-section that conforms to an arc of a circle, but for most rotational slides they tend to take a more elipse-like form. However, deep-seated slides may acquire even more complex forms, for whenever a line of weakness exists within the rock or soil, failure is likely to be associated with that line. Thus there is often considerable deviation from planar or circular failure surfaces due to structural or lithological factors, and this is one of the primary reasons for the existence of compound slides that are intermediate between the two main categories.

There are many other factors that affect the relationship between destabilising forces and resisting forces within slopes, as will be discussed more fully in Chapter 4, but one of the most significant is water. Water percolates into the ground and collects in fissures, discontinuities and the pore spaces between mineral grains where it exerts a pressure, known as **pore-water pressure**. The greater the head of water, the greater the pressure. In saturated soils, this pore-water pressure acts to separate the mineral grains, thereby reducing frictional strength. Thus the generation of landslides during and after prolonged or intense rainfall is rarely a result of saturation, waterlogging or lubrication as is frequently claimed in the media, but due to the creation of high pore pressures. The variations in water retention properties (permeability) with depth have therefore a profound influence on slope stability. In particular, the presence of clay strata is important because it acts as an aquiclude, inhibiting the downward migration of water. Similarly, interbedded sequences of porous sandstones and limestones separated by layers of impervious clay are known to be particularly susceptible to failure.

Because of the enormous variation in slope-forming materials and slope morphology, it is not surprising that slides have almost infinite variety of forms. The pragmatic approach adopted here (see earlier) is to consider slides on the basis of whether the basal shear surfaces are curved so as in cross-section they appear as an arc of a circle (**rotational** slides; *Figure 3.10*), planar (**translational** slides; *Figure 3.7*), or contain elements of both translational and rotational forms (**compound** or **translational (non-rotational)**: *Figure 3.12*).

(i) Rotational slides

Rotational landslides have received much attention in the literature, especially in the engineering and geotechnical fields, because they often occur in embankments, and the walls of cuttings through superficial deposits or rocks such as shales, mudstones and over-consolidated clays. This, in part, accounts for the prominence of this group of failures in the national landslide database *(Tables 3.1 and 3.2)*, where they account for 11% of all reported landslides, nearly 28% of landslides where type of failure is known and 31% of all specified failures, including those represented in complex landslides.

(a) single rotational slides

Single rotational landslides are considered by some to be the classic form of rotational failure but are often of relatively small size, hence the frequent use of the term "slumps". A total of 844 examples have been recorded in the landslide survey *(Table 3.2)*, the majority of which are in South-eastern England (207 or 25%), Wales (153 or 18%) and the Midlands (261 or 31%). Particularly large concentrations have been reported from Derbyshire (185), Mid Glamorgan (83), Essex (45), Kent (39) and Gwent (28), which merely serves to emphasise three points:

(i) the greatest concentrations of known landslides are almost certain to reflect the areas which have been searched/investigated most thoroughly;

(ii) single rotational failures tend to be widely developed where relatively homogeneous clay strata/superficial deposits occur, especially where these deposits are attacked by the sea (e.g. Essex, Kent);

(iii) single rotational failures also occur where sandstones and limestones are interbedded with clays as in the South Wales Coalfield and the Millstone Grit Series of the Pennines.

By extension, therefore, it must be assumed that single rotational failures are widely distributed throughout lowland Britain and much of Upland Britain *(see Figure 2.2c)* and that many more could exist than have yet been recorded.

In general, single rotational failures undergo one dominant phase of movement, often preceded by creep and followed by gentle settlement if the toe undergoes erosion[41]. They have most frequently been reported from excavated cuttings (e.g. the Sudbury Hill and Northolt slides of 1949 and 1955[42]), eroding river banks and coastal cliffs. Such slides are particularly well developed in coastal situations, especially the London Clay cliffs of Essex and Kent which account for 49 of the 146 recorded instances, the 'boulder clay' cliffs of Norfolk, Suffolk and Humberside (35) and the Lias cliffs of Dorset (14). Excellent descriptions have been provided of London Clay cliff failures, especially the Beacon Hill slide at Herne Bay[43], which rotated 25° indicating a displacement of 50m along the shear plane, and a series of deep-seated slides at Warden Point, Sheppey[44] *(Plate 5)*, where a number of houses have been lost due to rapid rate of cliff retreat.

The numerous inland occurrences of single rotational failures are largely made up of relatively small or un-noteworthy examples. Exceptions are the failures in glacial deposits at the site of the Selset Dam, Teesdale[45], and the large example that occurred at East Pentwyn, Gwent, on 22 January 1954[46].

(b) multiple rotational slides

Multiple rotational slides often develop as a result of retrogression of a single rotational slide. Well developed examples consist of a series of slipped, backtilted blocks, each underlain by a circular failure surface that merges tangentially to form a common basal shear surface *(Figure 3.11)*. In plan, they appear as a sequence of linear ridges, benches or terraces, arranged roughly as a staircase leading up to an arcuate rear scarp or backscar. Often the individual rotated blocks appear more disorganised but the general morphological characteristics are the same.

Multiple rotational landslides are generally large-scale features and are characteristic of sites where the sedimentary strata are horizontal or sub-horizontal. They are best developed where the underlying geology consists of a relatively thick layer of fissured clay or shale, underlain by a more competent rock and overlain by a strong cap rock[23]. The importance of the cap rock is that it probably limits the degradation of the scar produced by the initial rotational movements and thereby maintains the potentially unstable conditions at the head of the slope so that further rotational slides can develop. However, as the number of rotational movements grows, the potential for failure diminishes due to the passive support provided by previous failures; that is, until toe erosion at the base of the slope causes reactivation to retrogress upslope to the headscarp[47].

The most famous example is Folkestone Warren, Kent, where large rotational failures have developed in the Gault Clay and the overlying Chalk, above the more

3.7 The Landslip at Dunnose, 1810.

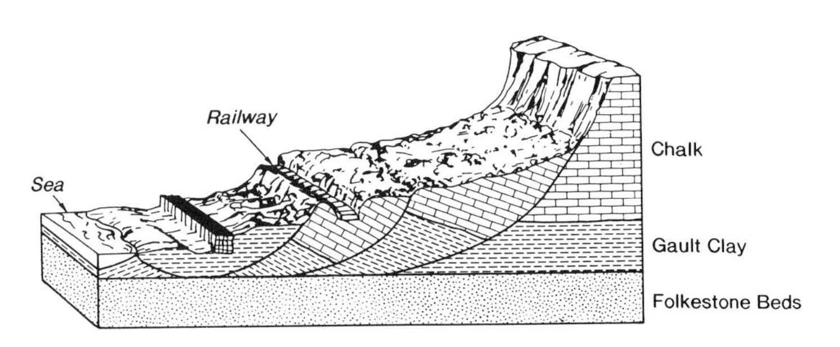

3.23 The multiple rotational landslide at Folkestone Warren, Kent (after Selby, 1982).

competent Folkestone Beds *(Figure 3.23)*. Landslide development is thought to have been the result of lateral expansion of the Gault Clay, accompanied by the generation landwards of a shear surface near the base of the clay[48] *(Figure 3.24)*. This lateral expansion created tension in the overlying Chalk which failed if there was insufficient support downslope, and continued until a situation was reached in which the support provided by the slipped blocks prevented further collapse of Chalk at the rear of the slide.

Large multiple rotational landslides have also been recognised along parts of the Isle of Wight Undercliff, particularly at The Landslip at Dunnose Point[49] *(Photo 3.7)*. Once again, the failure surfaces have developed in the Gault Clay and extended through the overlying Upper Greensand and Chalk. Graphic descriptions exist of the major retrogression of this landslide unit that took place in 1810 and 1818, extending over an area of 8ha[50].

Stonebarrow, Dorset, is a third excellent example *(Photo 2.5, Figures 2.13 and 3.25)* with Upper Greensand repeatedly failing over the Gault Clay. Two main sets of rotated blocks are presently visible, dating from the nineteenth century (pre-1887) and 14th May 1942, with previous examples displaced seawards and destroyed by other mass movements[47]. Investigations at this site indicate that it takes about a century for sufficient landslide debris to be removed for instability of the backscar to recommence.

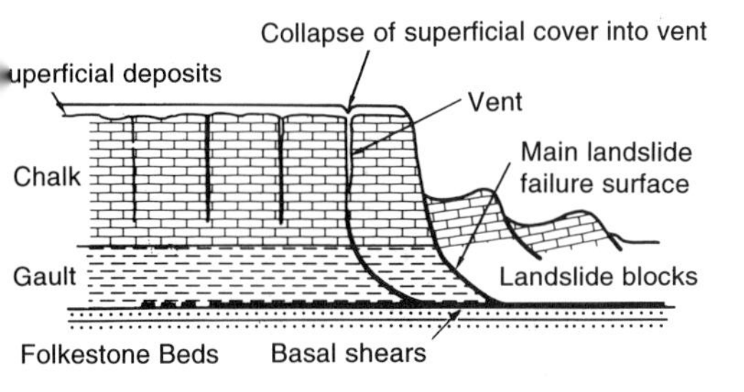

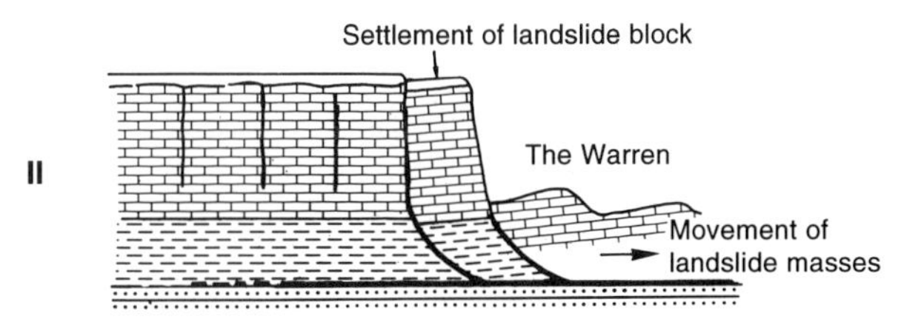

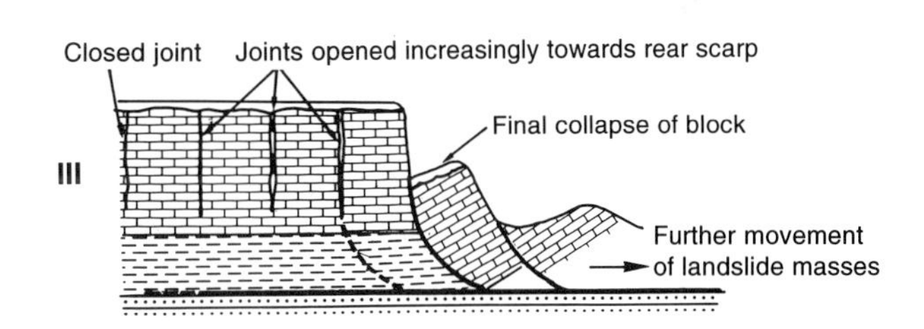

.24 Stages in the development of major slides at olkestone Warren, Kent (after Hutchinson, 1969).

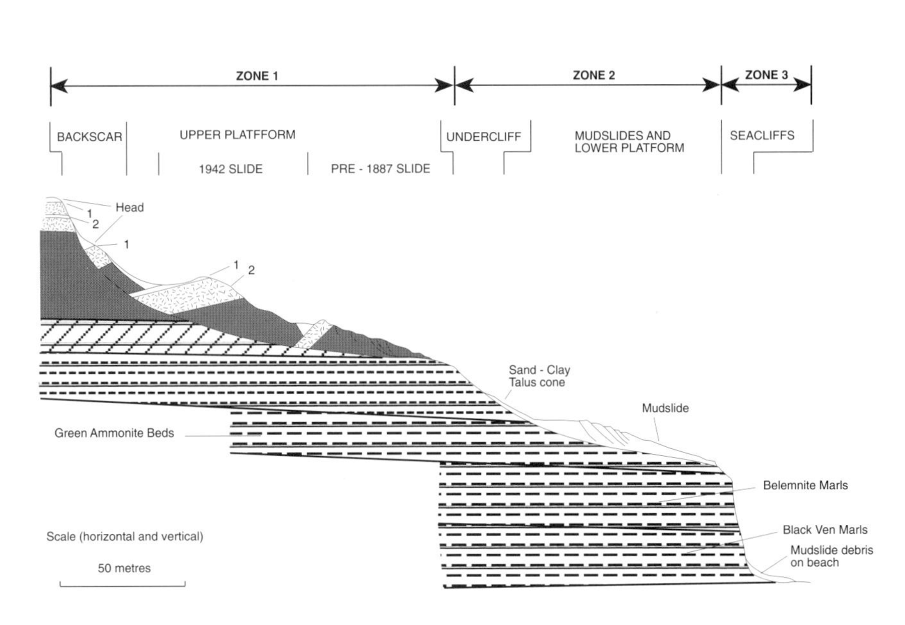

.25 Section through the tonebarrow landslide on the Dorset oast showing the relationship etween rotational blocks and udslide units (after Brunsden & ones, 1976).

Table 3.2 **The occurrence of rotational landslides in Great Britain**

County	Inland Rotational			Coastal Rotational			Combined Rotational		
	Single	Multiple	Successive	Single	Multiple	Successive	Single	Multiple	Successive
Avon	21	7	4	–	–	–	21	7	4
Bedfordshire	6	1	–	–	–	–	6	1	–
Berkshire	5	–	–	–	–	–	5	–	–
Borders	2	–	–	1	–	–	3	–	–
Buckinghamshire	4	–	–	–	–	–	4	–	–
Cambridgeshire	3	–	–	–	–	–	3	–	–
Central	1	–	–	–	–	–	1	–	–
Cheshire	1	–	–	–	–	–	1	–	–
Cleveland	–	–	–	–	1	–	–	1	–
Clwyd	4	–	–	–	–	–	4	–	–
Cornwall	–	–	–	3	1	–	3	1	–
Cumbria	10	1	–	–	–	–	10	1	–
Derbyshire	185	47	–	–	–	–	185	47	–
Devon	4	10	–	1	21	2	5	31	2
Dorset	5	19	21	14	30	2	19	49	23
Dumfries & Galloway	–	–	–	–	–	–	–	–	–
Durham	4	1	–	–	–	–	4	1	–
Dyfed	2	1	–	2	1	2	4	2	2
East Sussex	–	–	–	3	–	–	3	–	–
Essex	25	13	18	20	–	–	45	13	18
Fife	2	–	–	–	–	–	2	–	–
Gloucestershire	27	25	11	1	–	–	28	25	11
Grampian	–	–	–	1	–	–	1	–	–
Greater London	11	–	–	–	–	–	11	–	–
Greater Manchester	17	11	–	–	–	–	17	11	–
Gwent	28	5	4	–	–	–	28	5	4
Gwynedd	2	–	–	1	5	–	3	5	–
Hampshire	19	4	4	2	1	–	21	5	4
Hereford & Worcester	8	5	3	–	–	–	8	5	3
Hertfordshire	–	–	–	–	–	–	–	–	–
Highland	3	2	2	6	5	8	9	7	10
Humberside	2	2	–	9	4	–	11	6	–
Isle of Wight	5	–	–	8	12	–	13	12	–
Kent	10	–	4	29	7	2	39	7	6
Lancashire	3	2	–	–	–	–	3	2	–
Leicestershire	22	2	74	–	–	–	22	2	74
Lincolnshire	–	–	–	–	–	–	–	–	–
Lothian	1	–	–	–	–	–	1	–	–
Merseyside	1	–	–	6	–	1	7	–	1
Mid Glamorgan	82	10	7	1	–	–	83	10	7
Norfolk	2	1	–	10	–	1	12	1	1
North Yorkshire	24	18	–	10	9	3	34	27	3
Northamptonshire	22	2	28	–	–	–	22	2	28
Northumberland	2	1	–	–	–	–	2	1	–
Nottinghamshire	2	–	–	–	–	–	2	–	–

able 3.2 **The occurrence of rotational landslides in Great Britain (Cont'd)**

ounty	Inland Rotational			Coastal Rotational			Combined Rotational		
	Single	Multiple	Successive	Single	Multiple	Successive	Single	Multiple	Successive
rkney	–	–	–	–	–	–	–	–	–
xfordshire	1	–	–	–	–	–	1	–	–
owys	1	–	1	–	–	–	1	–	1
hetland	–	–	–	–	–	–	–	–	–
hropshire	20	9	1	–	–	–	20	9	1
omerset	2	10	–	–	–	–	2	10	–
outh Glamorgan	1	–	–	–	2	–	1	2	–
outh Yorkshire	3	–	–	–	–	–	3	–	–
taffordshire	–	1	–	–	–	–	–	1	–
trathclyde	8	–	–	1	4	–	9	4	–
uffolk	10	–	–	16	2	2	26	2	2
urrey	9	25	2	–	–	–	9	25	2
ayside	4	1	–	–	–	–	4	1	–
yne & Wear	–	1	–	1	–	–	1	1	–
Varwickshire	–	1	3	–	–	–	–	1	3
Vest Glamorgan	29	11	7	–	–	–	29	11	7
Vest Midlands	2	–	–	–	–	–	2	–	–
Vest Sussex	12	–	9	–	–	–	12	–	9
Vest Yorkshire	9	10	–	–	–	–	9	10	–
Vestern Isles	–	–	–	–	–	–	–	–	–
Viltshire	10	8	3	–	–	–	10	8	3
OTAL	**698**	**267**	**206**	**146**	**105**	**23**	**844**	**372**	**229**

In total 105 multiple rotational slides have been recorded around the coastline *(Table 3.2)*, including the two most extensive landslides in Great Britain, the Storr and Quirrang on the north-east coast of the Isle of Skye[51] (Plate 9; Figure 2.8), where the locally preserved geological sequence of Jurassic sediments partitioned by sheets of injected igneous rock (sills) and overlain by thick lavas, has proved particularly conducive to failure. However, these huge Scottish examples are something of an exception, for the geological conditions conducive to multiple rotational slide development are largely restricted to the south and east coasts, as is reflected in the statistics with the main concentrations of coastal examples in Dorset (30), Devon (21), the Isle of Wight (12), North Yorkshire (9; *Photo 3.8*) and Kent (7).

The presence of multiple rotational landslides on inland slopes usually gives rise to a much more subdued morphology. Rotated blocks tend to survive longer and therefore become broken up, reshaped and partly buried by the effects of other geomorphological processes. This was very clearly shown on Stonebarrow Down, Dorset, where the relatively gentle undulations on the sides of the Char Valley are the equivalent of the sharply defined features of the active coastal landslide complex[52] *(Figure 2.13)*. The Char Valley landslides are clearly of some antiquity for there are only a few signs of contemporary shallow instability. As similar slopes are widely developed in West Dorset, it must be assumed that multiple rotational

3.8 Multiple rotational landslide at Scarborough Castle, North Yorkshire (Cambridge University Collection: copyright reserved).

landsliding was widespread at certain times in the past. However, as only 19 inland examples have been recorded for Dorset, it must be assumed that many of these areas of ancient landsliding remain to be investigated and reported.

The total of 267 recorded instances of inland multiple rotational landsliding *(Table 3.2)* indicates both the widespread presence of this type of failure and the fact that certain areas have been thoroughly investigated. The large number recorded in Derbyshire (47) reflect their widespread development on the lithologically variable Millstone Grit Series of the southern Pennines. Sometimes these failures attain considerable size, as is illustrated by the Binns landslide[53] and they are by no means confined to the Derbyshire portion of the Pennines, as is reflected in the moderate numbers reported for West Yorkshire (10), Greater Manchester (11) and North Yorkshire (18). Other reported concentrations include the South Wales Coalfield[54] (26); Bredon Hill; the Cotswolds escarpment, especially at such locations as Cleeve Hill and Dovers Hill[55]; the Bath area at sites such as Beacon Hill, Bailbrook and North Stoke[56]; and the continuation of the Jurassic outcrop through Somerset and Dorset.

Numerous examples of multiple rotational landsliding have also been reported from the Lower Greensand Hythe Beds escarpment of the Weald, where large failures have developed in the underlying Atherfield and Weald clays. These were first reported in the westernmost part of the Weald where extremely large failures were recorded at Blackdown Hill and Telegraph Hill, overlooking the Vale of Fernhurst[57], and subsequently identified along the escarpment to the east through Surrey and Kent, to the abandoned sea cliffs behind Romney Marsh between Lympne and Hythe.

(c) successive rotational slides

It is apparent that some confusion exists as to the difference between multiple and successive rotational slides, in that multiple are sometimes referred to as successive because the failures occur 'in succession'. In reality, the two are very different, for successive rotational slides are usually small, shallow failures confined to the weathered mantle[23]. They are characteristic of degrading slopes developed in stiff fissured clays and are normally arranged 'head-to-toe' up the slope, thereby covering areas of considerable extent with regular steps across the slope *(Plate 18)*. However, irregular successive slips which form a mosaic pattern have also been reported.

The occurrence of successive rotational slides may be viewed as part of the long-term slope degradation process as established for abandoned cliffs of London Clay in Essex and Kent[58]. Free degradation in these clay slopes occurs in a number of stages involving the sequential development of shallow single rotational slides, successive rotational slides and undulations of the ground surface *(Figure 3.26)*. There is a close relationship between these different slide forms and slope angle, with the early stages of degradation achieved by single rotational slides. Initially, these slides encroach upon the cliff top but their size subsequently decreases as the slope angle is reduced until only a part of the slope is involved. Below a slope angle of about 13°, single rotational slides give way to successive rotational slips, which probably develop by retrogression from an initial failure on the lower parts of the slope. Successive rotational sliding proceeds until the slope has declined to about 8°, when the slides are converted into undulations and finally into a smooth slope at the long-term angle of stability against landsliding.

This free degradation process is accompanied by a build up of debris at the base of the slope in the form of an **accumulation zone** which protects the lower parts of the slope and progressively inhibits the development of widespread failure. At the same time, the upper parts of the slope, known as the **degradation zone**, declines in slope angle. Ultimately, free degradation results in a straight slope inclined at an angle

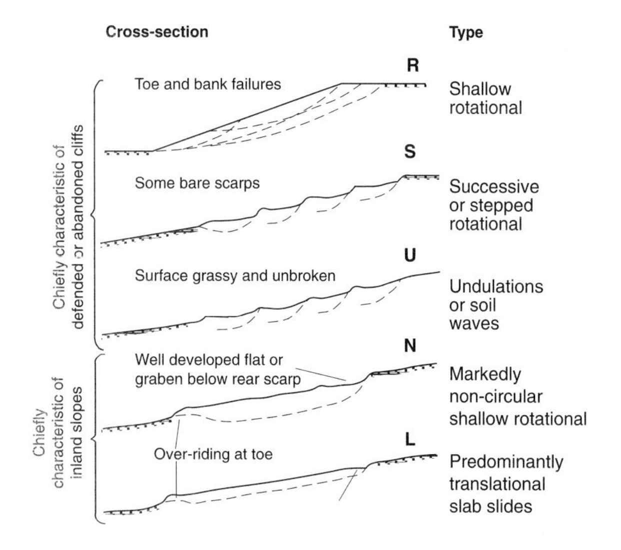

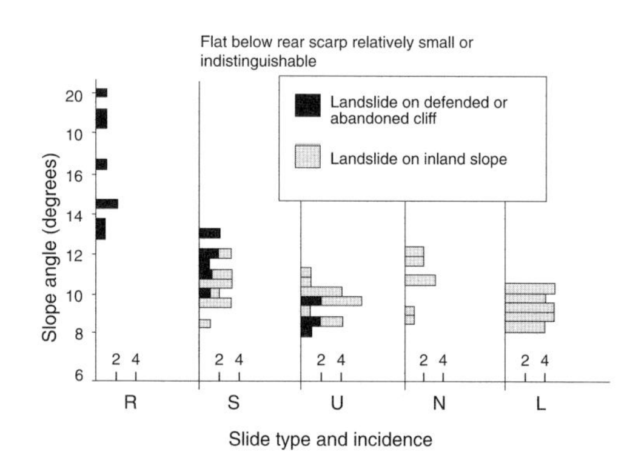

3.26 Main types of landslide observed on London Clay slopes (after Hutchinson, 1967).

corresponding to the residual strength of the material and the steady state groundwater conditions. This sequence has also been well described from the abandoned marine cliffs developed on the Hythe Beds, Atherfield Clay and Weald Clay below Lympne Castle, Kent[59] *(Plate 11)*.

Similar patterns of degradation have been recognised on other clay slopes, including the Upper Lias Clay of the Jurassic escarpment near Rockingham, Northamptonshire[60]. At this site successive rotational slides and translational failures are present on slopes of 8.5–9°, which is near to the ultimate angle of stability for the Upper Lias Clay, as observed at similar slopes in the Northamptonshire-Leicestershire area[61]. Along the Cotswolds escarpment successive rotational slides have been identified on lower slopes below large-scale rotational failures[55]. It appears that the initial multiple failures of the escarpment degrade through a sequence of progressively shallower, part-successive rotational failures and part-translational failures, until a stable angle is achieved *(Figure 3.27)*.

(ii) Translational slides (planar)

Stresses acting on slope-forming materials often result in their displacement downslope over a relatively planar shear surface located at shallow depth and oriented approximately parallel to the ground surface. This form of mass movement affects a great range of different materials – rock, clay, regolith, soil and even peat. As a consequence, it has come to be recognised as the most widely developed and common style of landsliding developed in temperate latitudes; a conclusion supported by the recent landslide survey which recorded 956 examples of translational slides (planar), representing over 20% of landslides for which type information is available *(Table 3.1)*. However, both these figures are likely to be underestimations, as this form of mass movement is often small-scale and therefore tends not to be recorded unless it occurs in an area that by chance happens to be subject to detailed investigation. The fact that evidence for 309 small examples was identified within an area of 25km^2 in central Devon as a consequence of such an

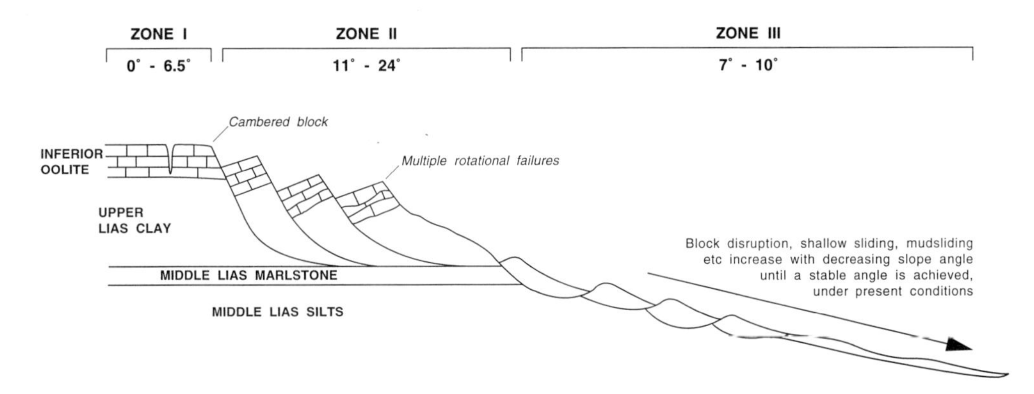

3.27 The degradation of valley slopes in the Cotswolds, showing the relationships between different landslide types (modified from Butler, 1983).

investigation[62], suggests that the actual number present on the surface of Great Britain could be huge*.

Because translational sliding develops in so many different materials, the resultant morphological characteristics and behaviour also vary greatly. Therefore, it is necessary to consider this movement mechanism under the following headings:

1. rock slides and wedge failures;
2. debris slides;
3. slab slides;
4. spreading failures;
5. mudslides;
6. peat slides;
7. block slides/block glides.

1. rock slides involve failure of masses of rock along planar shear surfaces which are usually related to discontinuities such as bedding planes, joints, cleavage or foliation planes. Therefore, the presence of such slides is not usually a reflection of lack of strength of the rock itself, but highlights instead the importance of internal weaknesses, i.e. the intact (free from any discontinuities) rock strength of the rock is of much less significance in determining rock mass strength than the frictional strength of its discontinuities.

One of the most important controls on rock slope failures is the orientation of the major discontinuities with respect to the ground surface. Most rock slides are associated with failure planes developed along naturally occurring major discontinuities oriented sub-parallel to the slope, but inclined downslope less steeply so that they intersect the ground surface (i.e. daylight). Thus the steeper the ground surface, the greater the probability that rock slides will occur. This, in turn, implies that rock slides will be most common on the 'hard rock' coastal cliffs of western and Northern Britain *(see Figure 2.2)*, on the steep sides of glacially overdeepened valleys in Wales, the Lake District and the Highlands of Scotland *(see Figure 2.5)*, on the flanks of river gorges cut through upstanding masses of resistant rock (e.g. the Wye Valley) and in excavations.

* In the National Landslide Databank these numerous small failures have been grouped together as 4 "areas of landsliding" to minimise the influence that such a detailed investigation of a very limited area has on any subsequent data analysis.

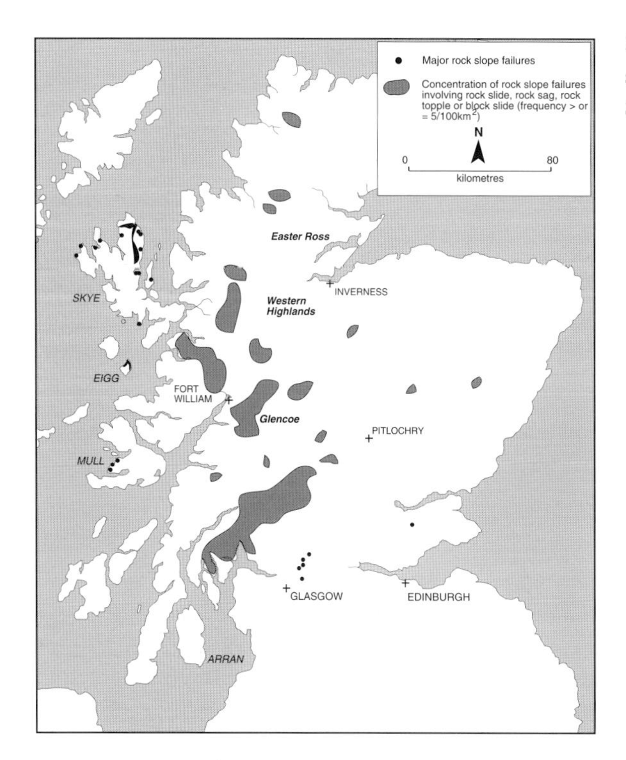

3.28 The distribution of rock slope failures in Scotland (after Smith, D.I., 1984).

Two main types of rock slide are common in Great Britain; **planar rock slides** *(Figure 3.6)* and **wedge failures** *(Figure 3.3)*. Planar rock slides are common where hard rocks have well developed discontinuities. For example, numerous large rock slides have been reported in the Scottish Highlands where their distribution is closely related to local geological and topographical conditions[63]. The majority of such failures *(Figure 3.28)* occur in areas of good topographic relief with steep slopes underlain by Moinean and Dalradian metamorphic rocks, particularly schists, which tend to have lower friction angles than other rocks, such as granites, and also possess pronounced foliation planes that form potential shear surfaces. It has been suggested that the dip of these planes relative to the valley slopes has a great influence on landslide type, with steeper valleyward dips being conducive to translational sliding whereas shallower dips favour failure by toppling.

One of the largest areas of rock sliding in Scotland occurs on the west side of Glen Ogle, 2km north of Lochearnhead, developed in mica-schist. The main rock slide extends 450m downslope by 300m wide, and comprises a broad frontal accumulation zone below an extensive area of instability crossed by numerous fissures up to 3m wide *(Plate 19)*. The railway line to Killin crossed the middle of the largest slide and construction resulted in the initiation of two small failures[64]. The railway was operational until 1965 when it was abandoned following a major slide *(see Chapter 7)*.

Approaching 200 rockslides have been recorded in the Highlands of Scotland, the majority of which appear to have occurred prior to 5,000 years ago. Indeed, a high proportion of the recorded rockslides are located within 600m of the mapped limits of the last minor ice advance[65], the Loch Lomond Advance or Loch Lomond Stadial

(11,000–10,500 years ago; *Figure 2.4*), indicating that the influence of ice erosion and the freezing of water in joints may have combined to generate instability*. However, the effects of repeated glaciation has been to create numerous steep rock slopes in conditions of critical conditional stability, so it seems probable that movements will continue into the future, especially if disturbed by excavation.

There are several examples of recent human-induced rocksliding. One of the most problematic was generated during improvements to the A40 near Monmouth in the early 1960s, when cutting slopes intersected steeply dipping, thinly bedded rocks which continued to unravel until stabilised by extensive, and expensive, remedial measures[66].

Often rocks are deformed to such an extent that discontinuities dip out of the slope at varying degrees to the horizontal. Under these circumstances, planar failure is replaced by the sliding of prismatic blocks of rock whose boundaries are defined by two sets of intersecting discontinuities, otherwise known as **wedge failures** *(see Figure 3.3)*. This type of failure is surprisingly common and the identification of potentially unstable wedges is an important element in the three-dimensional stability analysis of cut rock slopes[5]. Wedge failures in road cuts have been reported on low grade metamorphic rock in North Wales[19] and on the M5 motorway route through the Clevedon Hills[67], where they had to be stabilised by the use of rock bolts anchored in polyester resin.

2. **debris slides** generally involve shallow translational failure of the weathered materials and superficial deposits which mantle slopes *(Plate 20)*. Such slides are characterised by low depth to length ratios, often less than 0.05. The most common form of failure is slow movement along relatively shallow shear surfaces parallel to the ground surface. This type of failure is common on slopes where there is a marked change in strength and permeability with depth, as described for slopes in Exmoor[68]. Conditions for failure are at an optimum when the regolith is at residual strength *(see Chapter 4)*, where the water table is at, or close to, the ground surface and where water movement is parallel to the slope. Not surprisingly, such slides often occur when the ground is saturated after periods of heavy or prolonged rainfall.

It is difficult to generalise about debris slides because of the wide range of materials involved and the markedly differing slope angles. In general terms, the more clayey the medium and the gentler the slope (above the limiting angle for failure), the slower the movement. Normally the head is defined by a low arcuate scarp. The track of the slide is frequently only patchily covered by debris so that the basal shear surface may be visible. The depositional zone or toe area ranges in character from a low, asymmetrical mound (toe lobe with steepest slope facing downslope) crossed by shears in the case of clayey materials, to a jumbled mass of ridges, humps and hollows in the case of coarser deposits.

---

* It must be recognised that the limit of the Loch Lomond Advance has been the subject of considerable investigation in recent decades, so it is likely that rock slope failures may have been better reported in these areas than elsewhere.

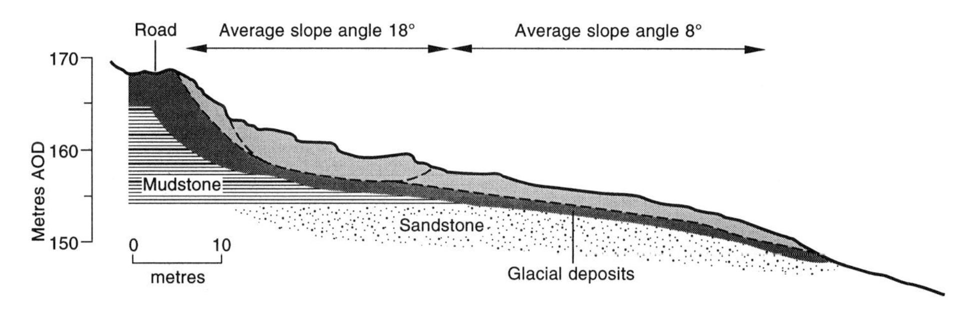

3.29 The shallow translational landslides at Glynrhigos Farm, South Wales (after Northmore & Forster, 1986).

It has been estimated that over 75% of the landslides recorded in the South Wales Coalfield are shallow translational failures in weathered material and solifluction deposits[69]. An excellent and well studied example is the Glynrhigos Farm landslide on the western side of the Dulais Valley to the north of Cilfrew. This is a complex of shallow debris slides *(Figure 3.29)* characteristic of many other landslides in the area, where movements have led to the creation of irregular lobes covering an area of 250m x 200m. Since detailed investigations by the British Geological Society began in 1979, several movements have occurred resulting in the creation of tension cracks, low scarps, bulging of banks and heaving of a minor road. The movements are wholly in the thin covering of superficial deposits that mantle the bedrocks in the area, and appear to be the reactivation of very ancient features produced under freeze-thaw conditions.

The existence of widespread active shallow mass movements in this area is not a new discovery, for a paper published in 1927[70] described a number of shallow translational slides in the South Wales valleys, including the more problematic examples at New Tredegar, Pentre and Blaina. The Blaina site has been investigated in recent years and it is now concluded that reactivation of this feature was partly the result of loading with colliery spoil[71]. Extensive research in the area has revealed that there is a close relationship between slope angle and landslide type, with planar failures in superficial deposits generally occurring on slopes of 17-19°[69]. As many slopes in the area are steeper than 17°, the potential for movements is clearly great. In addition, many ancient slides have been identified resting at angles above the limiting value, thereby highlighting the potential for reactivation through disturbance, as occurred during the construction of the M4 motorway near Port Talbot[72].

An important recent study of shallow translational debris slides focussed on the slope stability of Upper Carboniferous mudrocks in central Devon where such failures are now known to be surprisingly common[62]. Individual translational slides in this area are generally 10–20m wide, 15–40m long and about 2–4m deep, although multiple slides are quite common[73]. The slides are wholly developed in the weathered mantle, are slow moving, and often comprise a low arcuate scarp feature with a toe bulge which rarely exceeds 1m in height. However, many of the slides are combinations of shallow translational slides at their heads and flows downslope with no definite toe feature, and therefore resemble mudslides (see later). In general, sliding results from artesian pore-water pressures at the base of the soil. The geotechnical significance of this form of unstable and marginally stable ground has for long been appreciated in the site investigations for new road construction in the area[74], but several new housing estates in the suburbs of Exeter have encountered problems due to reactivated landsliding[73].

3. slab slides occur on clay slopes, especially where the clay has been weakened by fissuring and softened by water percolation. As these processes are more effective in the surface layers, they result in shallow rather than deep-seated failures, which are often very long and wide compared to their depth. They are, therefore, similar to debris slides developed in clayey deposits, except that the slab of moving material is much more extensive and behaves as a coherent unit.

Slab slides have been surprisingly widely reported from very different locations. Several examples have been described on freely degrading slopes developed in London Clay, where their incidence appears restricted to slopes of between 8° and 10.5°[58], thereby indicating that they are associated with the later stages of slope evolution following slope angle reduction by other mechanisms, especially rotational landsliding *(Figure 3.26)*.

Other well known examples include the Jackfield landslide which occurred in 1951–1953 in the Ironbridge Gorge on a 10.5° slope developed in Upper Carboniferous clays[75] *(see Photo 2.12 and Figure 3.30)* and a similar slide on a 9° slope of weathered Upper Lias Clay near Uppingham, Leicestershire[76]. More recently, two conspicuous slab slides were reported on cutting slopes associated with the A45 Daventry Bypass[77], also in Upper Lias Clay. Excavation works at the Fox Hill site initiated the failure of a 50m long slab of hillside, along a planar shear surface within a block of Upper Lias Clay which had been previously displaced by an earlier phase of landsliding *(Plate 21)*. Similar features were discovered during the excavation at nearby Newnham Hill *(Plate 22)*, where an 80m long lozenge-shaped slab failure occurred along a shear surface 7m below ground level.

30 The shallow translational landslide at Jackfield, onbridge Gorge, 1953 (after Henkel & Skempton, 1954).

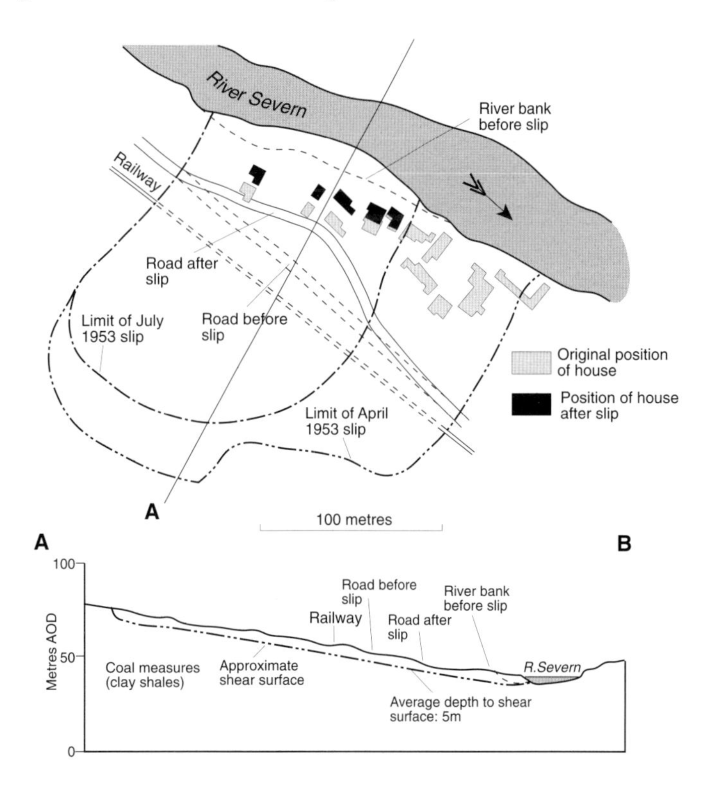

4. spreading failures are relatively large landslides characterised by a succession of cross-slope upstanding blocks (horsts) and intervening areas of lower ground that have subsided (grabens), produced by extension of the failed mass. This type of failure is very rare in Great Britain, only occurring in areas underlain by sensitive clays which suffer liquefaction (turn into a dense liquid) if exposed to failure or shock. If this happens, failure spreads rapidly through the clay strata causing overlying materials to move about as blocks or rafts and quickly producing a chaotic topography of small upstanding masses separated by depressions (horsts and grabens). Features akin to spreading failures have been reported during construction of the M4 motorway across the Gault Clay at Burderop, Wiltshire, and as a minor part of the 1987–1988 movements at Luccombe, Isle of Wight[78].

5. mudslides are highly distinctive mass movement features that have received considerable attention in the scientific literature following the careful analysis by Hutchinson and Bhandari which established the distinction between mudslides and mudflows[27]. They are now clearly defined as a slow moving form of mass movement in which softened clayey, silty and fine sandy debris moves downslope mainly by sliding over a basal shear surface[23, 79]. The resultant feature may be likened to a conveyor-belt *(Plate 23)*. The head is normally bowl-shaped, with an arcuate headscarp which collapses by slumping to feed the mudslide, although material can also be supplied by rotational slides, translational slides or debris slides. The track of the mudslide *(Figure 3.9)* is usually much longer than it is wide and has clearly defined margins marked by strong lateral shears. Long profiles vary, but many have minor undulations marked by cross-slope tension cracks or minor pressure ridges. Many mudslides feed into rivers, over cliffs or directly onto the shore, so that depositional features are often poorly developed, but if the moving material eventually comes to rest, the result is single or overlapping lobes which can attain enormous dimensions, as at Black Ven, Dorset *(Plate 7)*.

Mudslides are best developed in stiff, fissured clays which have been progressively softened, weathered or broken up by movement. However, examples have been recorded as developed in glacial tills, mudstones, siltstones and sandy clay debris. Movement regime is also variable. Most are perennial and have active lives that last for tens of years. However, activity is often seasonal or cyclical, with dormant periods during the summer months, dry periods or during times when debris supply to the source area is limited.

During active phases movement is rarely dramatic, perhaps 10–20 metres in a year, with pulses of a metre or so a day during very wet periods and occasional surges of 10–20 metres in a week. However, the rate of activity is extremely variable, both at a site and between sites, reflecting the supply of debris, the available moisture, gradient of track and size of the feature[79]. Thus the speeds recorded from the huge mudslides at Black Ven *(Plate 7)* will be significantly greater than those for the smaller features on nearby Stonebarrow *(Photo 2.5)* and orders of magnitude greater than the small, inland example on Bredon Hill, known as the Woollashill mudslide *(see Plate 13; Figure 2.12 and accompanying text)*.

Although mudslides vary through a spectrum of sizes, two characteristic groups are recognised on the basis of length of track; **elongate** and **lobate** mudslides. Typically elongate forms are long (800–2,000m) narrow features found on long slopes which are steeper than the limiting angle for movement. Examples of this type have been recorded on the Permian breccias in Torbay[80] and at Churchdown Hill, Gloucestershire, where they have developed in Middle Lias clays[81]. By way of contrast, lobate forms are generally short (20–80m) and commonly occur on short steep coastal slopes where the track is terminated by a cliff or the sea. Excellent examples have been reported from the London Clay cliffs of north Kent near Herne Bay[82] *(Photo 2.2)*, developed in Oligocene clays at Bouldner Cliff, Isle of Wight[83], from Stonebarrow, Dorset[52] *(Photo 2.5; Figure 3.25)* where they are developed in Middle Lias Clay, and at West Down Beacon on the east Devon[84] coast where they are developed in Permo-Triassic conglomerates and mudstones.

Without doubt the largest and most dramatic area of mudsliding in Great Britain is at Black Ven on the Dorset coast, east of Lyme Regis *(Plate 7)*. Here a vast amphitheatre contains numerous highly active mudslides which slide over a step-like sequence of lithologically controlled benches to feed three master slides which convey material to the sea[85]. The upper sections of the landslide complex are developed in Cretaceous Upper Greensand and Gault Clay, which overlie Lower Lias clays and thin limestone bands *(Figure 3.31)*. Much of the contemporary activity originates in the upper regions of the cliff, where rotational failures result in the accumulation of remoulded debris on the upper benches. This material is supplied with water from an aquifer in the Cretaceous strata and mudslides develop carrying the debris to the sea. The history of recent movements has been well documented[86] and Figure 3.32 presents a compilation of the available historical data. The threat to a small housing estate at Charmouth *(Figure 3.32)* due to mudslide reactivation was investigated in the early 1970s[87] and stabilisation measures emplaced. However it is the development of a major mudslide at the western end of the complex since the winter of 1986–7 that has proved the greatest hazard to property.

The localised development of lobate mudslides on the London Clay cliffs to the west of Herne Bay, Kent *(Photo 2.2)* has been the cause of debate. Originally it was argued[88] that such development occurred where the rate of marine erosion was broadly in balance with the weathering of the slope. At sites where the rate of marine erosion exceeds weathering, then there is a tendency for the slopes to fail by

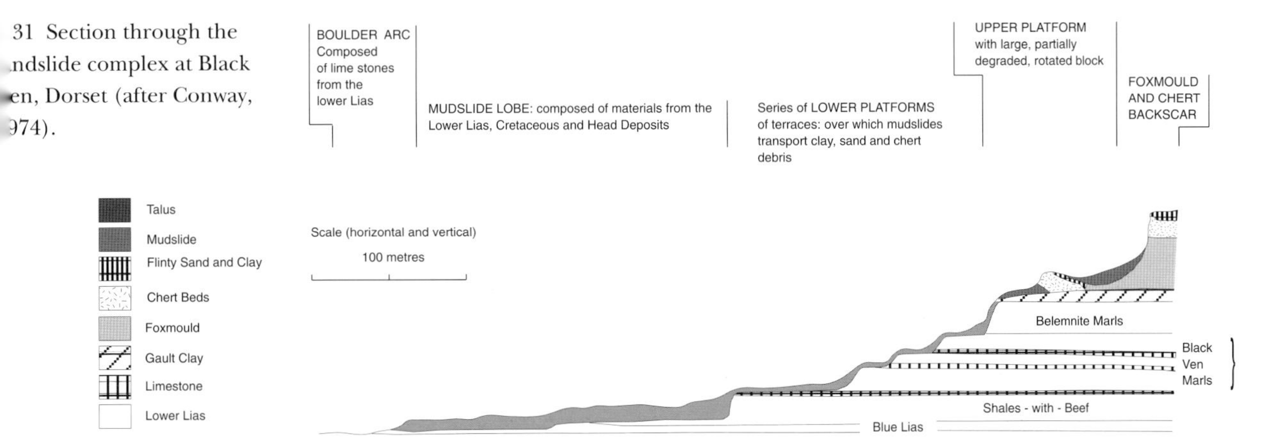

31 Section through the ndslide complex at Black en, Dorset (after Conway, 974).

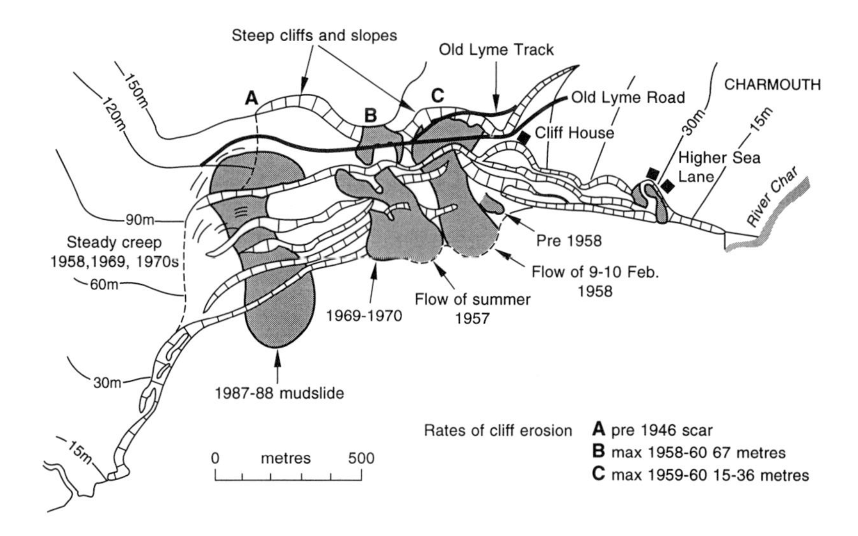

3.32 The evolution of the landslide complex at Black Ven (after Brunsden & Goudie, 1981).

deep-seated rotational slides. However, it has subsequently been suggested that the behaviour of cliffs developed in clays is also controlled by the groundwater conditions at the head of the slope[44]. This view is supported by work on the degradation of the Barton Clay cliffs at Hampshire, where lobate mudslides were identified at Highcliffe, Christchurch Bay[89]. The location of these mudslides is apparently controlled by the increased outflow of water at these sites, rather than variations in the rate of toe erosion. Similar conclusions were reached regarding the active mudslide systems at Eastcliffe, Lyme Regis, where recent instability has been caused entirely as a result of the introduction of water at the crest of the slope[90].

The significance of high pore water pressures at the head of a mudslide system is also demonstrated by the Lansdown mudslide developed in Fullers Earth clay in Bath[91]. This slide was triggered by the dumping of rubbish in an abandoned limestone quarry between 1963–1969, which blocked the drainage from springs issuing at the base of the Great Oolite limestone.

The majority of active mudslide systems have been reported from coastal cliffs developed in clays. Inland examples are few, but there is clear evidence that the presently active features *(e.g. the Woollashill mudslide on Bredon Hill, Plate 13; Figure 2.12)* are the surviving members of previously much more numerous features. Indeed, many inland clay slopes are mantled by solifluction deposits, otherwise known as head, a heterogeneous assemblage of weathered bedrock debris and superficial materials (e.g. glacial till, terrace gravels) often containing indistinct shear planes. Clay-rich solifluction deposits are now recognised as the products of widespread mudsliding under periglacial conditions, when snowmelt and the seasonal thaw of the upper levels of permafrost created a saturated mobile layer that slid and sludged downhill. As solifluction episodes occurred frequently during the later Pleistocene *(see Chapters 5 and 6)*, and affected the whole country, head deposits are ubiquitous. Although solifluction materials normally occur as sheets or fans of material, occasionally more morphologically distinctive features are encountered. Ancient mudslide lobes occur in places below landslide prone escarpments, the most famous example being just to the south of Sevenoaks, Kent, where inadvertent reactivation by bulldozers in 1966 resulted in the eventual realignment of the A21

Sevenoaks Bypass *(Figure 3.13)*. Detailed subsurface investigations revealed that these features had been stationary for 10,000 years prior to attempts to build the bypass[92].

6. peat slides; extensive areas of thick peat cover many upland parts of Wales, the Pennines, the Lake District, the southern Uplands and the Highlands of Scotland *(Figure 3.33)*. These organic deposits are subject to failure as are any ground-forming materials, although the nature of the materials determines that the characteristics of mass movement are slightly different.

Three types of failure have been recognised: bog bursts, bog flows and peat slides[93]. The limited number of records of such features makes it convenient to consider them together. Bog bursts and bog flows are usually confined to peat bogs in mountainous areas where heavy rainfall often causes the bog to swell until the crust ruptures and releases semi-fluid peat. There have been few records of such events in Great Britain, but descriptions do exist of events in Baldersdale, Upper Teesdale[94], and on the Isle of Lewis[95].

Peat slides are usually developed in areas of eroding peat. Observations indicate that the basal shear surface normally lies just below the peat-underlying clay interface. Several examples of peat slides have been recorded, usually following thunderstorm rainfall, including Stainmore on 18.6.1930[96], nearby Maldon Hill on 6.7.1963[97] and on the flanks of Noon Hill, Upper Teesdale, on 17.7.1983[98] *(Plate 24)*. Construction work also triggered a number of large peat slides during flood defence work at Borth, Dyfed[99], one of which was unusual in that it developed retrogressively suggesting that some liquefaction of the underlying clay might have been involved.

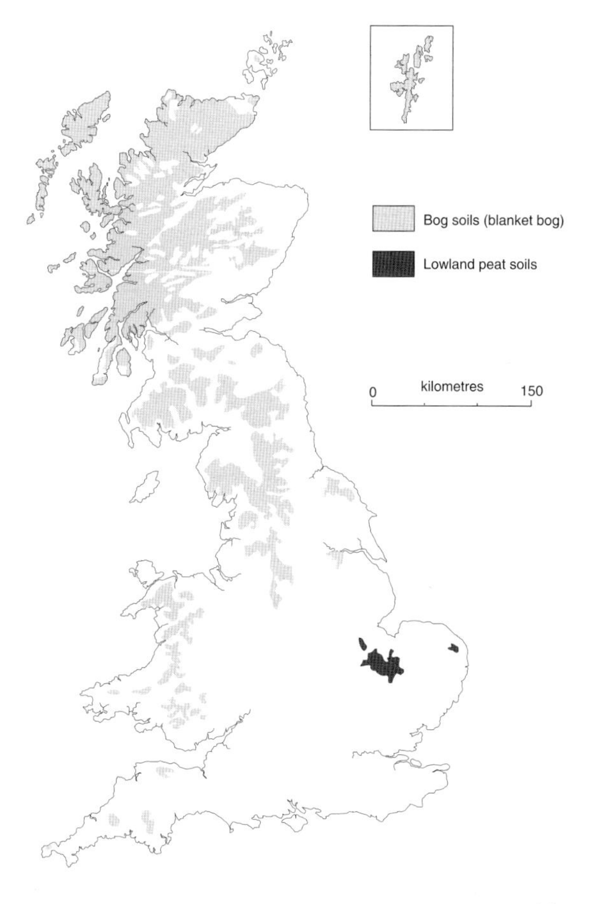

3.33 The distribution of peat soils in Great Britain.

7. **block slides** or **block glides;** the term block glide was originally coined to describe failures in which material remains largely undeformed as it moves over a shear surface. Blocks of material are detached from the main mass and displaced almost horizontally, with little or no tilt, leaving chasms or grabens in their rear. This kind of displacement is rarely observed and is mainly a small-scale feature associated with the disruption of soils and superficial deposits. The term block glide has also been associated with extremely large features where slabs of rock have moved over deformable clays. The classic example is the Bindon landslide on the Dorset coast *(Photos 2.6 and 2.7; Figure 2.7).*

(iii) Compound slides (translational non-rotational)

As was stressed earlier, mass movement features occur through a spectrum of failure types many of which do not fall conveniently within proposed classifications. There are many rotational slides which have a translational component, many translational slides that display signs of rotation and certain slides that appear to have equal amounts of both, thereby resulting in the occasional employment of unsatisfactory terms such as 'rotational translational slide' or 'transrotational slide'. As a consequence, there has been a movement to adopt an intermediate category of **translational (non-rotational)** or **compound slides** to describe features that display both translational and rotational characteristics.

Compound landslides are characterised by markedly non-circular shear surfaces formed by the combination of a steep curved or planar rearward (upslope) portion and a relatively flat sole *(Figure 3.12).* The position and form of the steeply inclined rearward portion of the shear surface is usually determined by discontinuities, such as stress release joints, and the two main components of the shear surface are normally linked by an element with a sharp radius of curvature. Displacement of material is therefore accompanied by some rotation, but this is limited to the headward areas.

Compound landslides are particularly common in stratified soft rocks and overconsolidated soils with sub-horizontal bedding[100]. The form of the failure clearly reflects structural control and the fact that bedding planes are usually the weakest zones within soil or rock masses. The lower strength along the bedding is sometimes due to previous shearing related to landsliding or geological movements, or differential softening as a result of seepage. In certain instances, such as the failures in Barton Clay at Highcliffe on the Hampshire coast[101], the position of the basal shear surface is determined by the elevation of weak bedding planes rather than the magnitude of shear stress.

In contrast to rotational and planar slides, in which movement can occur with minimal distortion of the failed mass, compound landslides are locked in place as a result of their unusual shear surface geometry and can only move when the slipped mass is divided by the development of internal shears[23]. The main internal dislocations generally develop in the rear portion of the slide, which often sinks down with little or no backward rotation to form a graben, whilst the remainder of the slide moves forward along the bedding *(Figure 3.12).*

3.9 The 1953 Miramar compound landslide developed in London Clay, Herne Bay (Acknowledgements to the Kent Messenger).

One of the best documented examples of non-rotational failure is the Miramar landslide, Herne Bay, Kent, which developed in London Clay on 4th February 1953. This landslide took place at the site of an earlier failure (c. 1883) which caused 20m of ground to be lost along a 100m section of cliff[102]. The 1953 landslide involved the settlement of a section of cliff between sharply dipping shear surfaces to form a near horizontal graben, whilst the original cliff face was pushed forward intact to form a ridge[43] *(Photo 3.9)*. Analysis of the failure[44] established that the rear of the slide had been displaced nearly 18m and that the position of the basal shear surface was controlled by the silty fine sands of the Oldhaven Beds[103] *(Figure 3.34)*.

.34 Sections through the 1953 ompound slide (the Miramar ndslide) in Herne Bay, Kent (after romhead, 1978a).

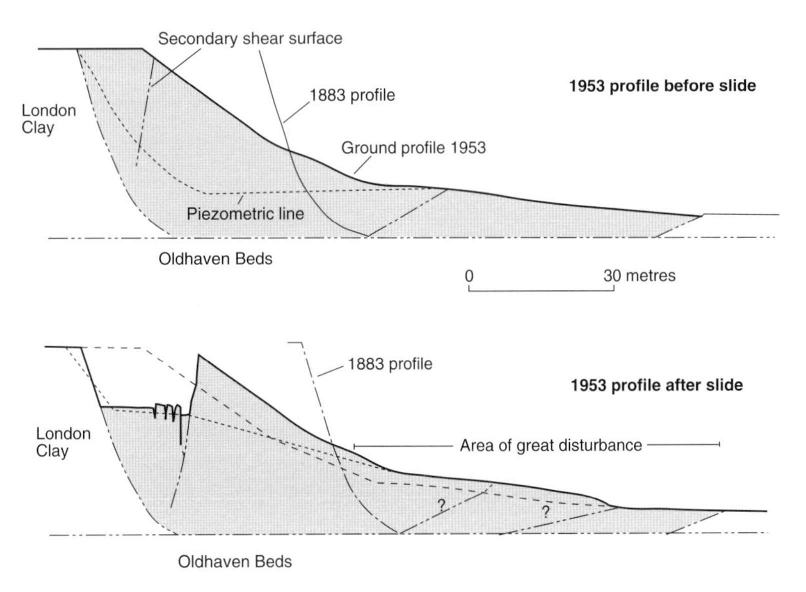

3.10 The St Catherines Point compound landslide, Isle of Wight (Cambridge University Collection: copyright reserved).

The famous Alport Castles landslide in North Derbyshire is also a compound landslide. This slide has produced highly irregular ground *(Photo 2.11)* covering 250ha, 1km in length and 2.5km wide. It is developed in a variable sequence of Upper Carboniferous sedimentary rocks comprising massive sandstones, mudrocks and thin sandstones of the Shale Grit and Mam Tor Beds, above the Edale Shales. The failure comprises a series of large displaced blocks, known locally as Alport Tower, Birchin Hats, The Tower and Little Moor, which lie below a main rear scarp up to 68m high[104]. Many of these blocks have been displaced downslope by 100–130m and subjected to backward rotation, with the overall non-rotational form of the slide controlled in part by the orientation of bedding planes and the jointing *(Figure 3.35)*.

A total of 537 compound slides have been identified so far *(Table 3.1)*. Prominent examples include the St Catherine's Point slide on the Isle of Wight *(Photo 3.10 and Figure 3.36)*, where translational movements of ridge "S" are thought to have been achieved by failure in clay layers within the Lower Greensand Sandrock, followed by

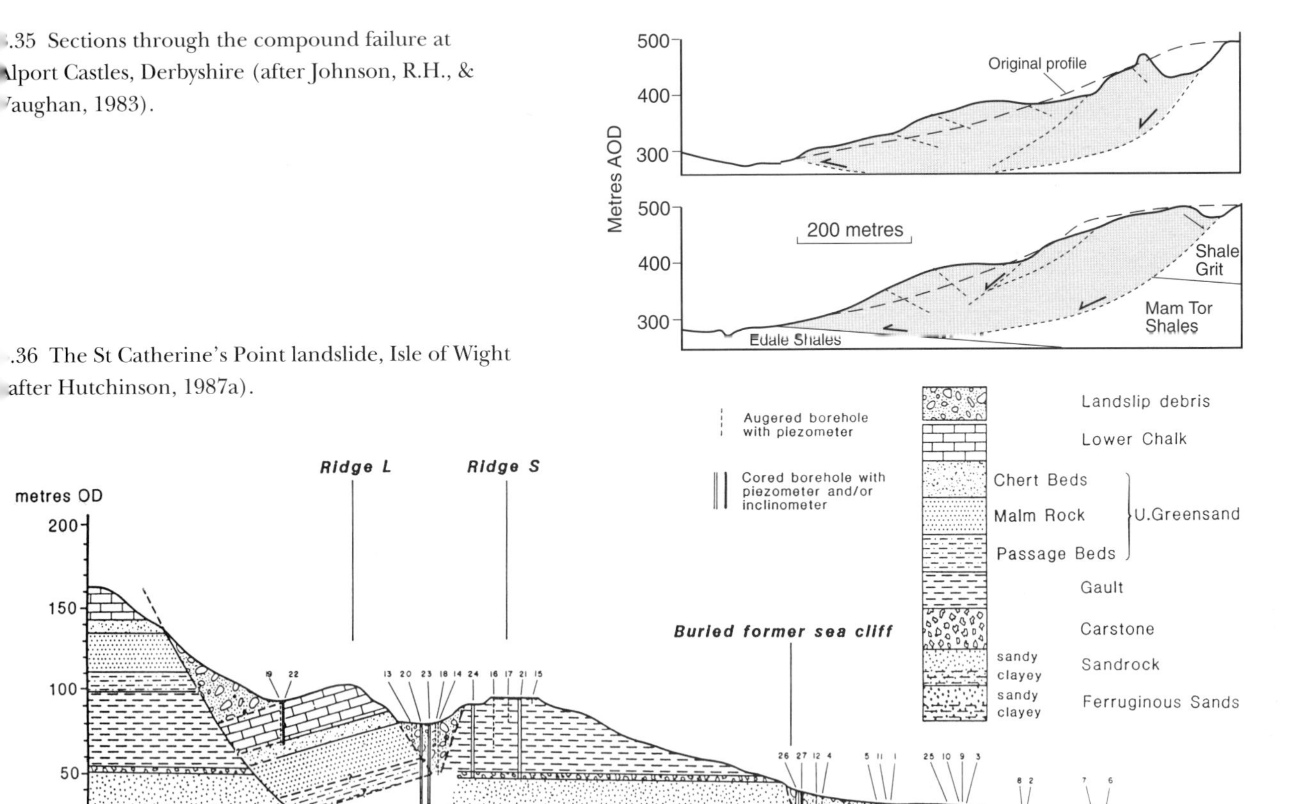

.35 Sections through the compound failure at lport Castles, Derbyshire (after Johnson, R.H., & 'aughan, 1983).

.36 The St Catherine's Point landslide, Isle of Wight after Hutchinson, 1987a).

the rotational failure of the landward ridge "L" [105] and the failure at Hawkley, Hampshire, in 1774, where Upper Greensand was displaced over a basal shear surface developed on one of the bedding planes in the underlying Gault Clay[106]. This latter landslide is of interest as it was described by the naturalist Gilbert White in "The Natural History of Selbourne" (1788).

Complex landslides

It has already been emphasised that mass movements occur as a spectrum of features often difficult to compartmentalise for classification purposes. In addition, different types of mass movement tend to exist in close association at sites of major instability. For this reason, it has proved necessary to develop the notion of **complex landslides,** where the displacement of particular masses of moving ground is achieved by the combination of different mechanisms in a downslope direction, eg fall to slide to flow, and **landslide complexes,** where extensive areas of instability consist of a patchwork of different mass movement mechanisms, all of which combine to transport material downslope.

The DOE commissioned survey identified 777 complex landslides *(Table 3.1)* or just over 21% of the landslides for which type information is available. For each of these complex slides a combination of separate mechanisms (e.g. falls, flows, translational slides, etc,) were recorded as part of the data collection exercise. In total, the descriptions of the 777 complex landslides included 1920 citations of component mechanisms, representing 41% of all the records of individual landslide mechanisms or types. These figures clearly illustrate the importance of complex landsliding in Great Britain.

Most of the extensive or conspicuous landslides can be described as 'complex'; indeed, nearly all landslide complexes consist of assemblages of complex landslides! The Bindon slide *(Photo 2.7)*, Black Ven *(Plate 7)*, the Storr *(Plate 9)* all contain more than one mass movement mechanism, although in all cases one mode of displacement is visually most dominant. It merely serves to illustrate this widespread characteristic by reference to two examples, one inland, the other on the coast.

The East Pentwyn landslide, Gwent, first occurred on 22 January 1954 when a deep-seated rotational slide developed at the base of a cliff in massive Upper Carboniferous Brithdir and Hughes Beds sandstones, underlain by the weak siltstones, claystones, sandstones and seatearths of the Rhondda Beds *(Figure 3.37)*. This failure initiated a large debris slide immediately downslope. In contrast to the primary failure, this translational slide was a relatively shallow feature with pronounced lateral shears, and had the appearance of a large boulder field. The debris slide, in turn, moved 300m downslope and reactivated a pre-existing solifluction sheet *(Figure 3.38)* which developed into a lobate debris apron sloping at around 13°. Subsequent detailed investigations have revealed that any successful stabilisation scheme will have to improve the stability of all elements of the landslide[107], for the effectiveness of any measures directed at a single element could be impaired by movements of other parts of the slide system. They also recognised that continued degradation of the backscar and scree accumulation at the head of the slide would cause movements of the displaced block, thereby ultimately affecting the stability of the debris slide and debris apron.

Stonebarrow, on the Dorset coast, is another example of a complex landslide that has been investigated in detail[47] *(Photo 2.5)*. Three morphological zones have been recognised, each characterised by a distinctive combination of mass movement processes *(Figures 3.25 & 3.39)*. The Backscar and Upper Platform form Zone I,

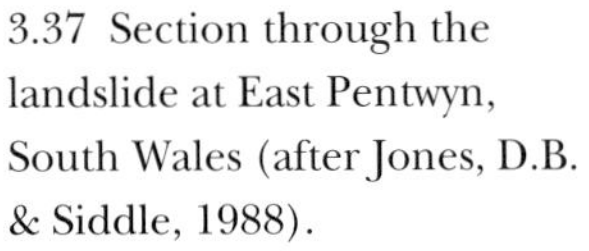

3.37 Section through the landslide at East Pentwyn, South Wales (after Jones, D.B. & Siddle, 1988).

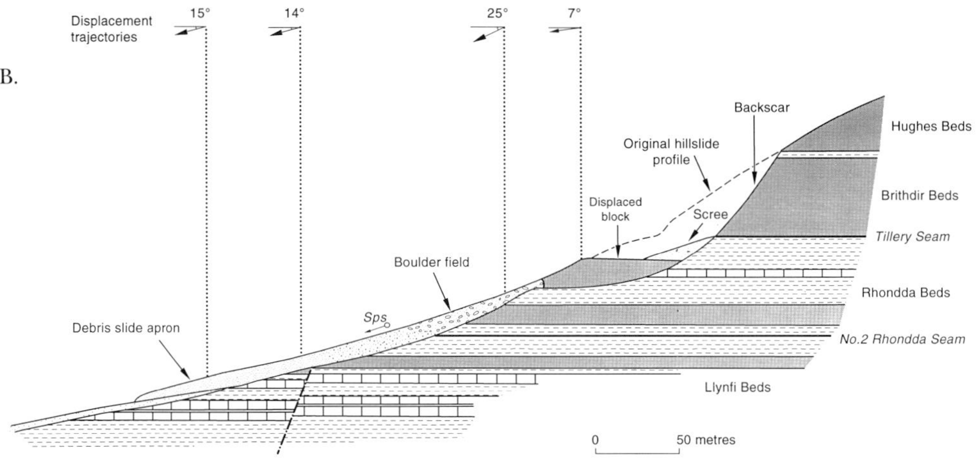

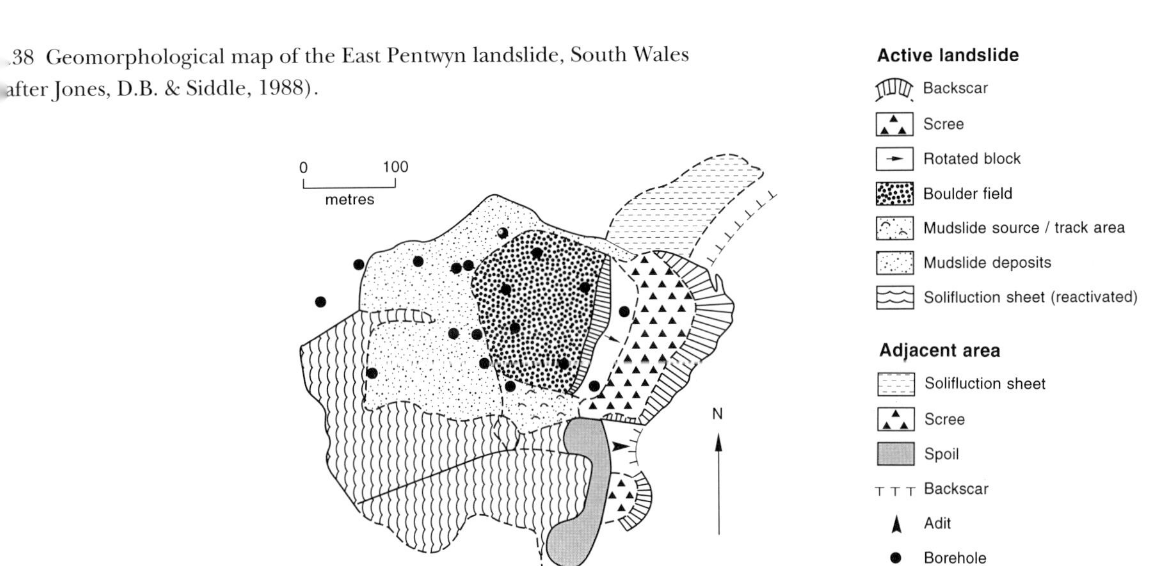

.38 Geomorphological map of the East Pentwyn landslide, South Wales
after Jones, D.B. & Siddle, 1988).

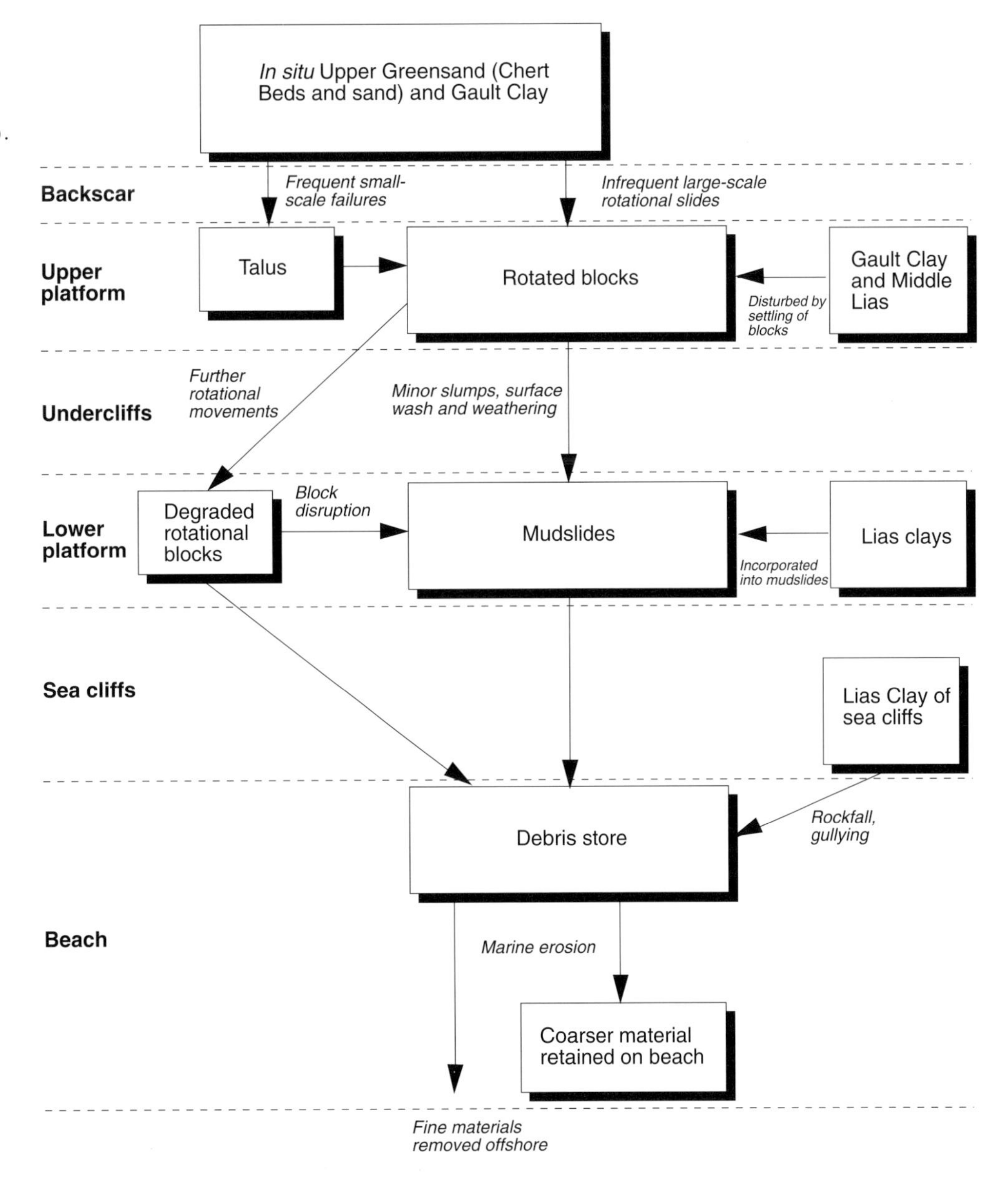

.39 Summary of the flow
f materials through the
tonebarrow landslide
omplex (after Bray, 1991).

where falling and multiple rotational failures predominate. The rotational blocks gradually become disrupted downslope towards Zone 2 (the Undercliff and Lower Platform) where a complicated sequence of mudslide units developed in the Middle Lias efficiently transports debris to the crest of the sea cliffs where it falls to the beach below. The near-vertical sea cliffs are developed in the Belemnite Marls and the Black Ven Marls (Zone 3) and fail by toppling failures and rock falls.

Landslide activity in the three zones at Stonebarrow is intimately related to, and ultimately controlled by, the rate of removal of material from the base of the sea cliffs by the sea. Large-scale movements in Zone 1 are followed by a period of block disruption when the rotated units (blocks) are gradually degraded and transported towards the beach via shallow mudslides and falls. Clearly the rate of supply of debris to the beach will increase over time as the block disruption process proceeds, with the result that debris supply may exceed removal by the sea, and the rate of cliff retreat diminish or temporarily cease.

However, the transport of material away from Zone 1 removes passive support at the base of the backscar, and eventually results in diminished debris supply to the beach. This, in turn, results in removal by waves exceeding supply from mudslides, so that the accumulations of material on the beach are removed and the cliffs re-exposed to wave attack. The renewal of cliff retreat causes further movements in the Undercliffs which eventually result in new large-scale failures in Zone 1 and so the cycle of activity begins again. This cycle is considered to take a little over a century although this figure may be modified in the light of current research[108].

Conclusion

The purpose of this chapter has been to illustrate the wide diversity of landsliding phenomena that affect Great Britain. It has sought to emphasise the range of mechanisms that affect both natural and artificial slopes, the ways in which these mechanisms frequently operate in varying combinations, the differences in scale and rates of process operation, and the variety of surface morphologies that result. Three important points are worthy of emphasis by way of conclusion.

First, while it is undoubtedly true that fast moving failures do occur (mainly falls and flows), as do displacements involving huge volumes of rock, such as the 1839 Bindon slide *(Photos 2.6.and 2.7)* and the 1915 rotational movements in Folkestone Warren *(Photo 2.3)*, such dramatic, conspicuous and newsworthy events represent a small minority of mass movements, most of which operate at relatively low speeds. Indeed, for many slides, much of the actual displacement is achieved by imperceptible deformation followed by brief pulses of movement that diminish with time, culminating in further gradual, imperceptible, movements as the slope materials attain a new stable condition. It is this insidious characteristic of landsliding that many find difficult to appreciate and which leads to an understandable underestimation of the extent and significance of mass movement.

Second, the character and intensity of landsliding is fundamentally a reflection of ground conditions, most especially materials, water and slope characteristics. These aspects will be considered more fully in Chapters 4 and 5. For the moment, however, it merely serves to note that Great Britain is well known for having topographic diversity shaped from a complex geological fabric. As a consequence, the distribution of the different types of slope failure, their associations and the resulting concentrations are complex. Thus, while it may be possible to make regional generalisations about landsliding, e.g. the Lower Greensand escarpment of the Weald is prone to landsliding as is testified by the failures encountered during road construction at Sevenoaks *(see Figure 3.13)*, in reality local conditions are of paramount importance in determining the nature and scale of landsliding activity. This is graphically illustrated on the Dorset coast where the two major landslide complexes of Black Ven *(Plate 7, Figure 3.31)* and Stonebarrow *(Photo 2.5)*, occur on very similar sequences of rocks and yet display markedly differing characteristics.

Finally, frequent mention has been made of certain types of inland failure being less active today than they were in the past (e.g. translational slides and falls). This highlights the importance of climate change in determining landslide activity. On many occasions during the last 2 million years, the climate of the British Isles was significantly colder than present with permanent snowfields, deeply frozen ground (permafrost) and extensive ice-sheets *(see Chapter 6)*. During these cold or arctic phases, snow avalanching would have been a significant mass movement process and some have claimed that the shallow linear depressions running down the faces of Chalk escarpments in southern England are the remnants of avalanche chutes. But of much greater importance was the widespread sliding of saturated surface layers every time the climate ameliorated from Glacial to Inter-glacial conditions and the permafrost melted from the surface downwards. Surface materials were repeatedly redistributed by a process of translational sliding known as **solifluction**, and gently inclined solifluction sheets and lobes are widely developed over Lowland Britain and in many parts of Highland Britain. But the effects of ground freezing were even more profound in those areas where valleys had been cut through rigid cap-rocks and into softer, clay layers below. Here the intense freezing often caused the rigid cap-rock layer to bend over and break into blocks, separated by great fissures (gulls), in a process known as **cambering,** while the clayey material in the valley floor often pushed upwards in **valley bulges.** These processes created large-scale structures, sometimes confused with the effects of faulting, which prepared the ground for later mass movements, as will be described in Chapter 5. Thus landsliding activity has displayed remarkable variation over time, thereby emphasising the significance of inheritance from the past.

NOTES

1. Sharpe, 1938
   Hutchinson, 1968
   Skempton & Hutchinson, 1969
   Coates, 1977
   Varnes, 1978
   Hansen, M.J., 1984
   Bromhead, 1986
   Hutchinson, 1988
   Zaruba & Mencl, 1976
2. Geomorphological Services Ltd, 1986-1987
3. Hutchinson, 1988
4. Whalley, 1974
5. Hoek & Bray, 1977
6. De Freitas & Watters, 1973
   Hoek & Bray, 1977
7. Hutchinson, 1972
8. Pitts, 1974
   Pitts & Brunsden, 1987
9. Varnes, 1958
10. Brunsden, & Jones, 1976
    Bray, 1990
    Bray, 1991

11. May, 1971
12. Jones, D.K.C., 1981
13. Brunsden & Jones, 1980
14. Duncan, 1984
15. Hutchinson, 1972
Hutchinson, 1980
16. Hutchinson, 1983a
Hutchinson, 1984a
17. Williams, & Davies, 1984
Williams, & Davies, 1987
Williams et al., 1991
Davies et al., 1991
18. Hutchinson, 1982
Hutchinson, et al., 1981a
19. Hawkins, T.R.W., 1985
20. Ballantyne & Kirkbride, 1987
21. De Freitas & Watters, 1973
22. Bromhead, 1986
23. Hutchinson, 1988
24. Holmes & Jarvis, 1985
25. Johnson, M A & Rodine, 1984
26. Rouse, 1984
27. Hutchinson & Bhandari, 1971
28. Statham, 1976
29. Addison, 1987
30. Baird & Lewis, 1957
31. Common, 1954
Strachan, 1976
Ballantyne, 1981
Innes, 1982
Innes, 1983
32. Innes, 1982
Innes, 1983
33. Common, 1954
34. Baird & Lewis, 1957
35. Jenkins et al., 1988
36. Bishop, 1973
Rouse, 1984
37. Bishop et al., 1969
38. Osman, 1917
39. Tankard, 1973
Hutchinson, 1984a
40. Lambe & Whitman, 1969
Carson, M A & Kirkby, 1972
41. Brunsden, 1979a
42. Skempton, 1964
43. Bromhead, 1978a
44. Bromhead, 1979
45. Skempton & Brown, 1961
Skempton, 1964
46. Gostelow, 1977
Jones, D.B. & Siddle, 1988
47. Brunsden & Jones, 1976
Brunsden & Jones, 1980
48. Hutchinson, 1969
49. Hutchinson, et al., 1981b
50. Webster, in Englefield, 1816
Brannon, 1825
51. Anderson & Dunham, 1966
52. Brunsden & Jones, 1972
Brunsden & Jones, 1976
53. Johnson R H, 1980
54. Conway et al., 1980
55. Butler, 1983
56. Kellaway & Taylor, 1968
Hawkins, A.B. & Privett, 1979
Forster et al., 1985
57. Wooldridge, 1950
Robinson, D.A. & Williams, 1984
Lee & Griffiths, 1989
58. Hutchinson, 1967
59. Hutchinson, et al., 1985b
60. Chandler, R.J., 1971
61. Chandler, R.J., 1970a
Chandler, R.J., 1970b
Chandler, R.J., 1970c
62. Grainger, 1983
63. Watters, 1972
Holmes, 1984
64. Smith, D.I., 1984
65. Holmes, 1984
66. Early & Jordan, 1985
67. Eyre, 1973
68. Carson, M A & Petley, 1970
69. Rouse, 1969
Conway et al., 1980, & Bridges, 1986
70. Knox, 1927
71. Gostelow, 1977
72. Newberry & Baker, 1981
73. Grainger & Harris, 1986
74. Burton, 1969
Dumbleton & West, 1970
75. Henkel & Skempton, 1954
76. Chandler, R.J., 1970a
77. Biczysko, 1981
78. Lee & Moore, 1989
79. Brunsden, 1984
80. Lee et al., 1988
81. Hutchinson et al., 1988
82. Hutchinson, 1973
83. Bhandari & Hutchinson, 1982
84. Grainger & Kalaugher 1987a
Grainger & Kalaugher 1987b
85. Denness, 1972
86. Arber, 1941
Arber, 1973
Brunsden, 1969
Conway, 1976

Chandler, J.H., 1989
Allison, R.J., 1990
87. Denness et al., 1975
88. Hutchinson, 1968
Hutchinson, 1973
89. Barton & Coles, 1984
90. Conway, 1979
Pitts, 1979
91. Cook, 1973
92. Weeks, 1969
Skempton & Weeks, 1976
93. Bower, 1959
94. Canden, 1722
95. Bowes, 1960
96. Hudleston, 1930
97. Crisp et al., 1964
98. Carling, 1986b
99. Ward, 1948
100. Barton, 1984
101. Barton, 1977
102. Hutchinson, 1965a
Hutchinson, 1965c
103. Hutchinson, 1965a
Hutchinson, & Hughes, 1968
104. Johnson, R H & Vaughan, 1983
105. Hutchinson et al., 1985a
Hutchinson et al., 1991
106. Barton, 1989
White, 1788
107. Jones, D.B. & Siddle, 1988
108. Brunsden, & Jones, 1980
Bray, 1991

CHAPTER 4

Causes of Landslides

The ultimate cause of all landsliding is the downward pull of gravity. However, the fact that all slopes of similar steepness or height do not show similar types and rates of landsliding, emphasises the existence of other factors that determine instability. Stated simply, if the force of gravity acting on a slope exceeds the strength of the materials forming the slope, then the slope will fail and movements occur, but if the materials forming the slope are capable of resisting the force of gravity, then there will be no movement unless and until changes occur that affect the balance of opposing forces. These changes are often divided, for the sake of convenience, into internal and external groups. External changes increase the stress placed on slope-forming materials, while internal changes reduce or weaken their resistance to movement. The majority of landslides are, therefore, the product of changing circumstances or alterations to the status quo.

The Strength of Materials

In order to appreciate the nature of these changing circumstances, attention must first be directed to a brief consideration of the nature of slope-forming materials and the stresses and resistances that occur within slopes.

A measure of the maximum shear stress* that a material can withstand before failure occurs is given by the **shear strength**. The amount of shear strength required to resist destabilising forces is known as the **mobilised shear strength** and is generally less than the total shear strength available. However, at that point in time when landsliding occurs, the shear strength is fully mobilised along the failure plane or shear surface.

Before proceeding further, it is necessary to recognise that soils and rocks display contrasting modes of behaviour, thereby giving rise to the two distinctive disciplines of soil mechanics and rock mechanics**. But what is the difference between soil and rock? Basically, this depends on the scientific perspective. To a pedologist, soil is a term reserved for the combination of mineral and organic material that supports life. Geomorphologists take a broader view and include all superficial deposits and disintegrated or fully decomposed rock. Geologists prefer the term 'regolith' to cover all materials overlying sound rock, with the term rock usually having an age connotation. However, to engineers, engineering geologists and geotechnicians, all materials are considered to be soils until they become so hard that they must be excavated by pneumatic drills or blasting. By this definition some really quite ancient materials, such as Lias Clays (c. 200 million years old) are considered 'soil', whereas the younger Great Oolite limestones are referred to as 'rock'.

The shear strength of both soils and rocks is determined by two fundamental parameters: cohesion and friction. Cohesion is produced by chemical bonding or

---

* **Stress** is defined as the force acting over a given area. **Shear stress** is the force acting to rupture a material so that it deforms by one part sliding over another.

** Comprehensive descriptions of soil mechanics and rock mechanics can be found in Lambe and Whitman (1969) and Hoek and Bray (1977) respectively.

Table 4.1 **Typical properties of soils and rocks**

		Material Cohesion	Unit weight saturated/ dry kN/m³	Friction angle degrees	KPa
Cohesionless	Sand	loose sand, uniform grain size	19/14	28–34	
		dense sand, uniform grain size	21/17	32–40	
		loose sand, mixed grain size	20/16	34–40	
		dense sand, mixed grain size	21/18	38–46	
	Gravel	gravel, uniform grain size	22/20	34–37	
		sand & gravel, mixed grain size	19/17	48–45	
	Broken rock	chalk	13/10	30–40	
		sandstone	17/13	35–45	
		limestone	19/16	35–40	
		shale	20/16	30–35	
		granite	20/17	45–50	
		basalt	22/17	40–50	
Cohesive	Clay	very soft clay, organic	14/6	12–16	10–30
		soft clay (e.g. lacustrine)	16/10	22–27	20–50
		soft glacial clay	17/12	27–32	30–70
		stiff glacial clay	20/17	30–32	70–150
		glacial till (mixed grain size)	23/20	32–35	150–250
	Rock	soft sedimentary rock e.g. sandstone, shale, chalk	17–23	25–35	1,000–20,000
		hard sedimentary rocks, e.g. limestone, older sandstones	23–28	35–45	10,000–30,000
		metamorphic rocks e.g. gneiss. slate	25–28	30–40	20,000–40,000
		hard igneous rocks e.g. granite, basalt	25–30	35–45	35,000–55,000

cementation of particles, and can result in enormous forces in certain kinds of rocks (*Table 4.1*). Even fine-grained sedimentary strata and superficial deposits sometimes display strong attraction between clay and silt particles which can give rise to significant cohesive strength, particula rly in the case of stiff clays and glacial tills. However, it is important to emphasise that engineering soils rely chiefly on their frictional strength for shear resistance.

Friction is the result of the compressive forces which hold particles together. The frictional resistance of a material is measured by a parameter known as the **angle of friction**, which can be explained with reference to the diagram reproduced as Figure 4.1:

(i) in Figure 4.1(a) the downward force (W), exerted by the mass of the block lying on a horizontal surface under the influence of gravity, is matched by an equal and opposite resisting force (N). In this instance there is no tangential shear force and the block is immobile;

(ii) in Figure 4.1(b) a horizontal force (H) is applied to the side of the block. As a consequence, the resisting force can no longer act to equal and oppose the

A simple explanation of the angle of friction.

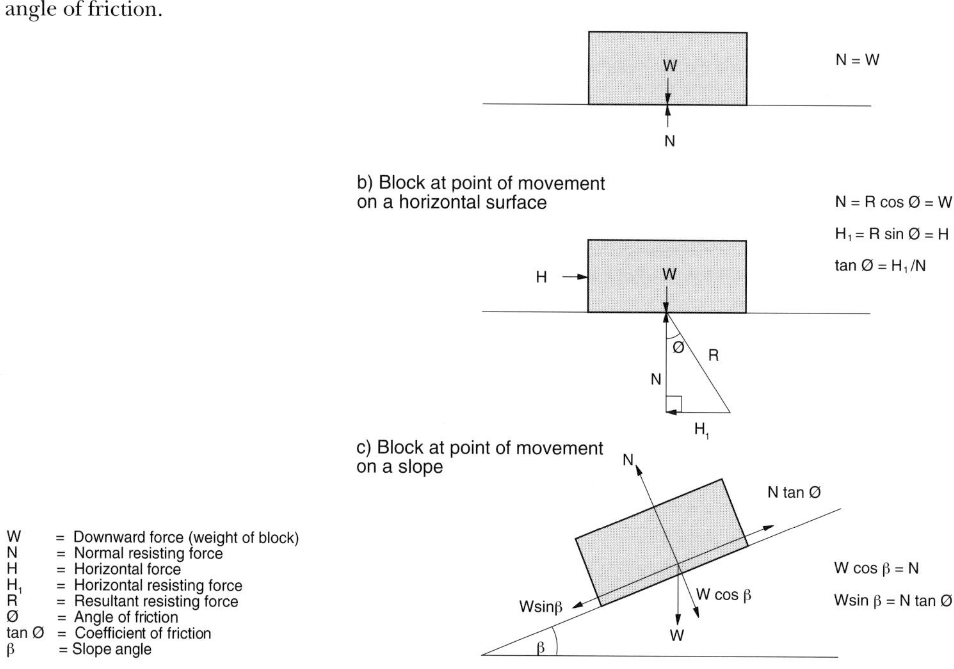

downward force (W), but must also act to oppose the horizontal force H. It therefore has two components, one acting vertically (N) and one horizontally (H_1), the resultant of which is the force R, which can be considered to act at an angle to the vertical. If force H increases, so too must resisting force H_1, with the result that the resultant force R becomes greater and acts with a greater deviation from the vertical. Eventually, the applied force (H) equals the fully mobilised horizontal (shear) resistance (H_1 max) and the block starts to move. At this point the resisting force R is at its maximum angle of deviation from the vertical. This is the **angle of friction**, otherwise known as the **angle of plane static friction** or **angle of shearing resistance**, and the ratio H_1/N is the **coefficient of friction**. It should be noted that the shear resistance (H_1 max) = NtanØ or normal stress x tangent of angle of friction;

(iii) in Figure 4.1(c) the block is shown lying on a slope so that the shearing force is provided by a component of the weight of overlying material resolved parallel to the inclined plane, i.e. W sin β . The angle of slope at the point at which the block starts to move downslope (i.e at **limiting equilibrium**) can be defined as when the downslope force just exceeds the upslope resistance generated by the product of angle of internal friction and the stress normal to the slope

i.e. $W \sin \beta \geq W \cos \beta \tan Ø$

or: $\tan \beta = \tan Ø$

In other words, in the situation where it is friction alone that determine stability, the block will move when the slope angle equals the angle of internal friction.

The importance of friction and cohesion in determining the strength of material was first recognised by Coulomb, a French engineer, in 1776. He was able to show that the shear strength (s) could be defined by the coefficient of friction (tan Ø) multiplied by the amount of stress applied at right angles to a plane (σ, or the **normal stress**), plus the cohesion (c), or:

$$s = c + \sigma \ (\tan Ø).$$

This simple Coulomb equation highlights the important fact that frictional resistance varies with the level of applied normal stress, i.e. load, whereas cohesion exists independent of compressive forces.

The strength of materials is not, however, merely the result of the interactions between mineral particles. The presence of water within pores or fissures can reduce the shear strength. This is not because of 'lubrication', as the media would often have us believe, but due to the fact that water in the ground exerts its own pressure which serves to reduce the amount of particle on particle contact and, hence, the frictional component of shear strength. The frictional resistance to movement depends, therefore, on the difference between the applied **total normal stress** and the pore-water pressure. This difference, or that part of the normal stress which is effective in generating shear resistance, is known as the **effective stress** and can be defined as:

effective stress (σ ′) = total applied normal stress (σ) – pore water pressure (u)

The Coulomb equation can, therefore, be modified to take effective stress into account, and becomes:

$$s = c' + (\sigma - u) \tan Ø'$$

in which c′ and Ø′ are modified parameters with respect to effective stress (i.e. **effective** cohesion*; **effective** angle of friction*).

When stress, in the form of a load, is applied to rock or soil, deformation occurs although the relationship between stress and strain** is non-linear. If the stress is relatively small, then the response may only be of an elastic nature. But once the applied stress exceeds a threshold value, large irreversible deformations take place in the material. Unlike steel and concrete, whose material properties are known or closely controlled, soils and rocks display remarkable diversity with material properties that vary greatly from site to site as well as with depth. However, for practical purposes, these very varied materials can be classified into three broad groups: cohesionless soils, cohesive soils and rocks.

Stress-strain Behaviour of Cohesionless Soils

For **granular soils** such as gravels, sands and non-plastic silts, shear strength is wholly dependent on the friction between particles, i.e. they are cohesionless. Frictional strength is in turn dependent on the size distribution of particles, grains or clasts, their arrangement, resistance to crushing, and the size, distribution and proportion of inter-granular spaces (voids). The last mentioned is known as packing and has a

* The shear strength parameters defined in this way have no physical meaning, i.e. the angle Ø′ is not a true angle of friction; c′ merely represents that part of the shear strength that is independent of the normal stress and not true cohesion.

** Strain is the deformation suffered by a material or body as a result of stress.

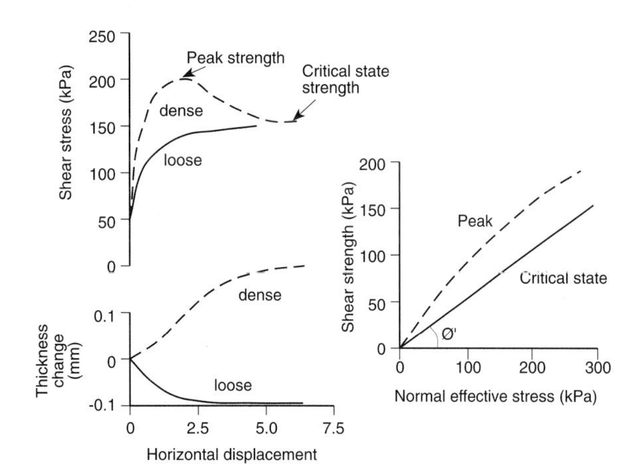

Typical shear stress/strain behaviour for sands (after Kenney, 1984).

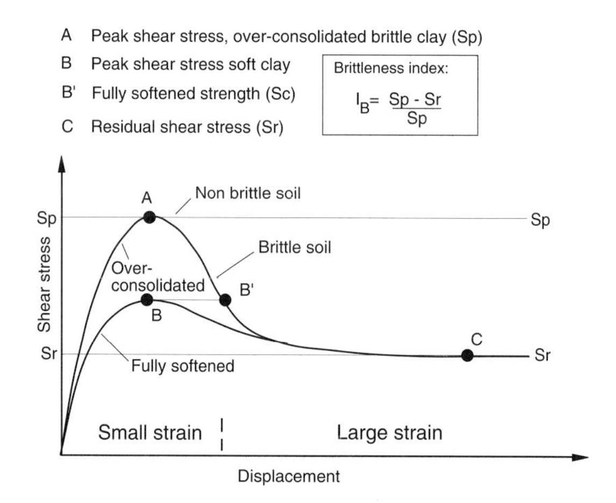

Typical shear stress/strain behaviour for cohesive soils (after Kenney, 1984).

significant influence on strength as illustrated in Figure 4.2 which highlights the important differences in stress-strain behaviour between loose and dense sands. When dense sands are sheared, they dilate initially (i.e. increase in volume) and this gives a high strength (**peak strength**). However, if shearing is continued, the strength drops and eventually the sand may be sheared at a constant volume (**critical state condition**). By contrast, loose sands contract during the initial shearing and as the packing becomes closer, the strength gradually increases with further deformation until it shears at a constant volume.

Stress-strain Behaviour of Cohesive Soils

Soils containing significant amounts of clay and silt are known as cohesive soils. There are two important differences between cohesive and granular materials. First, deposits rich in clays and silts are much less permeable* than sands and gravels, because of the small size of particles, chemical bonding and dense packing. However, they can and do contain water** but the low permeability inhibits the movement of this water if there is a tendency to change volume. As a result, it may take many years after a change in loading for the effective stresses to reach equilibrium. It is important to recognise, therefore, that shear behaviour is different

* **Permeability** is the measure of the ability of water to move through a soil or rock mass, including discontinuities. It must not be confused with **porosity** which measures the amount of water that can be held in the void spaces within a rock mass.

** The porosity and permeability of selected rocks and soils is shown below:

	Porosity (%)	**Permeability m/day**	**Av. permeability compared with clay**
Igneous rocks	10^{-5} – 10	10^{-7} – 10^{-2}	–
Limestone	1 – 10	10^{-2} – 10	–
Sandstone	5 – 20	10^{-2} – 100	–
Gravel	25 – 40	100 – 1000000	100000 – 1 billion
Sand	30 – 40	10 – 10000	100 – 100000
Silt	20 – 50	10^{-3} – 10	1 – 100
Clay	45 – 60	10^{-6} – 10^{-4}	–

These values clearly reveal that clays are the most porous of surface materials, i.e. can contain the greatest proportion of water per unit volume, but have permeabilities comparable with igneous rocks, i.e. the water can only move exceedingly slowly.

under **drained** conditions, where the excess water has been given time to flow out of the medium, as compared with the **undrained** state.

The other important difference concerns the shape of the particles; sandy grains are spherical, clay particles are plate-like. Thus, when a shear surface has developed in a clay, the adjacent clay particles tend to align themselves parallel to this surface. This realignment reduces friction and means that the shear strength along this surface can be significantly lower than that of the clay mass as a whole; this is known as the **residual strength.**

When cohesive soils are sheared in a **drained** condition (so that there is time for dissipation of pore pressures caused by the shearing), their behaviour is characterised by three important shear strength values which are mobilised under different stress conditions (*Figure 4.3*):

(i) a **peak** strength obtained by the rapid application of stress to previously unsheared brittle material;

(ii) a reduced critical state (fully softened) strength which occurs in deposits that have been altered by weathering or where the stress is applied very slowly, thereby allowing progressive deformation or strain softening; and

(iii) a minimum shear strength achieved at large displacements (concentrated in narrow shear zones) known as the **residual strength**, produced by the re-orientation of clay particles. Previously sheared ground (e.g. landslipped) usually displays residual strength unless stability has been improved by other means.

There are, however, important differences between types of clay that affect their shear behaviour. One of the most important factors is "brittleness" which is a measure of the difference between peak and residual strength values.

Soft*, low-plasticity silty-clays often show little brittleness, whereas soft plastic clays can be quite brittle. Most stiff* clays are very brittle, especially if they are plastic. As will be described later, this property of brittleness has an important bearing on the rate and scale of movement during initial slope failure.

Stiff clays often contain discontinuities such as fissures, which have a significant effect on shear strength. It has been shown[1] that the strength along such fissures is very similar to soft specimens of the same clay (i.e. the fully softened strength in Figure 4.3). In effect, the formation of discontinuities creates weaker zones within the soil mass which can have a major influence on landsliding.

When saturated clays and silts are confined and then sheared in an **undrained** condition (so that movement of pore-water is restricted), they appear to be purely cohesive with no frictional strength. In this instance, increases in normal stress are

* Soils are commonly described as either soft or stiff, referring to whether they are in a normally-consolidated or overconsolidated state, respectively. Normally-consolidated soils have not been subject to a greater vertical effective stress than they are carrying at present. Conversely, if a soil has experienced a higher load in the past (e.g. as a result of burial beneath other strata (overburden) or thick ice sheets) which has since been removed, they are said to be overconsolidated. In general, normally-consolidated soils are weaker and more compressible than over-consolidated soils.

wholly supported by pore-water which has not been allowed to drain away. Reference to the modified Coulomb equation: $s = c' + (\sigma - u) \tan \emptyset$ reveals that if the pore pressure (u) is equal to the normal stress (σ), then the frictional resistance will be zero. As a result, the shear strength will remain at a constant value, equivalent to the apparent cohesion. This situation is of importance in low permeability materials, where rapid load changes can result in the build up of very high pore pressures and the initiation of landslide movements on very low angled slopes (see **undrained loading**).

Stress-strain Behaviour of Rocks

It is important to make a clear distinction between **intact rock** and a **rock mass**. Intact rock is a body of rock which has no continuous discontinuities through it. As indicated in Table 4.1, intact rock can be extremely strong, largely as a result of considerable cementing and/or chemical bonding between particles (cohesion). However, most rock slopes are dissected by discontinuities such as joints, faults, foliations or schistocity, bedding planes etc, which have an overriding influence on rock slope stability.

When rock is sheared it behaves in a highly brittle fashion, characterised by **peak** strength and a much lower **residual** strength. The corresponding peak shear strength values along discontinuities are much lower (*Figure 4.4*), and they are, therefore, planes of weakness within a rock mass.

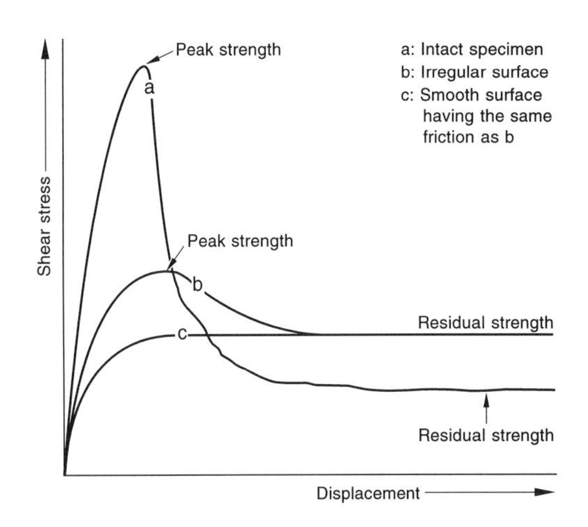

Typical shear stress/strain behaviour for rocks (after Petley, 1984).

The most important factor in controlling rock slope stability is the orientation of the discontinuities in relation to a rock face, which determines whether parts of the mass are free to fall or slide *(Chapter 5)*.

Slope stability

A **stable slope** is one where the resisting stresses are greater than the destabilising stresses and, therefore, can be considered to have a **margin of stability**. By contrast, a slope at the point of failure has no margin of stability, for the resisting and destabilising forces are approximately equal. The quantitative comparison of these opposing forces gives rise to a ratio known as the 'Factor of Safety' (F):

$$\text{Factor of Safety} = \frac{\text{Resisting stresses}}{\text{Destabilising stresses}} = \frac{\text{Shear strength}}{\text{Shear stress}}$$

The Factor of Safety of a slope at the point of movement is assumed to be 1. On slopes of similar materials progressively higher values represent more and more stable situations with greater margins of stability. In other words, the higher the value the greater the ability of slope-forming materials to accommodate change before failure occurs.

However, a slope is only as strong as its weakest horizon. Therefore, in order to accurately assess the stability of any slope it is necessary to establish the shear strength available at various depths and along any pre-existing shear surfaces. On this basis it is possible to identify a range of conditions to describe landslide slopes[2] *(Table 4.2)*.

First-time failures occur in previously unsheared ground and may involve the **mobilisation** of the peak strength. Once peak strength has been mobilised the soil must shed some load if it is to be strained further. This load shedding may, in turn, cause adjacent soil elements to mobilise all their strength and leads to the development of a shear surface. As a result, the soil mass must move to a new position so that equilibrium can be re-established between the destabilising forces and the residual strength along the shear surface. This response can involve a variety of mechanisms (falls, slides and flows) and will result in a change in slope geometry, a dissipation of pore-water pressures and a redistribution of the destabilising and resisting stresses. Such first-time slides are often characterised by large

Table 4.2 **A geotechnical classification of landslides, based on shear strength mobilised during failure (based on Hutchinson, 1988).**

1. **First–time slides** in previously unsheared ground; soil fabric tends to be random. Shear strength parameters are at or close to:

    1.1 **Peak strength;**

    1.2 **Fully softened strength.**

2. **Reactivation of earlier landslides**; soil fabric along the shear surface is highly orientated in the slip direction. Shear strength parameters are at or close to the residual value.

3. **Initiation of landsliding on discontinuous pre-existing shears**, produced by processes other than landsliding, e.g.:

    3.1 Tectonics: tensional and compressional forces generated by crustal movements;

    3.2 Glacitectonics; disturbance of rocks and superficial sediments by movement of ice sheets and glaciers, or due to their weight;

    3.3 Solifluction; seasonal downslope movement by flow or creep in periglacial environments;

    3.4 Rebound: rise in elevation of earth's surface following removal of weight by the melting of ice sheets;

    3.5 Non-uniform swelling; the product of differential volume changes due to varying water retention properties.

Landslides are likely to involve the mobilisation of both **peak to residual strengths** along different sections of the shear surface.

displacements, particularly if there are large differences between peak and residual strength values (brittleness).

First-time failures can occur anywhere due to progressively changing circumstances, but they are most dramatic where the rate or scale of change is relatively great. The under-cutting of slopes by rivers and the sea can produce first-time failures, often successively at the same location (**repeated failures**), including such major displacements as the great Bindon slide of 1839 (*see Chapter 2*). This slide is believed to have been the result of a block slide along the bedding in the Rhaetic Shales of the Westbury Formation, followed by subsidence of masses of Chalk and Upper Greensand into the resulting chasm or graben (*Figure 2.7*). Analysis of the slide[3] indicates that the shear strength mobilised was the peak strength of the Westbury Formation, although subsequent movements would have been at the residual strength. Similarly, the production of over-steepened valley sides by glacial erosion followed by the removal of support when the glaciers/ice sheets melted, resulted in numerous rockslides in the Highlands of Scotland and in Wales, including the Taren slide in Taff Vale, South Wales (*see Figure 2.10*). Finally, excavations are a major cause of first-time failures, although in this case the features are usually relatively small.

Failures involving the mobilisation of peak strength are, however, not common and most first-time failures in clay slopes actually occur at the **fully-softened strength** and not peak strength. It appears that limited slow movements may precede a slide, although the displacements are not sufficient to reduce the strength to the residual value. Examples of this type of behaviour, known as **progressive failure**, have been recorded on cut slopes in London Clay[4] and Upper Lias Clay[5].

By way of contrast, ground containing pre-existing shear surfaces usually results in failures that are slow moving with relatively limited displacements associated with each phase of movement. This is because the material along the shear surface(s) is at or near to its residual strength. The resultant failures may be **the reactivation of earlier landslides**, where part or all of a previous landslide body is involved in new movements, or new failures utilising shear surfaces created by other processes, as listed in Table 4.2.

The importance of such pre-existing shears within slope-forming materials cannot be over-emphasised. Probably the most important process is the development of bedding plane slip (flexural shearing) during the folding of sequences of hard rocks and clay-rich strata. Indeed, recent studies have suggested that shears can be expected to be present in the clay beds even at dips as low as 1°[2]. However, it is important to bear in mind that pre-existing non-landslide shear surfaces are unlikely to be continuous, and a landslide shear surface involving these features will undoubtedly have had to pass through some previously unsheared material.

There have been a number of well publicised examples of reactivated failures in recent decades including the translational failures near Swindon in 1969 and on the Daventry Bypass in the late 1970s[6] (*Plates 21 and 22*). Probably the most significant was the failure caused by the construction of an embankment to carry the M6

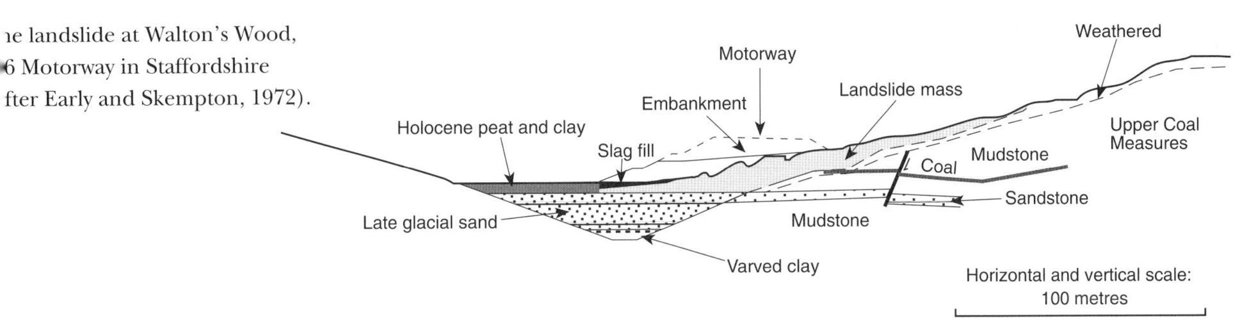

ıe landslide at Walton's Wood, 6 Motorway in Staffordshire fter Early and Skempton, 1972).

motorway at Walton's Wood, Staffordshire, which inadvertently reactivated a late-glacial landslide developed in mudstones, sandstones and thin coal seams of the Upper Coal Measures (*Figure 4.5*). During the winter of 1961–62 large parts of the embankment moved downslope at a maximum rate of c.5cm per month, with an overall vertical displacement of 60cm[7]. In addition to causing considerable delay and expenditure, this landslide was of great significance because of the discovery of continuous slip surfaces with reoriented clay particles which led to the development of the theory of residual strength as outlined by Skempton in 1964[8].

Temporal Variations in Slope Stability

If slope movements are the result of changing circumstances that affect the relationship between destabilising forces and resisting forces, then the simple two-fold division into stable and unstable slopes represents an oversimplification of reality. A better measure of slope stability is to judge a slope on its ability to withstand potential changes, or fluctuations, in conditions that have influence in terms of increasing stress or decreasing strength. Consequently, a three-fold classification of states is now widely employed[9]:

(i) **stable;** where the margin of stability is sufficiently high to withstand all transient forces in the short to medium term (i.e. hundreds of years), excluding excessive alteration by human activity;

(ii) **marginally stable;** where the balance of forces is such that the slope will fail at some time in the future in response to transient forces attaining a certain level of activity; and

(iii) **actively unstable slopes;** where transient forces produce continuous or intermittent movement.

Causal Factors

The recognition of these three stability states provides a useful framework for evaluating slope stability, but it is helpful to consider briefly the stable slope category in a little more detail. No landform is immortal; eventually weathering and erosion processes will combine to disintegrate rocks and remodel landshape. It therefore follows that there can be no such thing as absolute stability over long time-scales measured in many millions of years. Steeply inclined stable slopes will be ultimately converted to marginally stable slopes, given sufficient time, and eventually failure will take place unless remodelling is accomplished by other processes. The three stability states must, therefore, be seen to be part of a continuum, with the probability of failure being minute at the stable end of the spectrum, but increasing through the marginally stable range to reach certainty in the actively unstable category (*Figure 4.6*). Following failure, the hillslopes may become too gently

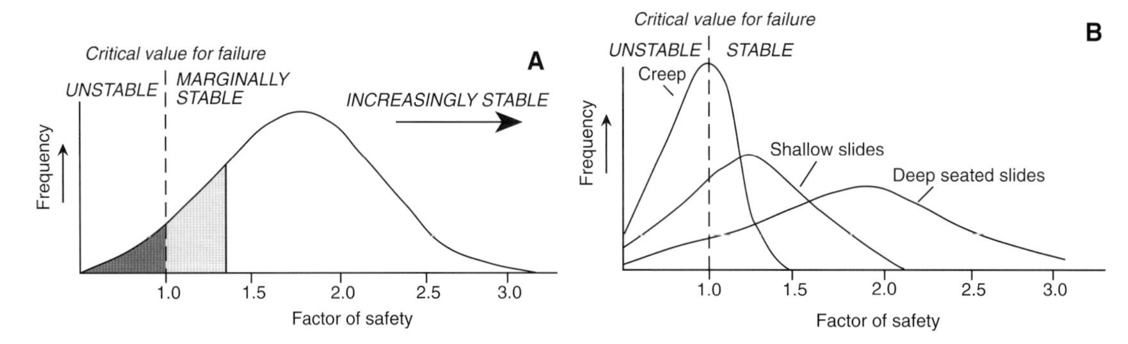

Simple Probability distribution curves for the stability of slopes: (a) for any set of slopes in a given environment, (b) for different mass movement processes acting on all slopes in a given environment. the message conveyed in these diagrams is that usually only a small percentage of slopes will be affected by mass movement processes at any point in time and the percentage decreases with increasing size of failures (adapted from Brunsden, 1979).

inclined for landsliding with relatively high Factors of Safety, but once again, environmental change may eventually result in new phases of erosion which will stimulate the generation of instability.

This perspective makes it possible to recognise that the work of destabilising influences can be apportioned between two categories of factors on the basis of their role in promoting slope failure. These two categories are (*Figure 4.7*):

(i) **preparatory factors** which work to make the slope increasingly susceptible to failure without actually initiating it, i.e. cause the slope to move from a stable state to a marginally stable state, eventually resulting in a relatively low Factor of Safety;

(ii) **triggering factors** which actually initiate movement, i.e. shift the slope from a marginally stable state to an actively unstable state.

This assessment of the role of destabilising factors is important because it emphasises:

(i) the effects of the long-term evolution of slopes in the development of instability;

(ii) that individual factors may have both long-term and short-term influences on slope stability; and

(iii) that instability is commonly the result of the variable interaction of a number of destabilising factors working over reasonable time spans.

Having established this conceptual framework, it is now possible to examine the actual causes of landsliding. It should come as no surprise that relative simplicity is immediately replaced by complexity, for the great diversity of landslide types described in Chapter 3 points to an equally wide range of causal factors.

Factors involved in the decline in stability of a slope (after Crozier, 1986).

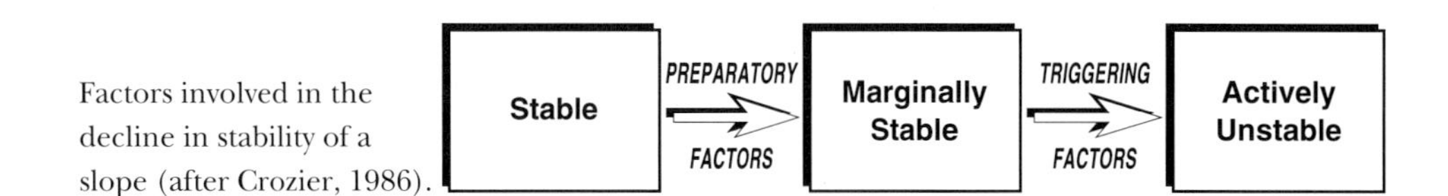

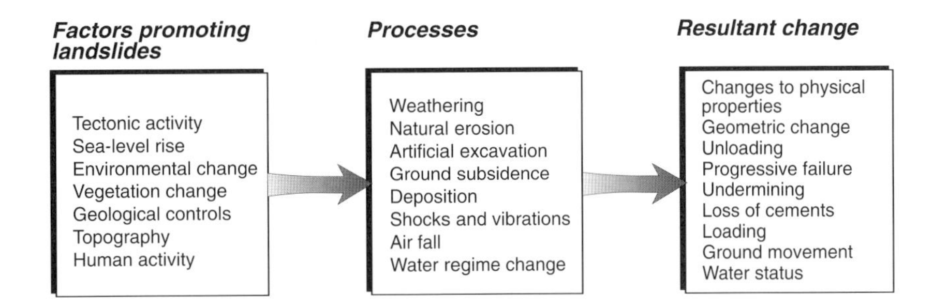

A simple sub-division of the causes of landslides.

At the outset it is worth noting that the scene for slope instability is in fact set by the external controls that determine the very nature of the physical environment. These include the geological framework, climate, environmental changes, topographic development, sea-level change and human activity (**factors promoting landslides**; *Figure 4.8*). A discussion of how these factors have influenced past and present landslide activity in Great Britain forms the basis for Chapters 5 and 6. The external controls influence the operation of **processes** such as weathering, erosion and water regime changes, which can lead to landsliding. These processes, however, do not actually cause landslides but may result in changes that affect the stability of slopes (**resultant changes**; *Figure 4.8*). Erosion, for example, can lead to increased rates of weathering, changed pore-water pressures, the removal of support and increases in slope angle or height. As was stated earlier, the actual initiation of slope failure is due to the stresses acting on a slope exceeding the available shear strength of the materials. It is, thus, those changes which lead to either **an increase in shear stress** (external factors) or a **decrease in shear strength** (internal factors) that are the direct cause of landslides (*Figure 4.9*).

Numerous excellent discussions of the causes of mass movement have been published elsewhere[10] and there is little point of repeating them here. The following sections are merely intended to provide a general introduction to each of the major external and internal causes, paying specific attention to their significance in Great Britain. The supporting statistics (*Table 4.3*) have been compiled as part of the DOE commissioned 'Review of Research on Landsliding in Great Britain'[11] and require a

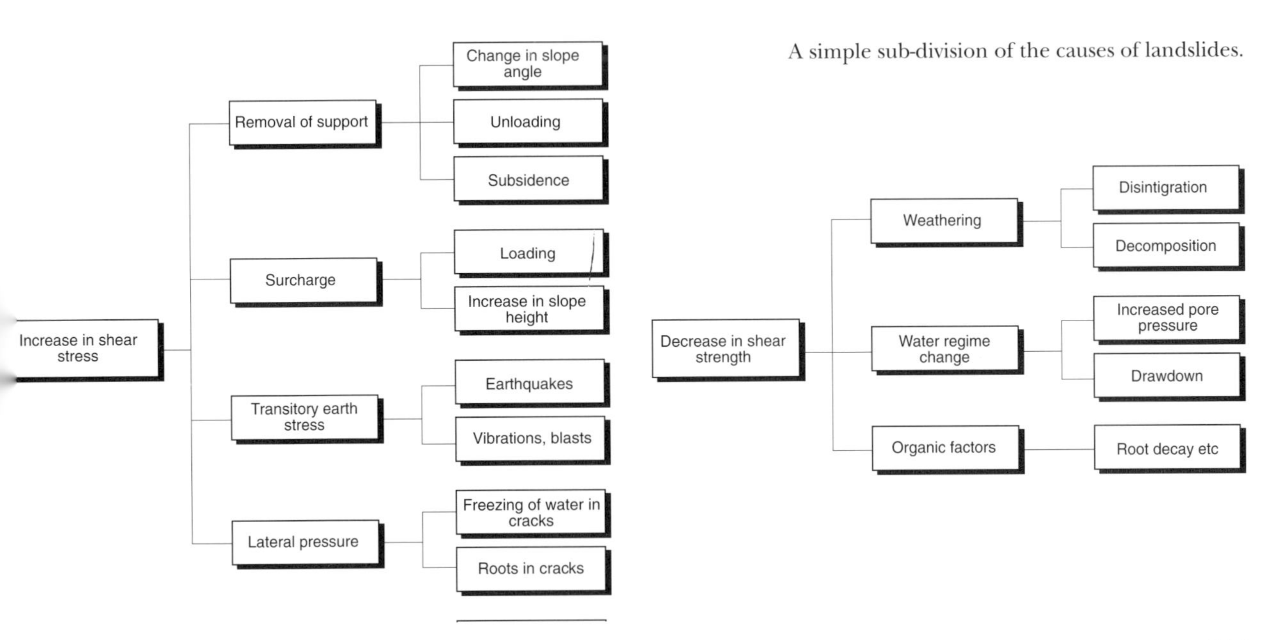

A simple sub-division of the causes of landslides.

Table 4.3 **The causes of landslides in Great Britain, as revealed by the DOE survey.**

Causal Feature (Process)	Number	Percentage of Causal Attributes	Percentage of Process Attributes
Weathering			
General	130	1.3	3.4
Physical	229	2.4	6.1
Chemical	29	0.3	0.7
Freeze-thaw	229	2.4	5.9
Dessication	5	0.05	0.1
Solution	23	0.2	0.6
Leaching	2	0.02	0.05
	647	**6.7**	**17.0**
Natural Erosion			
Undefined	11	0.1	0.3
General slope erosion	160	1.7	4.2
Basal erosion	1110	11.5	29.2
Subsidence	16	0.2	0.4
	1297	**13.5**	**34.1**
Artificial Erosion			
Excavation	256	2.7	6.7
Mining	9	0.1	0.2
Opencast working	11	0.1	0.3
	276	**2.9**	**7.2**
Ground Subsidence – Removal of Support			
Seepage erosion	103	1.0	2.7
Subsurface burning	0	–	–
	103	**1.0**	**2.7**
Undefined removal of support	39	0.4	1.0
Deposition			
Undefined	12	0.1	0.3
Natural	39	0.4	1.0
Artificial	75	0.8	2.0
	126	**1.3**	**3.3**
Shocks and Vibrations – Seismic			
Earthquake	0	–	–
Fault movement	10	0.1	0.3
Subsidence	63	0.7	1.6
	73	**0.8**	**1.9**
Shocks and Vibrations – Man-Induced			
Explosions	25	0.3	0.7
Machinery	5	0.05	0.1
Undefined	7	0.1	0.2
	37	**0.4**	**1.0**
Air Fall			
Loess air fall	0	–	–
Tephra air fall	0	–	–
General air fall	0	–	–
	0	**0.0**	**0.0**

Table 4.3 **The causes of landslides in Great Britain, as revealed by the DOE survey** *(continued).*

Causal Feature (Process)	Number	Percentage of Causal Attributes	Percentage of Process Attributes
Water Regime Change			
Intense precipitation	870	9.0	22.9
Snow melt	14	0.1	0.4
Dam Collapse	4	0.04	0.1
Flood conditions	55	0.6	1.4
Blocked drainage	46	0.5	1.2
Vegetation change	13	0.1	0.4
Other	85	0.9	2.3
General water regime change	117	1.2	3.0
	1204	**12.5**	**31.7**

Table 4.3 **The causes of landslides in Great Britain, as revealed by the DOE survey** *(continued).*

Causal Feature (Resultant change)	Number	Percentage of Causal Attributes	Percentage Resultant Change Attributes
Change to Physical Properties			
Pore pressure	203	2.1	3.5
Voids	1	0.01	0.02
Permeability	31	0.3	0.5
Fissures	41	0.3	0.7
Undefined	17	0.2	0.3
	293	**3.0**	**5.0**
Geometric Change			
Undefined	164	1.7	2.8
Gradient	1219	12.7	20.9
Height	88	0.9	1.5
	1471	**15.3**	**25.3**
Unloading			
Natural	121	1.3	2.1
Man-induced	41	0.4	0.7
General	50	0.5	0.9
Lateral expansion	103	1.1	1.8
Cracking	14	0.1	0.2
Other	7	0.1	0.1
	336	**3.5**	**5.8**
Progressive Failure Creep			
General progressive failure	224	2.3	3.9
Lateral expansion	125	1.3	2.1
Fissuring	79	0.8	1.3
Strain	45	0.5	0.8
Softening	284	2.9	4.9
Stress concentration	10	0.1	0.2
Other	2	0.02	0.03
Undefined	39	0.4	0.7
	808	**8.4**	**13.9**

Table 4.3 **The causes of landslides in Great Britain, as revealed by the DOE survey** *(continued)*.

Causal Feature (Resultant change)	Number	Percentage of Causal Attributes	Percentage Resultant Change Attributes
Undermining			
Removal of fines	55	0.6	0.9
Piping	3	0.03	0.05
	58	**0.6**	**1.0**
Loss of Cements			
Destruction	16	0.2	0.3
Removal	3	0.03	0.05
Leaching	1	0.01	0.02
Solution	16	0.2	0.3
	36	**0.4**	**0.6**
Loading			
General loading	12	0.1	0.2
Rapid depositoin	11	0.1	0.2
Addition of fill	65	0.7	1.1
Geometric change	8	0.1	0.1
Other	9	0.1	0.2
	105	**1.1**	**1.8**
Ground Movement			
Single	11	0.1	0.2
Multiple	73	0.8	1.2
Undefined	5	0.05	0.1
	89	**0.9**	**1.5**
Physical Effects			
Slope mantling	0	–	–
Change of soil grading	0	–	–
Perched water table	4	0.04	0.1
Other	0	–	–
	4	**0.04**	**0.1**
Water Status			
Saturation	1214	12.6	20.8
Rise in water table	341	3.5	5.9
Excess pore pressures	954	10.0	16.4
Drawdown	5	0.05	0.1
General	113	1.2	1.9
	2627	**27.3**	**45.1**

Notes: 2499 landslides in the GSL databank have one or more causal attributes specified.
Total number of causal attributes = 9629
Total number of process related causal attributes = 3802
Total number of resultant change related causal attributes = 5827.

little explanation. Although attempts were made to collect information on the cause of failure for each of the 8835 landslides recorded in the survey, this proved impossible in many cases where either no information was provided in the published literature, or the only indication of the existence of a landslide is contained on a map. By contrast, many of the identified landslides were complex features and reportedly caused by a number of different processes. As a result the 8835 reported landslides gave rise to 9629 separate records of 'cause' which were sub-divided into **causal processes** and the **resultant changes** that can occur (*Table 4.3*). In reality causal processes could only be identified in 3802 cases and resultant changes in 5827 instances.

External Factors

Removal of support

An obvious cause of slope instability is the removal of support provided by material in contact with, or part of, the slope. In reality, four kinds of support removal can be identified as important to the generation of slope instability, although the distinctions are somewhat blurred and arbitrary: the removal of **lateral support** provided by adjacent masses of material as a consequence of either overall changes in slope gradient (steepening) or the discontinued presence of some adjacent supporting mass; the removal of **restraining support** provided by a mass of material on the lower parts of the slope whose presence effectively inhibits movement; the removal of **underlying support** as a consequence of subsidence* following the collapse of subsurface voids created both naturally and artificially; and the decrease of **internal support** produced through chemical alteration and the physical removal of fine-grained material from within slope-forming deposits. The last mentioned acts mainly to decrease shear strength and will, therefore, be considered later under internal factors, but the other three constitute important external factors. Indeed, the recent DOE survey revealed that removal of support was identified in over 40% of those landslides for which information is available (*Table 4.3*) and is generally associated with undercutting by coastal, fluvial or glacial processes.

(i) Slope steepening

The most widely appreciated cause of slope instability is slope steepening, sometimes referred to as oversteepening. The logic that increasing slope gradient will facilitate mass movement stems from the reality that all the types of mass movement described in Chapter 3 require that material be displaced downslope to lower elevations. It follows, therefore, that the steeper the slope the greater the potential for mass movement, a relationship apparently confirmed by the close association that exists between landsliding and mountainous areas or coastal clifflines.

The generation of instability through slope steepening is partly a function of unloading, as was explained in Chapter 3 (*see discussion of sliding*). Simply stated, the compressional stresses on a horizontal plane surface at depth beneath a level plateau formed of uniform materials are everywhere equal. However, along the sloping boundaries of the level ground (e.g. valley sides or escarpments), the vertical compressional forces diminish with decreasing thickness of overlying material. As a

---

* Subsidence may be defined as the vertical downward movement of portions of the earth's surface due to changes in the nature of underlying materials or tectonic forces. Lateral movements are restricted to the margins of the area affected. Introductory discussions of subsidence can be found in Jones (1985), its impact in Britain is described in Wallwork (1960) and its significance to planning outlined in Planning Policy Guidance Note 14 : Development on Unstable Land (DOE, 1990).

consequence, the direction of principal stress ceases to be vertical and bends towards the sloping ground (*see Figure 3.22b*). The steeper the slope the sharper the alteration in principal stress direction, the greater the generated shear stresses and the lower the gravity-induced shear strength. Put another way, the steeper the slope the larger the proportion of material weight employed in generating shear stress parallel to the slope surface.

This situation can be clearly illustrated using the simple method of stability analysis known as **infinite slope analysis.** This method is explained in Figure 4.10 and can be used to analyse any slide that is displaced over a shallow shear surface parallel to the ground surface. Using this model, it can be shown that the stability of a **dry cohesionless slope** is wholly dependent on frictional resistance and that an increase in slope angle will lower the Factor of Safety, with failure occurring when slope angle equals the angle of friction. The existence of groundwater will also lower the Factor of Safety by varying amounts depending on the height of the water-table above the shear surface. The same applies in the case of **cohesive materials**, such as clays, although in reality pore-water pressures attain even greater significance because of low permeability. It is of interest to note, therefore, that in the case of cohesive materials underlain by a previously developed shear surface, where cohesion is assumed to be zero, and with the groundwater level at the ground surface, the Factor of Safety will be 1.0 when the slope angle (β) is approximately half the angle of friction. This explains why very gentle natural slopes in clays, which have been affected by previous landsliding, may be unstable. The residual strength of Upper Lias Clay, for example, is around 15° and, therefore, landslide movements may occur on slopes as low as 8°. Indeed, under periglacial conditions in the past, seasonal melting would have produced such high pore pressures that movements occurred on very shallow slopes (<4°).

The significance of slope steepening as a cause of landsliding is most clearly displayed in those instances where pronounced alterations to slope geometry result in instability. Artificial erosion (excavation) is obviously a prime cause of slope failure, as a result of both removal of support and over-steepening. Quarrying, opencast mining, excavations for foundations and the creation of cuttings for roads and railways, all drastically alter slope geometry and inevitably lead to the reactivation of pre-existing failures and the generation of new failures. One of the best examples occurred during the construction of the Stromeferry Bypass in the Scottish Highlands, over the period 1968-1971[12], when excavation at the base of a steep hillside developed in granulites, mica-schist and gneiss, triggered a major slide in May 1969 which blocked the adjacent railway line for five months (*Plate 25*). The fact that only a small amount of excavation initiated such a large slide indicates that the slope had been in a state of marginal stability prior to the commencement of construction works.

Pronounced slope steepening is also produced by natural (geomorphological) processes and is best exemplified by coastal landsliding. In many cases, such as cliff falls, the main cause of instability is simply over-steepening due to basal undercutting, but in other instances where instability is perennial, slope steepening is just a preparatory factor. This is most dramatically seen at Stonebarrow, Dorset[13]

Failure is assumed to occur by sliding a slab of soil on a plane shear surface which is parallel to the ground surface.

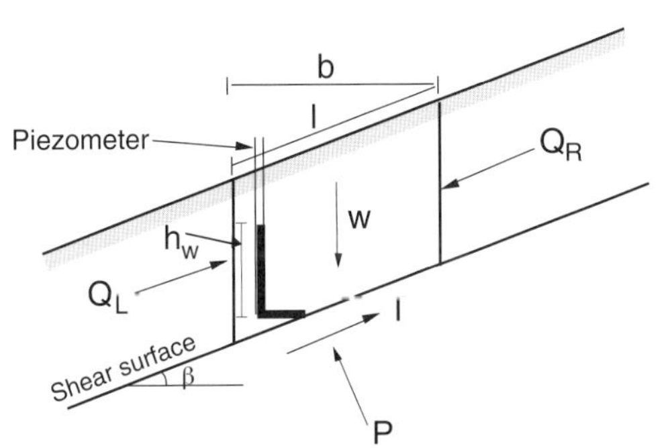

Soil properties: c',ø'

Unit weight soil: γ_z (assume equal to 2.0mg/m^3)

Unit weight water: γw (assume equal to 1.0mg/m^3)

Slab length (l) = b/cosß

Slab weight (w) = γzb

Pore pressure at base (u) = $\gamma_w h_w$

$r_u = u/\gamma z$

For slab shown: at base total normal stress σ, shear stress τ and pore pressure u

Since the slope is infinite, $Q_L = Q_R$

Resolving perpendicular to slope:

$$P = W\cos\beta = \sigma l$$

therefore $$\sigma = W/b \cos^2\beta$$

Resolving parallel to slope:

$$T = W\sin\beta = \tau l$$

therefore $$s = W/b \sin\beta\cos\beta$$

Failure criterion $s = c' + (\sigma - u) \tan ø'$

Mobilised shear strength $\tau = s/F$ where F is factor of safety

Hence $W/b\sin\beta\cos\beta = 1/F(c' + [W/b \cos^2\beta - u] \tan ø')$

$$\text{so } F = \frac{c' + [\gamma z\cos^2\beta - u] \tan ø'}{\gamma z\sin\beta\cos\beta}$$

$$\text{or } F = \frac{c'/\gamma z + [\cos^2\beta - r_u] \tan ø'}{\sin\beta\cos\beta}$$

Special cases:

If c' = 0 and groundwater at ground surface:

$$F = (1 - \gamma w/\gamma) \tan ø'/\tan\beta$$

therefore when F = 1

$$\beta \approx ø'/2$$

If c' = 0 and the slope is dry:

$$F = \tan ø'/\tan\beta$$

therefore when F = 1

$$\tan\beta = \tan ø'$$

The infinite slope analysis (after Nash, 1987).

(*Photo 2.5; Figure 4.11*) where erosion of landslide debris by the sea maintains mudslide activity, whereas removal of support to the rear cliffs by landslide movement downslope, stimulates rotational failures further inland (*Chapter 3*). Over-steepening of the sea cliffs leads to instability in the rear parts of landslide complex, which in turn, eventually causes increased debris supply to the shore. As this

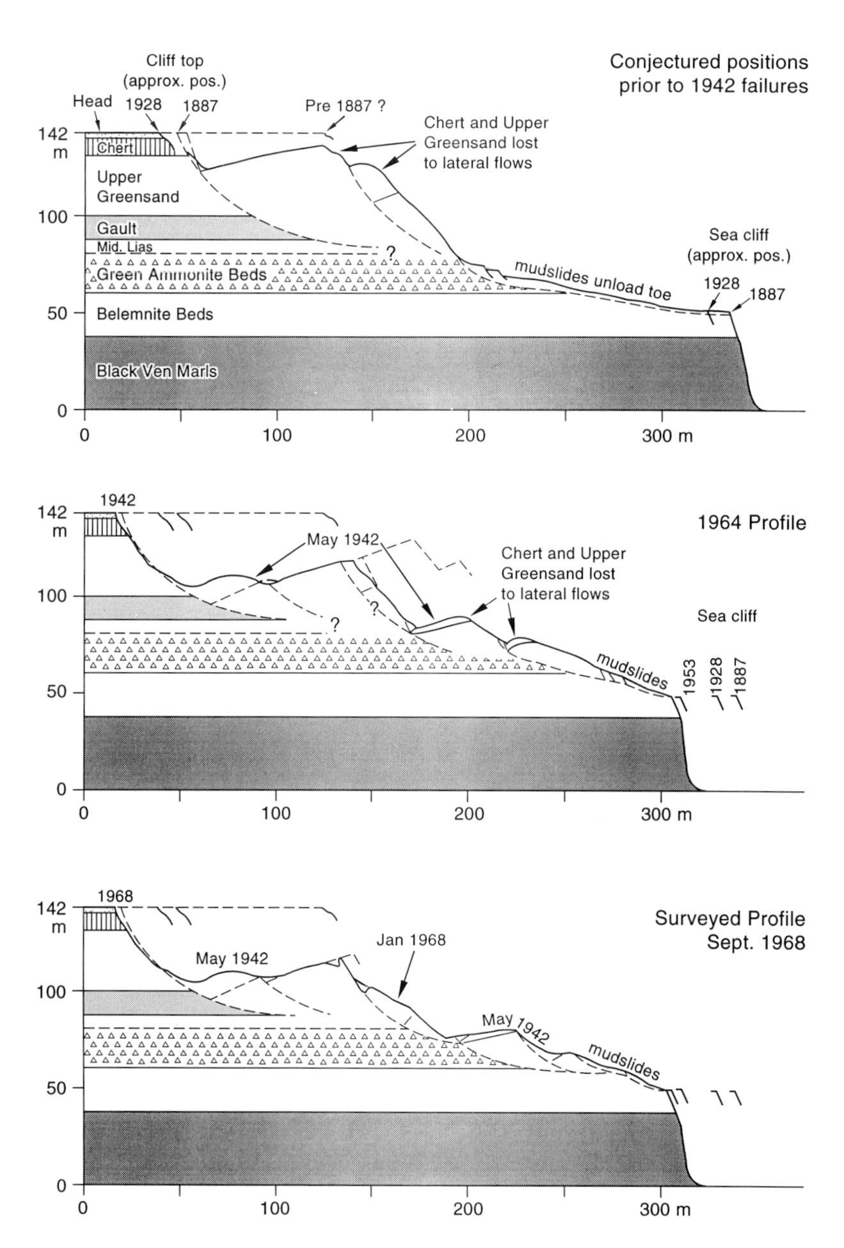

4.11 The cyclic evolution of the Stonebarrow (Fairy Dell) landslide complex, involving rotational sliding and subsequent block distribution (after Brunsden & Jones, 1976).

landslide debris has to be removed before the bedrock cliffs can be exposed to basal undermining, the result is cycles or pulses of landsliding activity generated by 'waves of aggression' transferred inland by each phase of cliff retreat.

The importance of the relationship between the rates of basal erosion and debris supply have long been recognised for the London Clay sea cliffs of north Kent and Essex. Three main modes of failure have been identified[14]:

(i) balance between rate of debris supply and debris removal results in a scalloped cliffline fashioned by shallow mudslide systems (*Photo 2.2*);

(ii) erosion-dominated sites result in straight, steep cliffs with large, deep-seated rotational landslides (e.g. Warden Point, Sheppey; *Plate 5; Figure 4.12*);

(iii) deposition-dominated situations result in the creation of stability through 'free degradation' (*see Chapter 3*) which may be permanent if erosion has ceased, as at Hadleigh Castle, Essex[15] (*Plate 10; Figure 2.9*).

In inland areas of Great Britain, the contemporary impact of slope steepening is much less dramatic, except in those localised instances where failures are produced by rivers progressively undercutting river cliffs (meander scars), bluffs and valley sides (e.g. the Jackfield landslide in the Ironbridge Gorge, *Photo 2.12; Figure 3.30*). Nevertheless, it is undoubtedly true that slope steepening produced by river incision has been a major contributory factor in the widespread creation of slope instability in the past. Long-term landscape dissection over the last 1–2 million years appears to have been considerable (*see Chapter 6*), with widespread river incision resulting in increased relative relief because of the differing durability of juxtaposed rocks. Progressively deepened and steepened valleys and the creation of pronounced escarpments frequently proved conducive sites for slope failure, sometimes resulting in extensive belts of slipped ground. Good examples include the Lower Greensand escarpment of Surrey and Kent, the Oolitic Limestone escarpment of the Cotswolds and the Millstone Grit escarpments of the Southern Pennines (*Appendix A*).

Valley development, especially valley deepening, can be envisaged to initiate a cycle of landsliding activity involving both shallow and deep landslides. The details will clearly vary from place to place depending on rates of incision, geological conditions and groundwater considerations, but river downcutting will create steep slopes and thereby initiate landsliding which, in turn, will work towards the creation of gentler slopes and in the process achieve valley widening. The best example of such a cycle is the three phase relationship between valley development and slope instability identified for the sandy-clay tills of County Durham[16] (*Figure 4.13*):

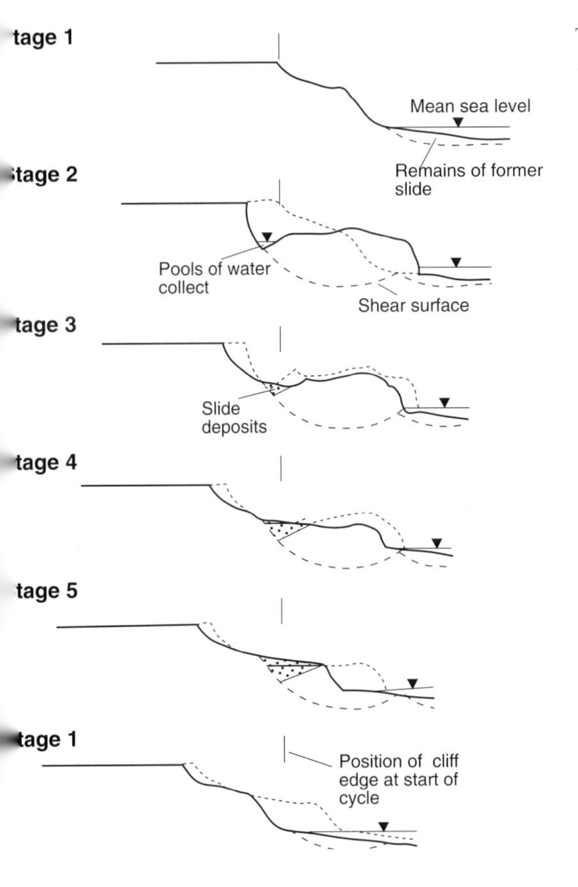

The evolution of deep-seated failures in London Clay cliffs (after Hutchinson, 1973).

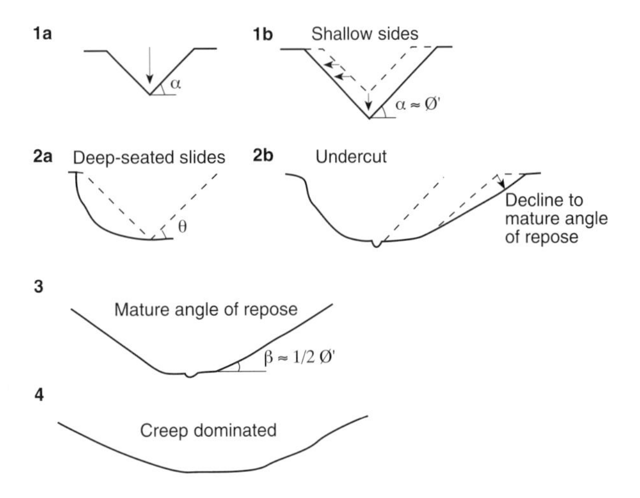

A simple model of valley-side evolution through landsliding (after Skempton 1953).

Stage 1; downcutting and undercutting by a river generates shallow surface slides, once critical slope angles have been attained.

Stage 2; the attainment of critical slope heights due to continued incision results in the development of deeper slides.

Stage 3; when downcutting slackens or ceases and/or the river ceases to erode the valley side-slopes, the slopes degrade through shallow landsliding to a relatively stable angle.

Finally, many landslides have been initiated or sustained by slope over-steepening due to glacial erosion. The resculpturing of pre-existing river valleys by glaciers frequently resulted in the creation of steep to vertical faces which collapsed when the supporting mass of ice was removed on deglaciation. The mechanism undoubtedly occurred in the deep valleys of the South Wales Coalfield (e.g. the Taren landslide, *Figure 2.10*), North Wales, the Lake District and the Highlands of Scotland. In many cases the resultant failures were determined by the development of pressure release joints parallel to the ground surface. These important discontinuities are produced by stress relief following the removal of load by the erosion that created the over-steepened slope. However, as they act to decrease shear strength, they represent an internal factor and, as such, will be considered later. They are merely mentioned here to indicate the close inter-relationship that often exists between internal and external factors. A similar complex relationship can be identified with regard to sediments laid down by meltwater adjacent to the decaying glaciers (pro-glacial deposits). These fluvio-glacial sediments frequently collapsed when the glaciers finally melted away and the underlying deposits were freed from the rigid binding of permafrost, as may have been the case in parts of the South Wales valleys (*Figure 4.14*). The removal of support by glacier ice is an external factor but the thawing of permafrost is an internal factor.

A simple model of slope development in the Taff Vale in the late glacial and Post-Glacial periods (after Fookes et al., 1975).

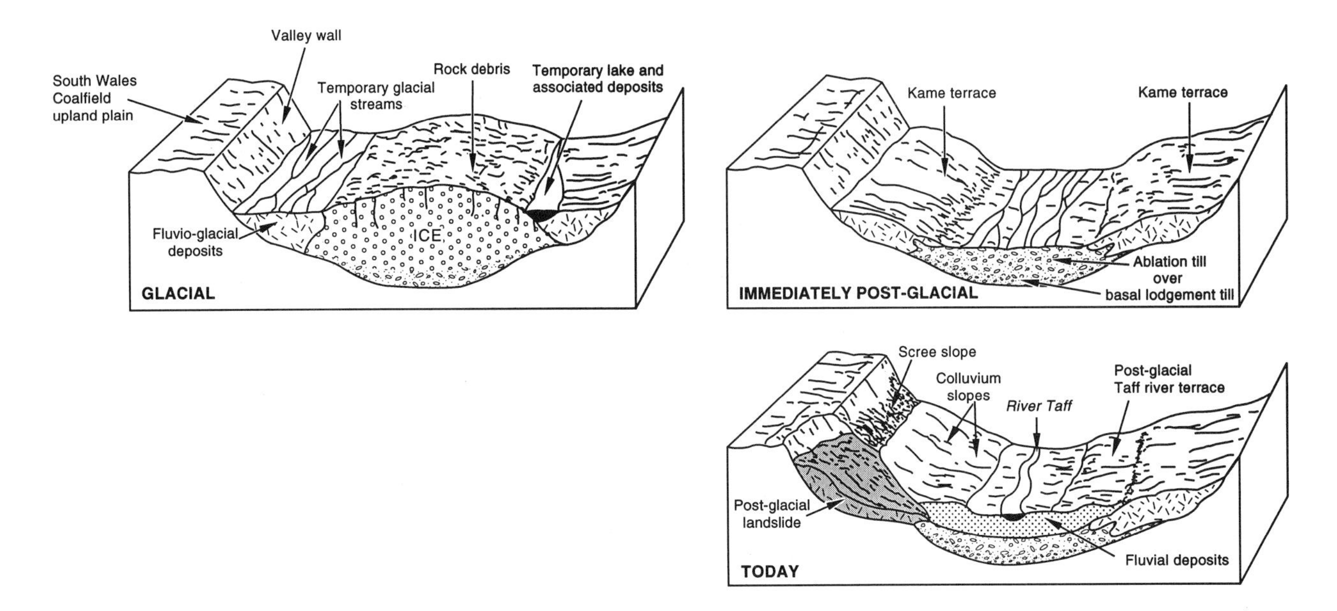

simple approach to
nalysing the stability of
rcular slips (after Fellenius,
918; Nash , 1987).

$\varnothing_u$ = O method

Failure is assumed to occur by rotation of a block of soil on a cylindrical slip surface on which the undrained strength may be mobilised.

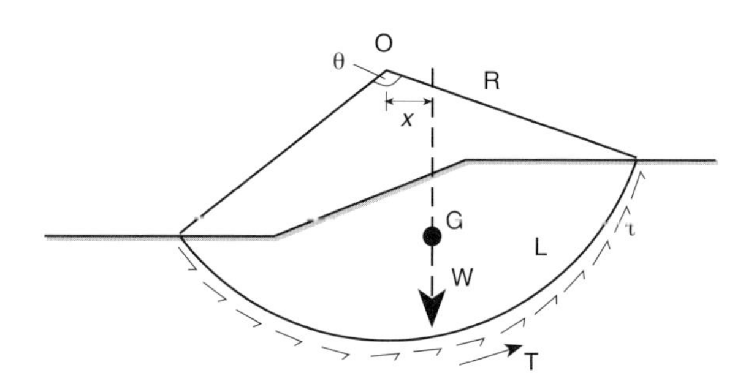

Shear strength (s) mobilised at failure = C_u (undrained shear strength)
Length of arc = L = Rθ
τ is average shear stress along L so T = τ.L
W is weight of soil block

Moments about O : Overturning moment = Wx
Restoring moment = TR

Failure criterion: s = C_u

Mobilised shear strength τ = s/F so τ = C_u/F where F is factor of safety
In equilibrium Wx = TR

Hence $Wx = \frac{C_u LR}{F}$

$$F = \frac{C_u LR}{Wx}$$

The analysis may easily be adapted to take account of varying shear strength, and the presence of surcharges and water at the toe.

(ii) Unloading

There are many instances where slope failure is initiated or reactivated through the removal of mass on the lower part of the slope, which has the effect of destabilising the slopes above. This unloading effect is sometimes referred to as **removal of restraining support** or **removal of passive support** and is most readily illustrated by the $\varnothing_u$=0 method of stability analysis. This is the simplest method available for analysing circular rotational slides in cohesive material (*Figure 4.15*). Failure is assumed to occur by the rotation of a block of soil on a cylindrical failure surface along which the **undrained shear strength** (C_u) is mobilised*.

Using this method it can be demonstrated that stability of a cohesive slope is controlled by:

(a) a **disturbing** or **overturning moment**; this is related to the relative difference between the driving and restraining forces imparted by the soil mass. The driving force is

* This approach is not usually reliable for estimating the stability of natural slopes, although it may be used to analyse construction of an embankment.

provided by the product of the soil mass and the distance between the centre of gravity (G) of the slide mass and the centre of rotation (o). On Figure 4.15 this is defined by W_x.

(b) a **restraining, resisting** or **restoring moment**, related to the available shear strength (Cu) along the shear surface (L) times the radius of curvature (R).

This simple model highlights the fact that the stability of slopes in cohesive soils is dependent on the weight of material (plus any other load) which acts as a driving force and the shear strength. Removal of material from the toe of the slope, i.e. the lower portion, has two possible effects: (i) it can reduce the length of the shear surface and thereby diminish the **resisting moment** and (ii) it causes the centre of gravity to move upslope, thereby increasing the destabilising force, i.e. the reduction in mass is more than offset by the change in the distance between the centre of gravity and the centre of rotation. It is readily apparent that small amounts of toe unloading can have significant repercussions. Such unloading can be achieved by natural erosion (river undercutting, wave attack, glacial trimming) or excavation, as occurred during construction of the Sevenoaks Bypass (*see Figure 3.13*).

Removal of support is best illustrated by reference to the problems encountered during regrading prior to house building along Marine Parade, Lyme Regis, in February 1962[17]. Around 20,000m^3 of material was removed from the slope, which was frequently affected by shallow landsliding. A few days after the earth moving operation had finished a large, deep-seated slide developed which moved several metres in a few minutes (*Plate 26*). It was established subsequently that the failure had occurred on a pre-existing landslide shear surface. The regrading, which was designed to improve stability against shallow landsliding, had in fact removed support from the toe of an unsuspected, potentially unstable, large slide.

Natural removal of support occurs most extensively in coastal situations where the stabilising effect provided by previously displaced landslide debris is negated by wave attack. This has already been identified in the discussion of Stonebarrow in the previous section, but is best examined by reference to Folkestone Warren, Kent, because this example also clearly illustrates how human intervention can exacerbate instability by interfering with the operation of 'natural processes'.

Studies into the evolution of Folkestone Warren (*Photo 4.1*) have revealed that the 130-155m high cliffs of Chalk overlying Gault Clay have experienced increased landsliding activity over the last 100 years[18]. This is attributed to decreased beach volume and the accelerated removal of the toes of previously displaced landslide blocks. Twelve major slips have been recorded since 1765, with a particularly dramatic failure in 1915 when virtually the whole of the Warren was involved in seaward movement, including one large block which displaced the Folkestone to Dover railway line by up to 50m (*Photo 2.3*). Although coastal landsliding would have been reactivated at this location some 6,000–7,000 years ago by the post-Glacial rise in sea-level, this most recent phase of pronounced failure was probably a consequence of the expansion of the Folkestone Harbour facilities over the years 1810–1905. In particular, the extension of the main pier resulted in the disruption of wave and current induced littoral (beach) drift of sand and shingle eastwards

4.1 The landslide complex at Folkestone Warren, Kent (Cambridge University Collection: copyright reserved).

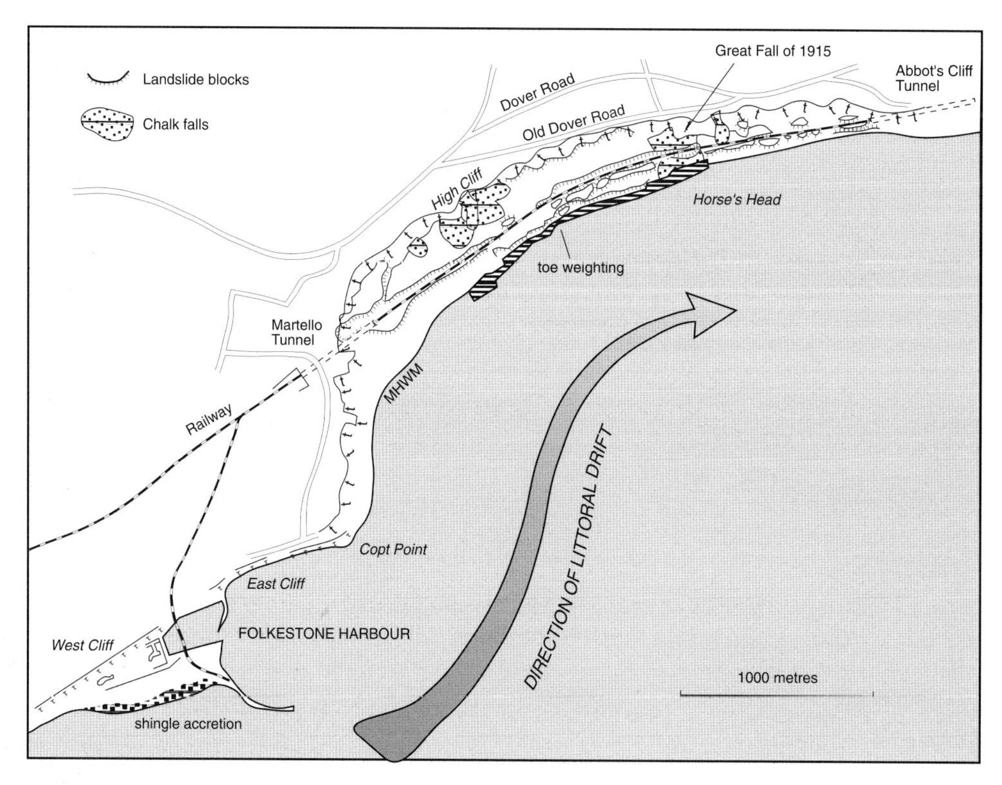

The influence of littoral drift on the stability of the landslide complex at Folkestone Warren, Kent (after Bromhead, 1986).

through the area, leading to the build up of trapped material west of the pier and beach shrinkage through undernourishment to the east at the foot of the Warren (*Figure 4.16*). Reduced beach volume would have led to increased wave erosion of the toes of pre-existing rotated blocks, which slid seawards thereby removing support from the base of the cliffs. The decrease in passive support would have created conditions of marginal stability, with actual failure taking place at times of high groundwater levels following rainfall. Today, stability is only maintained by the existence of major toe-weighting and erosion control structures (*Figure 4.16*) located over 200m seawards of the high failure scarp.

Finally, it is worth mentioning that basal erosion leading to unloading can result from the undramatic process of seepage erosion. Seepage erosion is a very important process on slopes where water-bearing sands overlie a relatively impermeable clay-rich stratum. In such settings, the drag of the groundwater emerging from the slope above the clay layers, may be large enough to dislodge and remove individual soil particles[19], thereby leading to localised slope steepening. This can lead, ultimately, to the undermining and failure of the overlying materials. The highest seepage discharge gradient and, hence, the most active erosion occurs after rapid incision or artificial excavation. In the latter case, it may lead to the problem of 'running sands' feared by many engineering contractors.

Along many stretches of coast, seepage erosion acts in close association with coastal erosion to create distinctive benched clifflines ('undercliffs') as at Chale Bay, Isle of Wight (*Plate 16; Figure 3.15*).

(iii) *Subsidence*

Removal of underlying support can be achieved by natural means through the collapse of subterranean cavities (voids) created in soluble rocks, mainly limestones, by solution weathering, or as a consequence of the underground extraction by humans of solids, liquids and gases. In both cases, it is the resultant vertical movements of rocks and soils, including the associated bending, fissuring and fracturing, that leads to slope instability. However, it must be recognised that some of these effects cause increasing shear stress (bending, surface depressions) while others lead to a decrease in shear strength (fracturing), so that subsidence can be considered both an **external** and **internal** causal factor. Nevertheless, it is convenient to consider it here.

In the context of Great Britain, natural subsidence is not particularly significant in the generation of slope instability. The creation of gorges following cavern collapse has resulted in the formation of cliffs prone to falls, and solution depressions do result in shallow surface movements in deformed surface materials, but the effects are, for the most part, localised and limited. By contrast, the impact of undermining for salt, ironstone and coals may be potentially more significant, especially in the case of coal mining where the destabilising effects are often superimposed on terrains that already have pre-existing landsliding because of topographical and lithological factors. This is especially true of those coalfield areas which were exploited by early, shallow mining techniques and where adits and inclined passageways (drifts) were driven into hillsides.

For many years there has been considerable debate as to the influence of underground coal mining on hillslope stability, especially in the South Wales Valleys[20]. It is generally assumed that the major deep-seated landslides in The Valleys had resulted from deglaciation when the support given by glaciers to the over-steepened valley sides was removed[21]. However, a number of major slides have occurred over the last 130 years, coinciding with the height of the industrial development of the area. A recent study, commissioned by the Department of the Environment, has investigated the relationship between landsliding and undermining[22], using field evidence from a number of slides, supported by a programme of physical and numerical modelling of slope behaviour. At East Pentwyn, for example, the method of mining involved leaving remnant pillars of coal to support the worked seams (the pillar and stall method). This method had a profound effect on the stresses within the hillside (*Figure 4.17*). Prior to mining,

4.17 Vertical stresses before and after mining of the Brithdir Coal seam at East Pentwyn, South Wales (after Jones, D.B. et al., 1991)

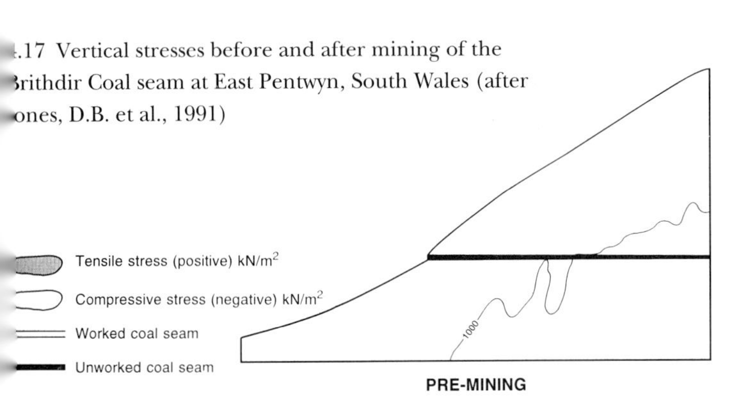

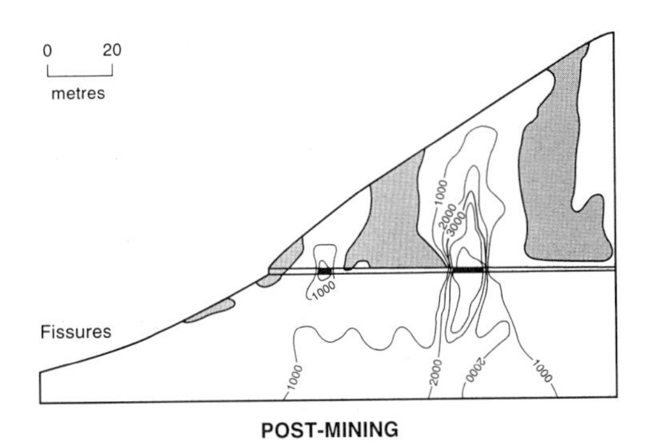

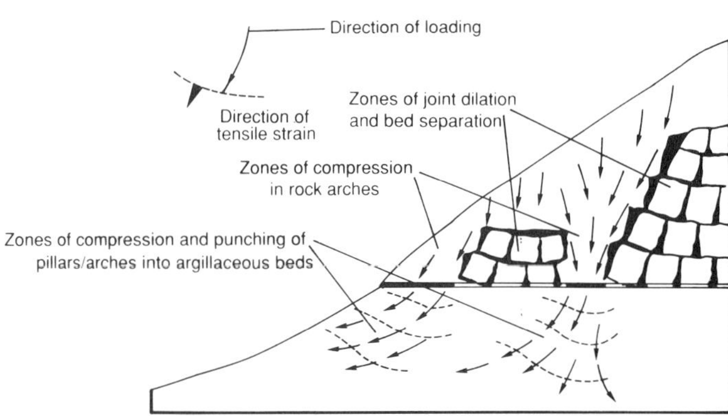

4.18 The effects of shallow mining on the hillside (after Jones D.B. et al., 1991).

stresses are compressional and approximately parallel with the hillside. After working of the coal seam, the stresses were probably concentrated about the pillars, where the stress values were more than doubled. As the stresses were passed into the underlying argillaceous beds, they would have resulted in lateral displacement towards the hillside. This, together with the vertical displacement of the overlying rocks, would have led to the development of fracture zones which subsequently acted as deep-seated failure surfaces (*Figure 4.18*). Nevertheless, the relationship between mining subsidence and landslide initiation/reactivation remains a highly contentious and sensitive issue, for numerous questions still remain as to the significance of mining-induced stresses compared with changes in groundwater conditions and the effects of surface tipping.

Surcharge

Surcharge can occur as a result of a change in the weight imposed on a slope following either natural processes (e.g. the sudden accumulation of rockfall debris in the source area of a slide) or human activity (e.g. construction of an embankment). These changes can either improve, lower or unalter the stability of a slope depending on both the nature of the materials and the positioning of the load. Alternatively, surcharge may simply follow from an increase in slope height caused by erosion or excavation.

In theory, on a slope developed in cohesionless material prone to shallow translational slides, an increase in weight or height will have no effect on the Factor of Safety which is dependent only on the angle of friction, the slope angle and the groundwater level (*Figure 4.10*). The potential influence of surcharge on a slope in cohesive materials prone to rotational slides is very different. Depending on the position of the load, increased weight will either reduce or increase the stability. In the simple model presented in Figure 4.15, an increase in weight at the head of the slope will move the centre of gravity upslope and, hence, increase the overturning moment (W_x), thereby increasing the potential for instability. Conversely, the addition of weight at the foot of the slope will have the opposite effect, and herein lies the basis of stabilisation through 'toe weighting'.

There are several ways in which shear stress can be increased by loading: saturation increases the weight of slope-forming materials but is usually less important than the effects of pore-water pressure in decreasing shear strength. Similarly, the weight of trees can apply a load but, generally, this is more than offset by the stabilising effects of roots. Snowfall, ashfall, loess fall and lava flow are all potential mechanisms of loading, but have little relevance in the context of Great Britain. The most important sources of loading are natural deposition through rockfall and scree accumulation, and artificial additions in the form of dumped waste, fills and structures. The relative size of different examples of surcharges is shown in Table 4.4.

Table 4.4 **Examples of surcharge (after Crozier, 1986)**

Source	Loading (kPa)
10 cm thick concrete slab	2.3
Spruce forest	2.4
1 metre of dry soil	10.0
1 metre of gravel[1]	10.8
1 metre of 30% moist soil[2]	13.0
One storey concrete building	18.5
1 metre of saturated soil	20.0

Note: [1] assuming a void ratio of 0.6 and specific gravity of solids at 2.7. [2] assuming porosity of 50%

Falls and slides apply surface loads to lower slopes which sometimes initiate further instability. An example is provided by the large rockfall at Gore Cliff, Isle of Wight, which occurred in July 1928 (*Photo 2.9*). The weight of this fall, estimated at between 100,000 and 200,000 tonnes, together with subsequent smaller falls, resulted in the initiation of major movement on the landslide slopes below in September 1928. Over 10ha was affected, with parts of the Undercliff below Gore Cliff reported as being "driven bodily seawards"[23]. The foreshore was forced up to form a 3–4m high pressure ridge which was later removed by the sea. A similar pattern is believed to have been partly responsible for the massive landslide at Folkestone Warren in December 1915 (*Photo 2.3*), which followed the sudden loading of inland portions of the landslide complex by a 1 million tonne Chalk fall from the rear scarp[24].

In both the Gore Cliff and Folkestone Warren examples, the movements were probably the result of **undrained loading.** This mechanism involves rapid loading with no dissipation of increased pore-water pressures and, hence, the mobilisation of the undrained shear strength (Cu). Such a mechanism can enable landsliding to occur on slopes considerably flatter than normally expected and, on steeper slopes, enhanced speeds of movement[25]. Undrained loading is of particular significance in the seasonal activity of coastal mudslides.

Surcharge can also be generated by the slow accumulation of loading debris but the cause-effect relationship is more difficult to prove. Movements of the East Pentwyn landslide in South Wales may have resulted in part from falls off the backscar[26], and surcharge by scree accumulation has been suggested for the upper parts of Stonebarrow[13] (*Figure 3.25*).

Water bodies, such as a lake or the sea, can often provide a surcharge at the base of a slope which improves the stability conditions. An interesting example has been recorded at Sandgate, on the Kent coast, where recorded landslide movements of around 1mm per day decelerate or cease at high tide level and accelerate at low tide

4.19 Cumulative graph of crack opening at Sandgate, Kent, showing the relationship between tide levels and ground movement (after Hutchinson, 1988).

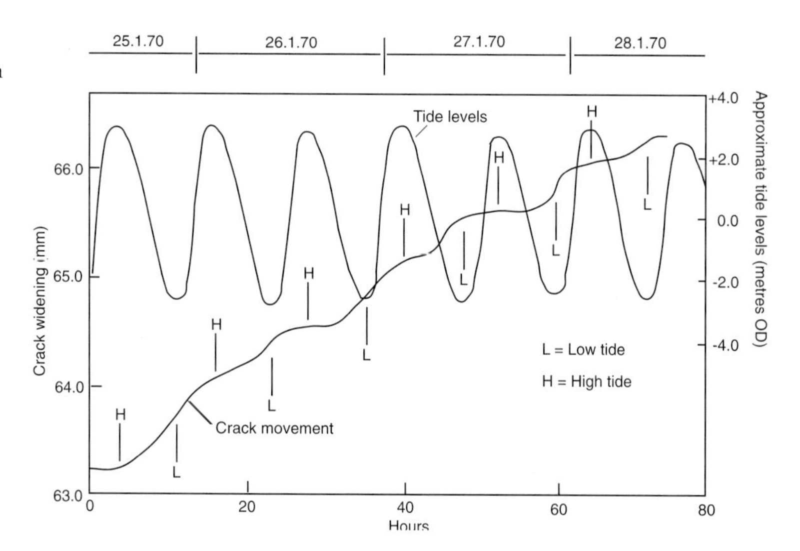

(*Figure 4.19*)[2]. However, the reverse also applies, in that the lowering of water levels can contribute to instability through unloading (see the subsequent section on **Water Regime Change**).

Landslides caused by loading are often associated with human activity. The dumping of colliery waste was a factor in the Aberfan disaster of 1966 and certainly contributed to the recent movements of the Blaina landslide[27]. The latter slide involved a grass-covered tip of colliery waste which failed during the early years of this century, destroying a row of six cottages and threatening the local school. The construction of fill embankments may also result in instability problems, as occurred at Walton's Wood on the M6[7] (*Figure 4.5*). More widespread earthworks can result in failure as occurred at Bury Hill, Staffordshire during the construction of a housing estate[28]. Investigations revealed that the tipping of colliery waste had caused a landslide to develop in the underlying Etruria Marls at some time between 1938 and 1955. Extensive cut and fill operations during 1959–1960 merely served to reactivate movements along pre-existing shear surfaces.

4.20 The influence of slope height on the stability of slopes developed in cohesive materials (after Crozier, 1986).

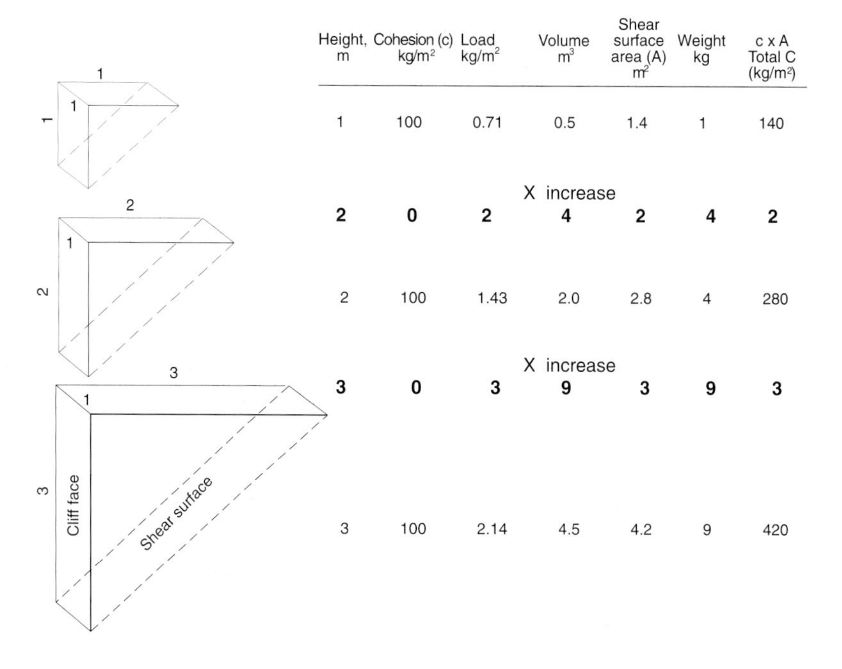

Height, m	Cohesion (c) kg/m²	Load kg/m²	Volume m³	Shear surface area (A) m²	Weight kg	c x A Total C (kg/m²)
1	100	0.71	0.5	1.4	1	140
			X increase			
2	**0**	**2**	**4**	**2**	**4**	**2**
2	100	1.43	2.0	2.8	4	280
			X increase			
3	**0**	**3**	**9**	**3**	**9**	**3**
3	100	2.14	4.5	4.2	9	420

A different form of surcharge is produced by increasing slope height due to natural erosion or excavation. This is because shear stress increases at a greater rate than the shear strength mobilised on a potential shear surface through the base of the slope. This situation is illustrated in Figure 4.20. When the height of a slope formed of homogeneous material is doubled (at a constant slope angle), the weight of material above the potential shear surface and, hence, the total stress increases four-fold. At the same time, the area of the shear surface is only doubled, thereby indicating that the available cohesion is only doubled. Clearly a doubling of height represents only a doubling of strength as against a four-fold increase in stress. Triple the height of the slope and the cohesion available to be mobilised on the potential shear surface rises to three times the original value, while the stress increases ninefold.

Transitory Earth Stress: Shocks and Vibrations

Shock waves are an important cause of slope failure in many areas of the world, as is testified by the widespread landsliding that often accompanies major earthquakes. Earthquake shocks can be considered to be **external factors** as they increase the shear stress along potential shear surfaces, whilst shear resistance is often unaltered. Considering Figure 4.15, the effect of an earthquake is to produce a horizontal acceleration which increases the overturning moment (W_x) and, hence, reduces the Factor of Safety.

Vibrations may also destabilise materials by literally shaking off portions of rock or soil slopes defined by joints and tension cracks, so as to yield falls and slides. A small number of failures are claimed to have been triggered by human-induced vibrations, which seems reasonable considering the magnitude of recorded explosions and the increase in transport generated vibration. Nevertheless, it is the seismic shock category that probably has the greatest significance for slope instability, despite the widespread view that Great Britain is relatively aseismic.

Seismic vibrations can be generated by the collapse of subterranean voids (both natural caverns and mine galleries) and crustal movements (earthquakes). Subsidence has been a widespread phenomena in areas of deep mining and numerous recordings exist of pseudo-earthquakes, 'goths' or 'bumps', although the actual mechanisms remain unclear. This phenomenon has been especially prevalent in north Staffordshire during the 1970s, with 11 recorded in 1975 and 59 noted in 1976–7, but the extent to which such minor shocks can stimulate slope instability has yet to be investigated in detail. However, subsidence was almost certainly the cause of instability in the past when primitive mining practices were employed at shallower depths. Wallwork[29] noted that 'subsidence in its more violent forms is now rarely experienced, the cataclysmic upheavals of the 19th and early 20th centuries have given way to a more insidious form of subsidence'. It is likely that the dramatic events referred to have contributed to slope instability, although there is little direct evidence.

Earthquake-generated shock waves present an even more significant potential cause of mass movement, although here again, there has been little research on this topic in the context of Great Britain. The British Isles do experience fairly frequent minor earth tremors (*Figure 4.21*). Indeed, there have been a number of shocks that must

4.21 Earthquake magnitudes and intensities in and around Great Britain.

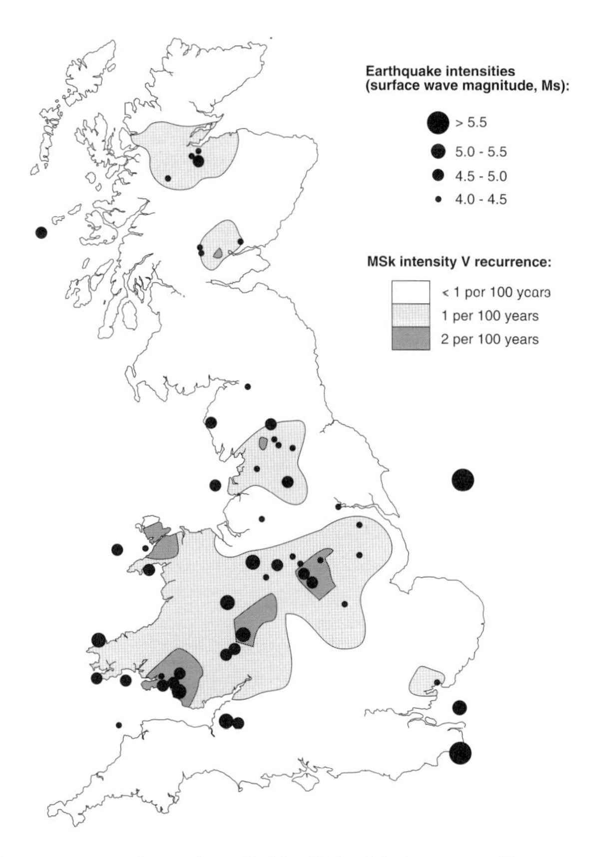

have caused significant damage*: the great earthquake of AD 974 which, according to Symeon of Durham "took place all over England"; the quakes of 1246 and 1382 located in the Straits of Dover, the latter having sufficient strength to damage Canterbury Cathedral; the very powerful Inverness shocks of 1769 and 1816; the Colchester earthquake of 1884, otherwise known as 'The Great English Earthquake', which damaged 1200 buildings over a radius of 200km and resulted in the deaths of four people; the Hereford earthquake of 1896; the Dogger Bank quake of 1932; and the Wrexham earthquake of 1990.

The scale of damage caused by some of these shocks which included the destruction of buildings and opening of fissures, indicates that they could have produced failures on marginally stable slopes and cliffs, and there is one source who claims that the 1932 shock caused coastal falls from Flamborough Head to the Isle of Wight[30].

Seismic shocks may also cause instability through the liquefaction of sands, silts and clays. The result is a dramatic reduction in shear strength and therefore must be

---

* Earthquakes are characterised by the **magnitude** of the measured surface wave (Ms values) and the **intensity** of the degree of shaking felt during an earthquake. The latter value is based on observations of earthquake effects which are then compared with a standard scale. In Europe the Medvedev-Sponheur-Karnik (MSK) scale is generally used, with earthquakes greater than intensity V recognised as significant events. Figure 4.21 shows the recurrence of earthquakes of greater than MSK Intensity V, which is defined as:

> "Doors open, slam shut. Felt by all indoors. Loose household articles may fall. Many people awoken, many alarmed, some may run outdoors. Animals disturbed, particularly at night. Field walls may fall. Very occasional Grade 1 damage (fine cracks in and fall of plaster, slight cracks in chimneys. Loose bricks, tiles and architectural ornaments fall. Windows break)".

classified as an internal factor, but for the sake of convenience it will be considered here. A loosely packed sand, for example, has strength because of contact between individual grains. If shaken, it will compact into a denser form with more inter-grain contacts and therefore greater strength. However, if the shaken sand is saturated the conversion to a denser form is restricted because the water cannot drain away. Consequently, the changing fabric of sand grains transfers some of its inter-granular effective stresses to the pore-water and pore-water pressures increase. In extreme cases, all the effective stresses are transferred to the water, the inter-particle frictional strength declined to zero and as the water has no resistance to shearing, the sand body deforms as a liquid. Sensitive clays behave in the same way, as do certain badly consolidated engineering fills. Uplifted, young marine clays are especially prone to this type of failure because the original high salt content tends to be removed by leaching, leaving the clay particles in their original disposition but without the bonding provided by the salt crystals.

The best example of liquefaction in Great Britain is the large failure that occurred in June 1975 during construction of the Portavadie Dry Dock in western Scotland[31]. The Dock was being excavated through a variable sequence of Quaternary deposits and recent marine laminated silts, when a slide occurred which caused slurried laminated silts to move up to 270m away from their original position on the excavated face (*Plate 27*). The slide is considered to have been caused by a chain reaction from an initial relatively small slip on the steep excavated face. This first failure produced a shock wave which passed through the soil mass causing loss of structure and large-scale rapid movement. The sensitive nature of the laminated silts is thought to have been the result of salt loss through leaching by freshwater.

Lateral Pressure

Weathering (see internal factors) generally results in a decrease in shear strength. However, a number of weathering processes operating upon a soil or rock mass may also contribute to the build up of lateral pressure and cause an increase in shear stress. These processes include root wedging, cleft-water pressure, ice wedging, salt crystal growth and clay swelling[9].

Internal Factors

Internal factors (*Figure 4.9*) may be less visible than external factors but they are no less important in the generation of instability. Their effects on the reduction of shear strength (shear resistance) can trigger slope failure, but they are particularly important in the longer term progressive alteration of slopes to conditions of marginal stability. Three main groups of internal factors are often recognised, although there is some disagreement in the literature as to the most desirable number and the correct classification of certain factors. Close inter-relationships exist between many internal factors, and between internal and external factors, thereby making classification problematic.

Weathering

Ground materials located at or close to the ground surface are subjected to a range of processes that work to achieve change by physical and chemical means. The nature of these weathering processes have been fully reviewed elsewhere[32] and it merely serves to note that they are normally divided into two groups: **physical weathering,** sometimes referred to as mechanical weathering, which cause

Table 4.5 **Classification of weathering processes (after Cooke and Doornkamp, 1990)**

I Processes of disintegration

a. **Crystalline processes**
Salt weathering (crystal growth, hydration, thermal expansion)
Frost weathering

b. **Temperature/pressure change processes**
Insolation weathering
Sheeting, unloading

c **Weathering by wetting and drying**
Moisture swelling
Alternate wetting and drying
Water-layer weathering

d. **Organic processes**
Root wedging
Colloidal plucking
Lichen activity

II Processes of decomposition

a. **Hydration and hydrolosis**

b. **Oxidation and reduction**

c. **Solution, carbonation, sulphation**

d. **Chelation**

e. **Biological chemical changes**
Micro-organism decay, bacteria, lichens.

breakdown without significant mineral alteration (**processes of disintegration**) and **chemical weathering** which achieve change by chemical alteration of minerals (**processes of decomposition**) (*Table 4.5*). The effects of physical and chemical weathering are complementary. However, physical processes tend to be more rapid and partition rock masses into progressively smaller pieces, thereby increasing surface area and facilitating accelerated rates of chemical weathering. Thus physical disintegration can be seen to act as a control on chemical decomposition[33].

The degree to which ground materials are altered and weakened by weathering processes has a profound influence on mass movement by reducing shearing resistance. This is determined both by the nature of the combination of weathering processes operating on particular masses of material and the extent to which the processes of change have been allowed to work. The terms 'degree of weathering' or 'engineering grade of weathering' (*Table 4.6*) are used to describe the extent to which a rock has been altered. This varies with (1) the nature of exposed rocks and superficial deposits, (2) the rate of operation of the various weathering processes (itself determined by temperature, rainfall, vegetation, groundwater conditions, etc.), (3) the duration of weathering activity , i.e. the length of time that weathering processes have been able to modify particular masses of rock and soil, and (4) the inter-relationship between rate of weathering penetration into the ground and rate of ground lowering by surface erosion. The varied character of the geological fabric 'able 4.6 **Engineering grade classification of weathered rock (after Fookes et al, 1971)**

rade	Degree of decomposition	Soils (i.e. soft rocks)	Rocks (i.e. hard rocks)
I	Soil	The original soil is completely changed to new structure and composition.	The rock is discoloured and is completely changed to a soil in which the original fabric of the rock is completely destroyed.
	Completely weathered	Soil is discoloured and altered with no trace of original structures.	The rock is discoloured and is changed to a soil, but the original fabric is mainly preserved. The properties of the soil depend in part on the nature of the parent rock.
/	Highly weathered	Soil is mainly altered with occasional small lithorelicts of original soil. Little trace of original structures.	The rock is discoloured; discontinuities may be open and have discoloured surfaces and the original fabric of the rock near the discontinuities is altered; alteration penetrates deeply inwards, corestones present.
I	Moderately weathered	Soil is composed of large discoloured lithorelicts of soil separated by altered material. Alteration penetrates inwards from the surfaces of discontinuities.	The rock is discoloured; discontinuities may be open and surfaces will have greater discolouration with the alteration penetrating inwards; the intact rock is noticeably weaker, as determined in the field, than the fresh rock.
	Slightly weathered	Material is composed of angular blocks of fresh soil, which may or may not be discoloured. Some altered material.	The rock may be slightly discoloured; particularly adjacent to discontinuities which may be open and have slightly discoloured surfaces.
	Fresh rock	Parent soil shows no discolouration, strength loss, or other effects due to weathering.	The parent rock shows no discolouration, loss of strength, or any other effects due to weathering.

of Great Britain (*Figure 2.2*) and the variable intensity of denudation, indicate that surface materials may be anything from completely fresh rock to wholly altered and much weakened material. The nature of contemporary weathering is also extremely variable and depends on latitude, altitude, proximity to the coast and nature of land use. Thus the harsh, sub-arctic, rock shattering weathering environment of elevated parts of the Scottish Highlands is very different from the warm temperate chemical rotting conditions prevailing in lowland Cornwall.

(i) *Processes of disintegration*

Six processes of disintegration are of relevance to mass movement in Great Britain: pressure release/unloading phenomena; strain softening; expansion and contraction due to variable moisture content; salt weathering; frost weathering and root wedging. These physical processes are significant in the context of mass movement in that they (i) create fissures and enlarge joints thereby reducing the strength of the ground-forming materials; (ii) create pathways (fissures etc.) for the increasingly efficient ingress of water into the ground materials, thereby aiding in the process of decomposition; and (iii) on occasions actually provide sufficient stresses to trigger failure.

Few surface materials have experienced the same conditions since formation. For the most part, they have suffered lengthy periods of burial beneath a cover of overlying rock strata and/or superficial sediments, with the result that they have become consolidated due to the weight of overlying material. In those cases where the strata were once very much more deeply buried than they are today, the effects of past loading persist and the materials are called overconsolidated. Erosion processes result in the dissection of rock sequences and the removal of surface materials so as to reveal progressively older strata and in so doing, cause the reduction of confining pressure as the buried rocks come close to the land surface and the weight of overburden diminishes. This leads to the expansion of rock due to stress relief, mainly through the formation of fissures and joints which partition the rock mass into small units, thereby facilitating the ingress of water and an associated increase in water content. This expansion is due to the release of strain energy stored in compaction by the bending and compression of clay minerals. As exhumed materials are free to expand in the vertical direction but not the horizontal, this leads to the development of high horizontal stresses. These horizontal stresses are released very slowly and result in the progressive reduction in the long-term stability of a slope. This process is likely to have been important in the initiation of cambering and valley bulging, major deformations of surface strata that took place during the colder phases of the Pleistocene (*see Chapter 5*).

These effects can be well illustrated to the Upper Lias Clay of the East Midlands and the London Clay of South-eastern England. Both are overconsolidated clays, the Lias having experienced burial beneath 1,000m of covering strata, the London Clay up to 400m. In each case, therefore, the clay particles are initially very closely packed resulting in low permeability and low water contents. The gradual removal of overburden by long-term denudation processes has resulted in the creation of fissures in the near surface layers which are altered by weathering to the appearance of a breccia extending to a depth of 10m or more, with nodules of the original clay material set in a matrix of remoulded clay. Many of the landslides in Lias Clay have been found to have failed at well below anticipated peak strength, including those along the ironstone escarpment at Rockingham[34], thereby indicating the significance of progressive failure.

Progressive failure is also an important factor when considering the stability of cuttings in clays. When a slope is overstressed, because it is either too steep or too high, elements of clay must shed part of the load they are supporting to neighbouring elements. As a result the available shear strength decreases as parts of the soil deform under increasing strain or displacement. Thus, if a failure occurs the average mobilised strength will be close to the fully softened strength. A good example of progressive failure is provided by the movements on the face of an Oxford clay pit at Saxon Pit, Cambridgeshire[35], where a block of material slid gradually about 0.2m along a shear zone developed as a result of this mechanism. Progressive failure was also reported to have been one of the factors that caused the collapse of Carsington Dam, Derbyshire in June 1984[36]. This embankment dam had been designed on the basis of peak shear strengths for the foundation and fill materials. However, progressive failure and strain softening (*see below*) were shown to have reduced the apparent Factor of Safety by over 20% in a 2 year period. An unusual feature of the slide, which occurred where fill was being added to the dam, was that a relatively small initial failure (50m long) spread 500m along the length of the dam over the next two days. Research into the mechanics of the failure has suggested that the initial slide resulted in the removal of support to adjacent sections (lateral load transfer). This caused progressive failure of sections of the dam that originally had a significant margin of stability, and led to the spread of the failure across the valley.

Many natural slopes (e.g. river bluffs) and excavations may stand for a period of time before suffering failure. Skempton[37] found that the timing of failure in slopes excavated in London Clay (mainly railway cuttings) varied with slope steepness, with 25° slopes failing in 10–20 years and 18° slopes after 50 years. Specific examples of such **long-term** or **delayed** failures are the slides at Wembley Hill in 1918 (14 years delay[37]), Kensal Green (29 year delay[38]) and Upper Holloway (81 year delay[39]). Similar long-term failures have been recorded for other clay strata, including a slide in Weald Clay near Sevenoaks in 1939, 70 years after excavation[40].

A number of alternative explanations have been advanced to explain these failures. In addition to the progressive failure mechanism already described, other possible reasons include strain **softening** and **swelling**[41]. Softening involves the reduction in the available shear strength resulting from an increase in the void ratio (the ratio of the volume of pore spaces to the volume of solids) under constant effective stress. This is a time-dependent process and results in the gradual reduction of available shear strength to the fully softened strength (*see Figure 4.3*).

Excavation of overconsolidated clay slopes often leads to a short-term depression of pore-water pressure. For example, in a London Clay cutting at Edgewarebury, on the M1 motorway, the pore-water pressure was substantially depressed even after a 20 year period[4]. Similar observations have been made in coastal landslides such as Folkestone Warren[24], where movements in 1940 caused unloading and depressed pore pressures. The recovery of pore pressures generally involves a long time period during which swelling, as a result of stress relief, accompanied by an increase in water content, are probably the most important factors.

The above discussion has concentrated on effects developed in clay strata. Intense erosion of hard rocks frequently results in the development of pronounced joints parallel to the ground surface which, in the case of marine cliffs or glacially overdeepened valley-sides and troughs, means steeply inclined to vertical discontinuities that partition the rock mass into unstable sheets prone to falls, topples and rockslides. These pronounced discontinuities are known as stress relief joints, dilatation joints or exfoliation joints, and are thought to be responsible for the development of many of the recorded rock-slope failures in the Scottish Highlands (*Figure 3.28*).

Salt weathering can also attain local significance in the context of Great Britain. Salts are deposited as a consequence of atmospheric pollution (acid deposition) and sea spray. Both crystal growth and hydration can result in the development of high pressures in confined spaces, thereby leading to fissure development and enlargement. Sodium Chloride (NaCl) and Calcium Sulphate ($CaSO_4$) are common in coastal areas, and although not particularly aggressive, they can produce pitting on rock faces (tafoni) and assist in crack enlargement. Root wedging is important on steep slopes in hard rock terrains and greatly assists in the opening of joints. Falls and topples are known to have been triggered by root growth or when the trees have been disturbed by windthrow.

The freezing of water is a well known disruptive influence. The cooling of water results in an increase in density to a maximum at 2°C, from which point density decreases with decreasing temperature. The change from liquid to solid form also increases specific volume by 9% and the lower the temperature the greater the pressure exerted by freezing until levels are reached far in excess of the tensile strength of most rocks. However, such pressures are not usually achieved outside the laboratory because freezing within fissures tends to result in ice being extruded from the fissure mouth. Nevertheless, the evidence of frost shattered blocks of rock and jagged debris in arctic and alpine environments indicates that freezing does disrupt rocks. Indeed, investigations into the development of talus slopes in Scotland have shown that rockfall activity during the last glaciation was over a hundred times that experienced in contemporary conditions[42].

The potency of frost weathering is determined by the nature of the rocks (tensile strength, porosity), their water content, the speed of temperature change and the frequency of freezing (number of diurnal or seasonal freeze-thaw cycles). Freezing was clearly an important factor in slope stability during the arctic conditions of the Pleistocene as is testified by the abundant accumulations of screes (*Plate 14*) and rockfall debris in the uplands of Wales, the Lake District and the Scottish Highlands (*Photos 3.3. and 3.4*). Frost action continues to be important in the uplands and on coastal cliffs, in the latter case probably working in conjunction with salt weathering. Investigation of the factors important in generating falls from coastal Chalk cliffs in Kent[43] (*Figure 4.22*) has indicated that freezing is of significance.

Finally, brief mention needs to be made of the ability of certain clay minerals to readily hydrate and dehydrate, with consequent marked changes in volume. This characteristic is particularly marked in the clay mineral **montmorillonite,** less developed in **illite** and negligible in **kaolinite.** Soils containing montmorillonite are

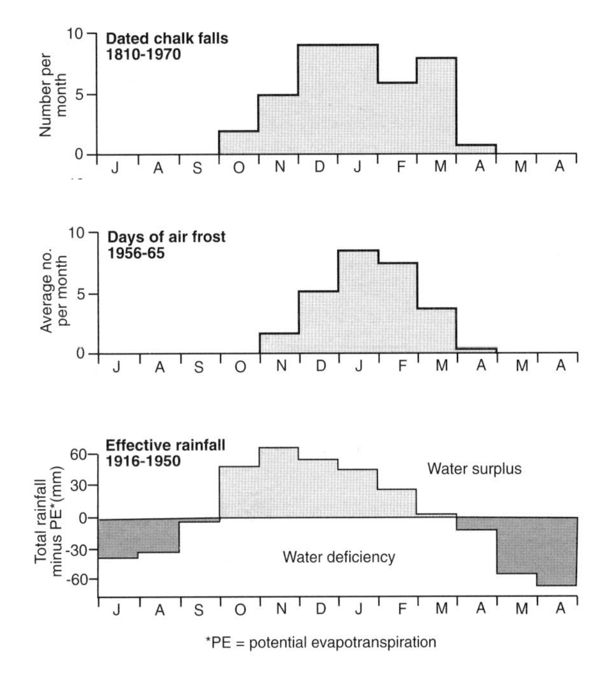

4.22 The incidence of chalk falls, frost and effective rainfall on the Kent coast (after Hutchinson, 1972).

therefore prone to expansion and contraction, thereby creating fissures, and can readily deform if they have high water contents. Such soils are well known for the problems they can cause to building foundations in dry years, such as the famous 1976 drought[44], but they can also assist in the development of shallow slides.

(ii) Processes of decomposition

A very broad range of chemical processes act to alter rock crystals, sediment grains or the cements that bind grains together to form sedimentary rocks. In certain cases, these changes result in altered volumes, in other cases new secondary minerals are produced. The net result, however, is that materials are progressively weakened as weathering grade increases, so that stability is changed to marginal stability as shear strength is reduced. Perhaps the most well-known example of the effects of chemical rotting is the conversion of tough, durable granite to a mixture of clay (kaolin) and sand capable of slumping and flow.

Weathering changes lead to a reduction in shear strength and increase in water content, as indicated in Table 4.7. Indeed, large reductions in the undrained strength, by as much as 50%, have been recorded. Peak strength also may decline from the non-weathered to the weathered state. This change usually involves a large drop in apparent cohesion with only a small change in the angle of friction.

The oxidation of pyrite (ferrous sulphide) is one of the most important chemical weathering processes in Great Britain. The processes involved can result in large volume increases[45], with resulting pressures of up to 500kNm^2. At Mam Tor in Derbyshire, for example (*Photo 2.10*), weathering of pyrite in the Edale Shales probably leads to a reduction in the residual shear strength[46]. This, and other studies[47], indicates that residual strength of clays is, to a large extent, determined by the type of clay minerals and the chemistry of the constituent pore-water. Significant

Table 4.7 **Engineering properties of selected fresh and weathered mudrocks and over-consolidated clays (after Taylor & Cripps, 1987)**

	Formation	London Clay	Lower Oxford Clay	Upper Lias Clay	Keuper Marl	Coal Measures Mudrock
	Geological age	Tertiary	Upper Jurassic	Lower Jurassic	Triassic	Carboniferous
Bulk density Mg/m^2	Fresh	1.92–2.04	1.84–2.05	1.87–2.09	2.20–2.50	2.15–2.76
	Weathered	1.70–2.00	–	1.79–1.96	1.80–2.30	1.86–2.18
Undrained shear strength kN/m^2	Fresh	80–800	96–12,000	40–1,200	130–2,800	9,000–103,000
	Weathered	40–190	52–93	20–180	70–200	15–335
Effective shear strength c' kN/m^2 Ø' Degrees	Fresh c'	31–252	10–216	27	>30	c' ≈ 131
	Ø'	20–29	23–40	24	>40	Ø'≈ 46
	Weathered c'	1–18	0–20	1–17	2–80	0–25
	Ø'	17–23	21.5–28	18–25	25–42	26–39

reductions in shear strength, and particularly in cohesion, can occur as a result of weathering through an increase in soil moisture, physical disruption and loss of bonding. Residual strength of the weathered mantle may also be modified by fluctuations in pore-water chemistry. Indeed, seasonal fluctuations in the concentration of salts in pore-water could be expected to modify the residual strength of natural clays. Residual strength may be lowest at low pore-water salt concentrations and highest for concentrated pore-water, because where there is a high salt concentration of high valency cations and little soil water, there will be a tendency for strong ionic bonds to form between and within particles, and cohesion will be strong. When the pore-water is dilute the bonds are chemically weakened and possibly exchanged. Diffuse double layers develop causing the swelling and dispersion of particles, and consequently a reduction in the friction between particles. Long-term changes in environmental chemistry can also be expected to cause similar modifications in shear strength and may provide an explanation for the long-term and progressive failure of clay slopes.

In the case of more durable bedrocks, weathering activity is largely confined to enlarging joints which, in turn, facilitates greater penetration of water into the rock mass. This process has been shown to work on Exmoor where shallow failures in Devonian Slates occur after prolonged periods of rainfall due to the gradual reduction in strength of the rock materials[48]. The strength of rocks and superficial deposits may be further reduced by the removal of cements in solution, most especially Calcium Carbonate ($CaCo_3$).

The ubiquitous nature and variable character of chemical weathering make it certain that it contributes, to some degree, in the majority of slope failures in Britain. However, it is a 'preparatory' or 'passive' factor and therefore is rarely seen as significant in generating instability. This explains why chemical weathering appears as an insignificant process in the causal data-set (*Table 4.3*) with only 29 citations (<1% of recorded attributes).

Water regime change

Changes in water content can quickly affect slope stability and have been responsible for triggering or reactivating more landslides in Great Britain than any other factor (*Table 4.3*). The most important effect of an increase in pore-water pressure is the reduction in effective stress and, hence, available shear strength (described earlier).

High pore-water pressures that contribute to the development of slope failures can normally be achieved in a number of ways:

(i) by intense rainfall;

(ii) by snowmelt;

(iii) by blocked surface drainage lines causing lateral seepage;

(iv) by the creation of impervious barriers to subsurface water flows (i.e. walls, deep foundations etc.);

(v) by poor surface water disposal (e.g. road drains);

(vi) by seepage from broken water pipes, sewer pipes, septic tanks, cesspits, swimming pools and sunken storage tanks; or

(vii) by combinations of the above.

Examples of landslides caused by high pore-water pressures are widespread throughout Great Britain. The following examples are merely intended to illustrate how high pore-water pressures can act as both preparatory and triggering factors.

Between November 1987 and January 1988 landslide movements occurred at Luccombe Village, Isle of Wight, resulting in severe damage to a number of properties (*Photo 4.2*). The village had been built in about 1930 on relatively gentle slopes above a 80m high sea-cliff developed in alternating sandstones and clays of the Lower Greensand. The slopes upon which the village was built are now known to form part of an ancient landslide system developed in the Gault Clay and overlying Upper Greensand, and thus the recent movements have involved failure along pre-existing shear surfaces at, or close to, residual strength[49] (*Figure 4.23*).

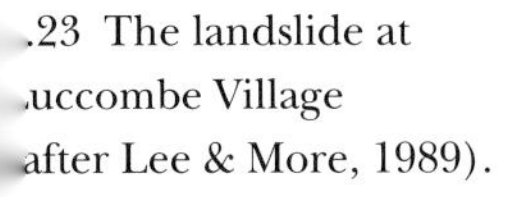

.23 The landslide at
.uccombe Village
after Lee & More, 1989).

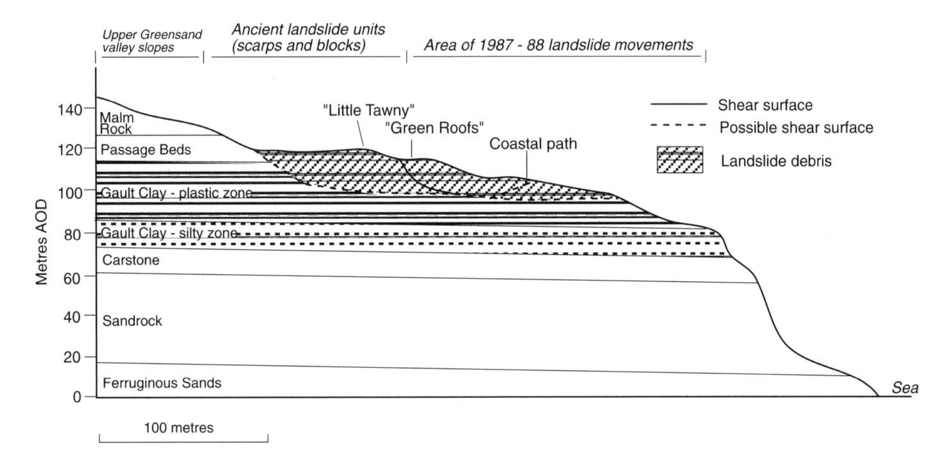

4.2 Severely back-tilted property, Luccombe, Isle of Wight.

Retreat of the sea cliffs is estimated at 0.3m per year. This has had the effect of reducing support to the landslide slopes underlying the village as the coastline retreated inland of the original position of the landslide toe. As a result of this long-term process, the stability of the inland slopes has gradually deteriorated. In addition, the development of housing in the area has contributed to a reduction in stability through the disruption of the natural drainage and artificial groundwater recharge by leakage from septic tanks and water supply pipes. It is clear that human activity, together with long-term coastal erosion, have acted as preparatory factors, and have produced a situation whereby the slopes within the village have become increasingly susceptible to reactivation. However, the movements in 1987–1988 were triggered by a prolonged period of heavy rainfall during which 638mm fell between September 1987–January 1988 (the 4th largest wet phase since 1947).

In the case of Luccombe Village, a **transient** (fast changing) factor (rainfall) assumed an abnormal value resulting in pore-water pressures that rapidly lowered the factor of safety, itself having suffered gradual decline over time. Examples of such transient phenomena are common. For instance, during the night of 2–3 September 1983 heavy rainfall, with an intensity of 40mm/hour, fell over the Snowdonia area. This intense storm had a return period of 75 years[50] and triggered numerous debris flows, a number of which blocked the A5 trunk road to the south of Bethesda, trapping several vehicles. Instances of landslides being triggered by heavy rainfall have been reported from many parts of upland Britain including Exmoor[51] (the 1952 storm), the North York Moors[52], Teesdale and Weardale[53] and the Ochil Hills of Central Scotland[54]. A similar combination of circumstances led to the dramatic movements at Blackgang, Isle of Wight, in January 1994 (*Plate 28*).

The relationships between rainfall arriving at the ground surface, the generation of pore-water pressures within ground composed of variable materials, and the creation of instability, are extremely complex. A slope experiencing a general, long-term reduction in stability may withstand the same pore-water pressure on many occasions without displaying signs of stress until a point is reached when a transient event of similar size reduces the Factor of Safety to below 1 and failure occurs. Similarly, modification to slopes through the creation of foundations, retaining walls etc, may impede water movements, thereby causing saturation so that the same rainfall events that were harmless in the past now create high pore-water pressures that may lead to failure. Patterns of water supply to the ground can vary considerably over time and this may mean that the significance of a specific rainfall episode is increased because of the amount of water already in the ground prior to the event. Moisture status prior to a specific event is known as **antecedent moisture condition** and is affected by the pattern of previous rainfall episodes (antecedent precipitation), seepage from water bodies such as ponds and leakage of water pipes, storm sewers etc. Such antecedent conditions may be extremely variable and can display both short-term fluctuations and long-term trends. As a consequence, the relationships between perceived cause (rainfall) and response (instability) may vary considerably both in terms of the size of the generating event and the interval between 'cause' and 'effect' (the lag).

The effects of variable moisture status have already been referred to in the discussion of instability at Luccombe (*see above*), but these can be exemplified by consideration of the movements at the nearby town of Ventnor, Isle of Wight. Ventnor is a small town built within a large coastal landslide complex, known as the 'Undercliff', (*Photo 2.4*) developed in Upper Greensand, Gault Clay and Lower Greensand Strata. A range of contemporary movements are known to occur, although the scale of movement is only slight (*Figure 4.24*). However, as they take

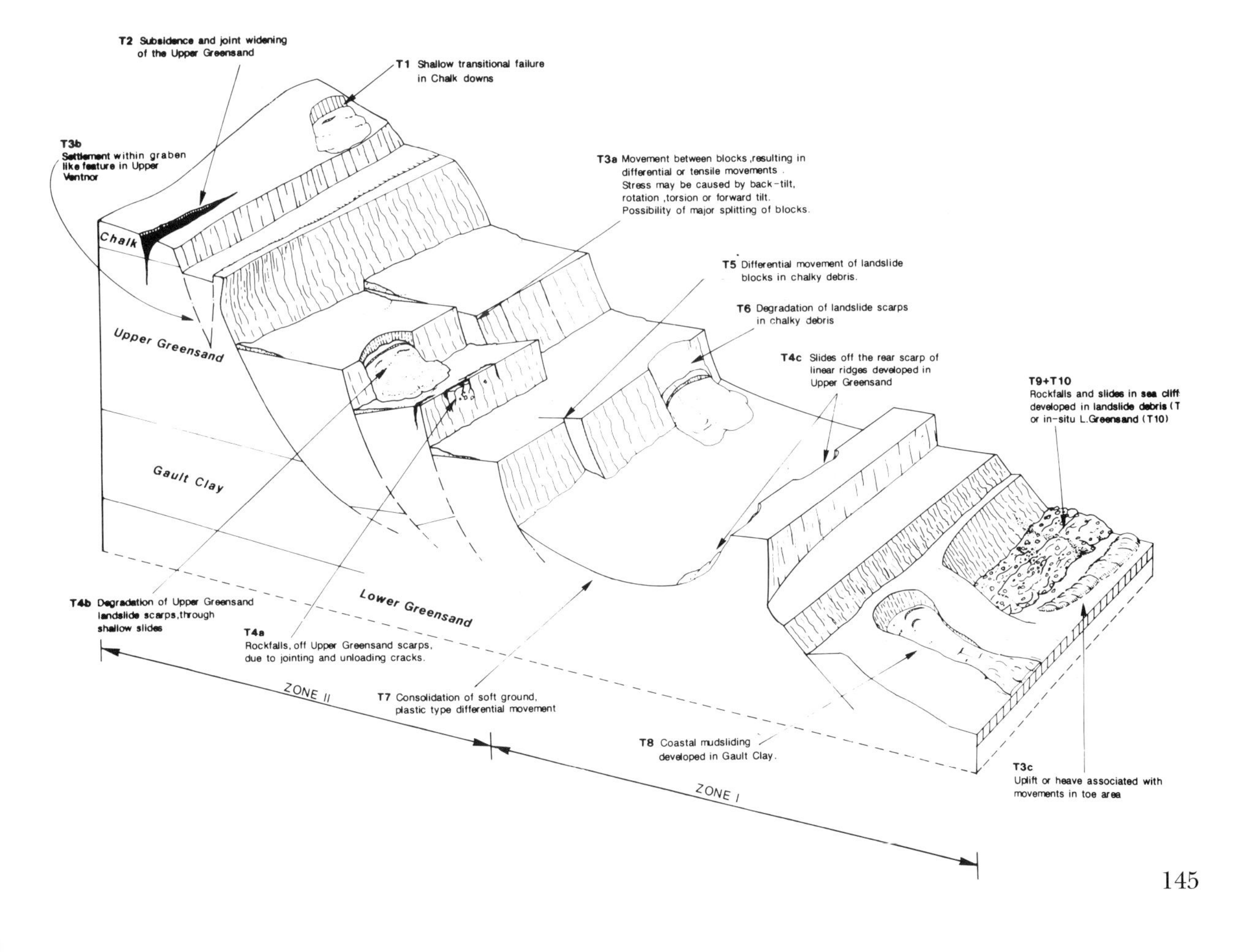

4.25 The relationship between antecedent rainfall and landslide activity in Ventnor, Isle of White (after GSL, 1991; Lee et al., 1991a(.

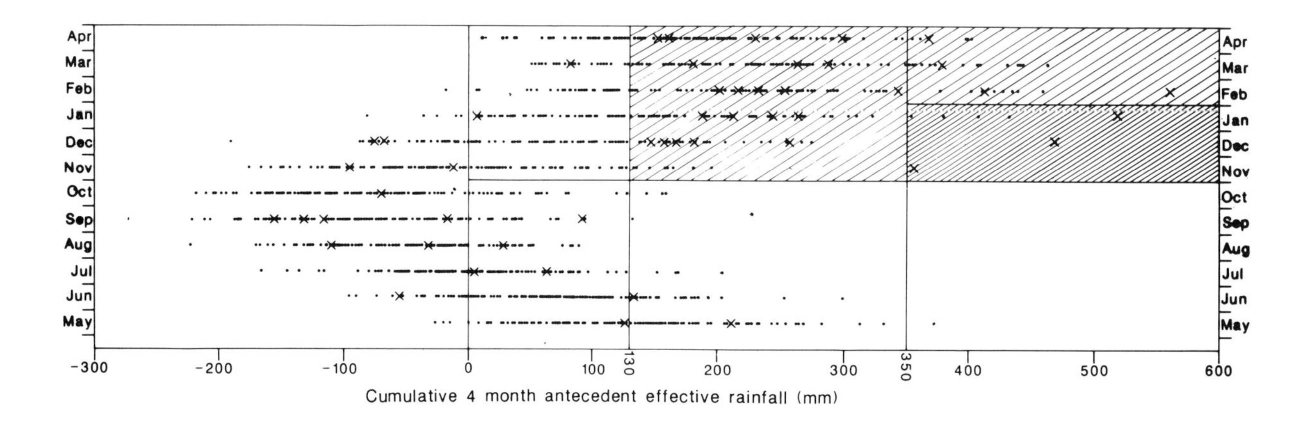

place within an urban area with a permanent population of over 6,000, the cumulative damage to roads, buildings and services in parts of the town has been significant[55]. Since 1800 there have been eight main periods of reported movement; 1873–1879, 1910–1916, 1921–1928, 1932–1941, 1954–1955, 1960–1961, 1976–1980 and 1988. These movement episodes show a coincidence with periods of high rainfall. In addition, there is evidence of a marked increase in annual rainfall at the site since 1920 and this has been reflected in a rise in the number of reported incidents of movement. The relationship between movement in the town and the proceeding four month rainfall is shown in Figure 4.25 and demonstrates that the higher the antecedent rainfall conditions the greater the probability of movement occurring. Indeed, the most damaging period of movement which took place in the winter of 1960–61, followed the highest winter rainfall total since records began in 1839. However, it must also be recognised that the pattern of movement reflects two other factors: the development of water supply and sewage disposal infrastructure over the period in question and the fact that greater awareness of landsliding may have increased the number of reported incidents.

4.26 The changes in pore pressure and factor of safety during and after the excavation of a cutting in clay (after Bishop &Bjerrum, 1960).

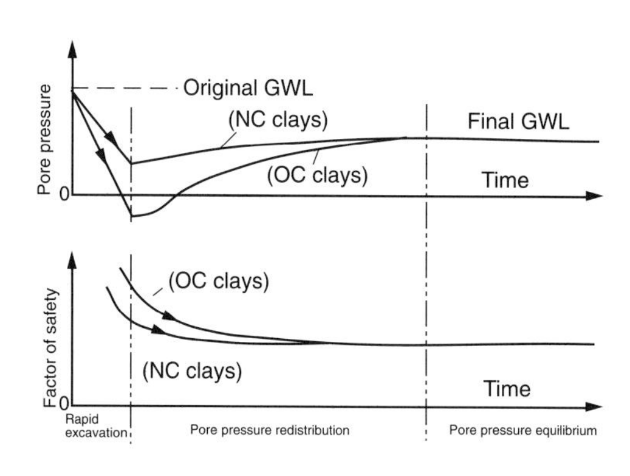

The importance of pore-water pressure changes during slope excavation have already been discussed with reference to the delayed failure of London Clay cuttings (*see weathering*). Figure 4.26 shows the changes in pore pressure and Factor of Safety when a cutting is made in clay. Initially there is a reduction in pore pressure and an increase in stability. This is known as the **short-term** or **undrained** case where pore pressures have not achieved a balance with the changing conditions. Although cuttings are generally more stable in the short-term, failure could still occur if the new slope is too steep or too high. Well known examples include the failures in Upper Lias Clays at Tinwell and Empingham, Leicestershire[56], and the movements of London Clay at Bradwell, Essex [57]. In the **long-term** pore pressures will increase to an equilibrium value (i.e. the **drained** condition), thereby reducing the effective stress and the stability of the slope, and possibly leading to delayed failures. The opposite situation will be true in the case of loading (e.g. an embankment) where the short-term stability is critical because of the build up of high pore pressures which then gradually dissipate.

Where bodies of water, in the form of rivers, lakes and reservoirs abut slopes, groundwater levels tend to be raised to correspond to the free-standing water surface in the adjacent water body. Should the water-level be reduced quickly, the effects of loss of passive support can be compounded by high pore-water pressures within the newly exposed slopes if groundwater levels cannot be reduced quickly enough by drainage. Under these circumstances, failure can occur. This cause of instability, known as drawdown, is restricted to riverbanks and the margins of reservoirs. Examples include failures at Llyn Peris in Gwynedd[58] and at Cod Beck Reservoir in North Yorkshire[59].

Finally, mention must be made of the decrease in material strength brought about by the flushing away of fines from within the soil or superficial deposits by subsurface flows of water. These fines may be either matrix material or released by the destruction of cements or produced as weathering by-products. Their removal results in the creation of voids within the material fabric which has the effect of increasing permeability and reducing resistance to shearing. The process is akin to seepage erosion, as discussed under external factors, but occurs wholly within the ground.

Under certain circumstances, desiccation cracking can provide routeways for storm infiltration to saturate a relatively permeable layer within the soil. Lateral (downslope) seepage then occurs within the soil (subsurface flow) which may be fast enough to move soil particles. The subsurface flow continues to follow fissures and voids until reaching the surface lower down the slope. Once such a routeway is established, successive flows will form a small tunnel (pipe) which will become enlarged and may eventually collapse to form a gully. Requirements for piping development are: a soil liable to cracking in dry periods; a relatively impermeable layer in the soil profile; the existence of a hydraulic gradient; and a dispersible soil layer. The existence of piping networks within the soil naturally result in weakening and can assist in the development of shallow slides. Indeed, at Plynlimon, North Wales, soil slides have been found to be in close association with pipes[60].

Organic factors

Plant root systems play an important role in influencing the stability of hillslopes. It is generally accepted that roots contribute to soil strength by providing artificial cohesion. As a result the Coulomb equation may be modified to:

$$s = (c' + \blacktriangle c) + (\sigma - u) \tan \varnothing'$$

where $\blacktriangle c$ is artificial cohesion. This additional cohesion may be as much as $10kN/m^2$ [61] and can have a significant effect on reducing the likelihood of shallow landslides. However, should the vegetation be cleared or trees harvested, this additional strength may be rapidly lost because of root decay. There are many examples of accelerated landslide activity after timber harvesting in the United States and New Zealand[61], but no reported instances in Great Britain.

Conclusions

The preceding discussion is merely intended as introductory and by no means covers the entire range of causes. Fuller treatments are readily available in the various references cited. What this chapter has sought to show is that the many forms of instability described in Chapter 3 are the product of an equally broad range of causal factors that interact variably over space and through time. Some trigger instability through their short-term fluctuations or transient behaviour, while others act in an insidious fashion to progressively reduce the resistance of slopes to failure, thereby lowering the threshold at which materials will move.

Mass movement is not an exceptional or unusual phenomenon but a perfectly normal geomorphological process which operates to reduce the elevation of the earth's surface. The need for denudation is also perfectly normal, due to the presence of forces within the earth that cause the establishment of relief through uplift and subsidence of the crust. Thus the distribution of mass movements owes much to the details of geological history and geomorphological evolution, which are inherent in the shape of the ground surface (topography) and the nature, disposition and relationships of the underlying materials (stratigraphy).

But it is also true that many of the causal factors are greatly influenced by changes that occur over time and that this results in temporal variations in landslide activity. For example, mass movement resulting from human activity could only have occurred during the brief span of time that humans have had a significant influence on geomorphological evolution – say 8000 years in southern Britain. Previous to that time, other influences become dominant which are usually grouped under the heading **environmental change.** Of particular significance within this group is **climatic change.** Pronounced variations in temperature and rainfall obviously have serious implications for slope stability, especially as regards the operation of particular causal factors and the resultant changes in types of failure.

These spatial and temporal aspects of mass movement are considered in the next two chapters beginning with the variable influence of ground conditions.

NOTES

1. Skempton & Petley, 1967
2. Hutchinson, 1988
3. Pitts & Brunsden, 1987
4. Skempton, 1977
5. Chandler, R.J. & Skempton, 1974
6. Biczysko, 1981
7. Early & Skempton, 1972
8. Skempton, 1964
9. Crozier, 1986
10. Terzaghi, 1950
    Brunsden, 1979a
    Crozier, 1986
11. Geomorphological Services Ltd, 1986–87
12. Smith, D.I., 1984
13. Brunsden & Jones, 1976
    Brunsden & Jones, 1980
14. Hutchinson, 1973
15. Hutchinson & Gostelow, 1976
16. Skempton, 1953
17. Hutchinson, 1962
    Hutchinson, 1984b
18. Hutchinson, 1969
    Hutchinson et al., 1980
19. Hutchinson, 1982
20. Knox, 1927
    Franks, C.A.M., 1986
    Franks, C.A.M., et al ., 1986
21. Higgenbottom & Fookes, 1970
22. Halcrow, 1989
    Jones, D.B. et al., 1991
23. Anon, 1928a
    Anon, 1928b
    Colenutt, 1928
    Jackson, J F, 1928
24. Hutchinson et al., 1980
25. Hutchinson & Bhandari, 1971
26. Jones, D.B. & Siddle, 1988
27. Knox, 1927
    Gostelow, 1977
28. Hutchinson et al., 1973
29. Wallwork, 1960
30. Anon, 1931
31. Clark, A.R. et al., 1979
32. Ollier, 1984
    Brunsden, 1979b
33. Taylor & Spears, 1970
34. Chandler, R.J. 1971
    Chandler, R.J. 1972
35. Burland et al., 1977
36. Skempton, 1985
    Middleboe, 1986
    Coxon, 1986
    Potts et al., 1990
    Vaughan, 1991
37. Skempton, 1948
38. Skempton, 1964
39. De Lory, 1956
40. Toms, 1948
41. Chandler, R.J., 1984
42. Ballantyne & Kirkbride, 1987
43. Hutchinson, 1972
44. Clark, M.J., 1980
45. Taylor & Cripps, 1987
46. Steward & Cripps, 1983
47. Moore, 1988
    Moore, 1991
48. Carson, M A & Petley, 1970
49. Lee & Moore, 1989
    Moore et al., 1991a
50. Addison, 1987
51. Gifford, 1953
52. Bevan et al., 1978
53. Carling, 1986a
54. Jenkins et al., 1988
55. Geomorphological Services Ltd., 1991
    Lee et al., 1991a
    Lee et al., 1991b
    Moore et al., 1991b
56. Chandler, R.J., 1984
57. Skempton & La Rochelle, 1965
58. De Jorge, 1982
59. Vaughan, 1965
60. Newson, 1975
61. Sidle et al., 1985

CHAPTER 5

The Pattern of Landsliding

Introduction

The preceding discussions on types and causes of mass movements have focussed attention on the fundamental truth that not all ground materials are equally prone to failure. This fact is well illustrated in the discussion of coastal landsliding contained in Chapter 2 and is an underlying reason for the variations in rates of erosion along the coastline (*Figure 2.6*). It has also emerged that not all ground materials fail in the same ways and that certain types of mass movement tend to be associated with particular types of material. For example, the investigation into the distribution of debris flows in Scotland[1] (*Figure 3.20*) revealed that they are particularly widespread on Torridon Sandstone, granite and other coarse-grained crystalline rocks of the Highlands which weather to produce cohesionless coarse regoliths, but are much less frequently developed on adjacent areas of schist which breakdown to form much finer, silt-rich debris. This same consideration may explain why debris flows appear rare in the Southern Uplands where slopes are often steep but the regolith is predominantly fine-grained.

The nature of ground materials is, therefore, a fundamental influence on the distribution and character of landsliding. It is one of four important general **location factors**, along with topographical characteristics, climate and groundwater conditions (hydrogeology). The variable interaction amongst specific aspects of these locational factors helps to determine the precise siting of instability (**site factors**), the timing of failure (**triggering factors**) and to dictate the condition of displacement in terms of the form, rate and duration of movement (**controlling** or **perpetuating factors**)[2]. These four will, if taken together, account for much of the pattern of recorded failures in Great Britain (*Figure 2.14*) and therefore provide the main focus of this chapter.

However, consideration of prevailing characteristics will only explain contemporary landsliding. As many of the recorded landslides have turned out to be relics produced under very different environmental conditions in the past, a full explanation of landsliding in Great Britain requires consideration of temporal variability as well as spatial variation. These longer-term temporal factors include sea-level change, climatic change, crustal movements, landscape evolution and human intervention, all of which display longer trends and cycles upon which are superimposed the short-term fluctuations (transient behaviour) and pulses of activity already considered under the heading of triggering factors. These temporal aspects will be discussed in Chapter 6, except for a group of features known as **cambered strata** and **valley bulges**.

Sometimes, where valleys have been cut through rigid cap-rocks and into softer, clayey layers below, the effects of intense freezing under harsh periglacial climates often caused the cap-rock layer to bend over and break into blocks, separated by great fissures, while the clayey material exposed in the valley floors pushed upwards to form distinct ridges. These geomorphological processes created large-scale structures, sometimes confused with the faulting produced by tectonic activity, which prepared the ground for later mass movements. Because the results (cambered strata and valley bulging) so considerably disturbed ground materials, they are best considered in this chapter along with other features of rock structure.

The distribution of the 8835 presently recorded landslides in Great Britain is shown on Figure 2.14 and the major known concentrations are described in Appendix A. These concentrations indicate areas known to be relatively 'landslide prone' although uncertainty exists regarding the limits of these areas and the extent to which the intervening areas with few data represent ignorance or prevailing stability *(see Chapter 2 for discussion of the deficiencies in current knowledge)*. In this chapter, six groups of explanatory variables will be examined: lithological factors, hydrogeology, structure, stratigraphy, topographical character and climate.

Lithology

The geotechnical properties of earth materials provide fundamental controls on both the susceptibility of slopes to failure and the resultant types of mass movement. The simple divisions between superficial deposits and bedrock, or between sedimentary strata and other types of rock or, for that matter, between rocks of differing antiquity, are only of interest and value once the differing character of materials is appreciated (*Figures 2.2 and 2.3*).

Rocks and soils are not uniform solids but aggregations of solids with significant voids, fissures and fractures filled with air, water, organic material, salts and other secondary materials. It is important, therefore, to distinguish between the basic strength of the materials and the much lower actual strength of the rock or soil mass due to the presence of weaknesses in the form of discontinuities, voids and secondary deposits. In the case of rocks, the former is referred to as **intact strength**, the latter as **rock mass strength**. Intact strength is a function of **cohesion** and **friction** as discussed in Chapter 4.

The values of friction and cohesion for a range of common rocks and superficial sediments is displayed in Table 4.1 and demonstrates the differences in shear strengths between **frictional** materials (e.g. sands and hard rocks like granite) and **cohesive** materials (clays and mudrocks). These differences are reflected in their contrasting susceptibilities to slope failure.

Generalised relationships have been established between grain size of constituent particles in sedimentary materials and the angle of internal friction (*Figure 5.1*). This demonstrates the difference in angle of internal friction, and hence resistance to shear, between granular materials and finer grained materials such as clays and silts.

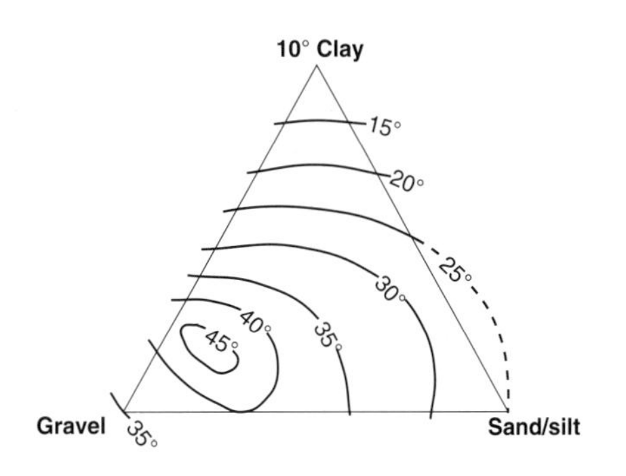

5.1 The relationship between clay content and angle of friction (after Kirkby, 1973).

Table 5.1 **The relationship between lithology and inland landslide type in Great Britain, as revealed by the DOE Survey**

Lithology	Number of recorded landslides (inland)	Unspecified	Rockfalls	Topples	Sagging failures	Single Rotational	Multiple Rotational	Successive Rotational	Compound	Translational	Flows	Complex	Cambered/foundered strata
clay	1795	1068	1	2	–	149	81	124	58	51	76	73	112
chert	3	2	–	–	–	–	–	–	–	–	–	–	–
shale	624	290	4	1	–	51	14	–	7	68	64	125	–
marl	66	49	1	–	–	5	2	1	2	2	1	3	–
siltstone/mudstone	579	339	2	1	–	45	7	3	7	54	11	103	7
sandstone	999	624	14	–	–	57	14	9	6	57	15	187	16
gritstone	261	137	1	–	–	39	46	–	9	4	–	25	–
conglomerate/breccia	40	18	4	–	–	–	–	–	1	13	3	1	–
quartzite	31	20	–	1	1	–	–	–	–	6	–	3	–
evaporites	7	7	–	–	–	–	–	–	–	–	–	–	–
dolomite	6	6	–	–	–	–	–	–	–	–	–	–	–
limestone	780	625	10	4	–	18	13	–	9	22	5	22	52
chalk	45	11	12	–	–	1	1	–	1	5	–	7	7
interbedded sedimentaries (argillaceous)	1109	903	–	–	–	24	37	5	9	56	8	54	13
interbedded sedimentaries (arenaceous)	900	468	2	–	–	175	32	104	12	18	6	42	41
interbedded sedimentaries (carbonaceous)	144	67	1	–	–	8	1	–	2	46	–	19	–
interbedded sedimentaries (calcareous)	347	313	–	1	–	2	4	1	7	1	–	10	8
slate	16	5	4	–	–	–	–	–	–	4	2	1	–
schist	449	325	3	15	25	1	–	–	4	44	–	32	–
gneiss	17	10	–	–	4	–	–	–	–	2	–	1	–
granite/granodiorite	38	27	4	–	2	1	–	–	–	–	1	3	–
gabbro/dolerite	41	36	2	–	–	1	–	–	–	1	–	1	–
basalt	125	114	–	–	–	2	2	2	–	–	–	5	–
ultrabasics	–	–	–	–	–	–	–	–	–	–	–	–	–
volcanic tuffs & pyroclastics	55	42	5	–	–	5	–	–	1	–	1	1	–
unknown lithology	93	49	17	–	–	5	2	–	1	6	4	9	–
Total	**8570**	**5555**	**87**	**25**	**32**	**589**	**256**	**249**	**136**	**460**	**197**	**728**	**256**

Notes: 6423 inland landslides in the GSL databank are recorded as involving one or more bedrock lithologies. 8570 bedrock attributes are recorded

In general, materials become weaker as the clay content increases and therefore more prone to failure. This relationship is borne out by the available data for Great Britain which reveal that failures in clayey bedrocks (clay/shale/mudrock/ interbedded argillaceous rocks) dominate both the inland and coastal data-sets (Tables 5.1 and 5.2). However, it has to be noted that low permeability and high water contents promote failure under the prevailing climatic regime, as compared with arid areas where clay slopes often stand at quite steep angles. Also, failures in clay tend to be quite deep-seated and conspicuous, thereby increasing the likelihood of their being recorded.

In the case of reported landsliding in superficial deposits (*Table 5.3*), the association between mass movement and clay materials is not immediately obvious. Instead, the *able 5.2* **The relationship between lithology and coastal landslide type in Great Britain, as revealed by the DOE Survey**

ithology	Number of recorded landslides (coastal)	Unspecified	Rockfalls	Topples	Sagging failures	Single Rotational	Multiple Rotational	Successive Rotational	Compound	Translational	Flows	Complex	Cambered/foundered strata
ay	301	79	16	2	–	66	13	4	3	6	41	71	–
hert	3	–	–	–	–	–	–	–	–	–	–	3	–
hale	107	34	15	1	–	3	3	1	5	33	1	11	–
narl	34	12	3	1	–	1	1	–	–	1	3	12	–
iltstone/mudstone	44	6	7	2	–	–	–	–	2	16	–	11	–
andstone	146	48	23	9	–	5	7	–	2	8	6	38	–
ritstone	39	2	11	2	–	–	–	–	1	21	–	2	–
onglomerate/breccia	84	3	52	2	–	–	–	–	5	19	–	3	–
uartzite	–	–	–	–	–	–	–	–	–	–	–	–	–
vaporites	–	–	–	–	–	–	–	–	–	–	–	–	–
olomite	–	–	–	–	–	–	–	–	–	–	–	–	–
mestone	143	31	74	2	–	–	3	–	4	20	1	8	–
halk	71	14	37	1	–	4	4	–	1	6	1	3	–
nterbedded sedimentaries (argillaceous)	31	16	2	1	–	–	–	2	–	1	2	7	–
nterbedded sedimentaries (arenaceous)	91	31	9	2	1	9	4	2	1	6	4	22	–
nterbedded sedimentaries (carbonaceous)	14	8	1	1	–	–	–	–	–	2	–	1	1
nterbedded sedimentaries (calcareous)	23	8	4	–	–	2	1	2	–	–	1	5	–
ate	74	7	25	2	–	–	–	–	4	32	–	4	–
chist	28	19	–	2	1	–	1	–	1	3	–	1	–
neiss	7	4	–	–	–	–	1	–	–	1	–	1	–
ranite/granodiorite	18	16	–	–	–	–	–	–	–	–	–	2	–
abbro/dolerite	25	20	3	–	–	–	1	–	–	–	–	1	–
asalt	46	28	3	–	–	2	3	7	–	–	–	3	–
ltrabasics	3	2	–	–	–	1	–	–	–	–	–	–	–
olcanic tuffs & pyroclastics	6	4	–	–	–	–	–	–	–	2	–	–	–
nknown lithology	71	37	21	–	–	1	4	–	–	1	–	7	–
otal	**1409**	**429**	**306**	**30**	**2**	**94**	**46**	**18**	**29**	**178**	**60**	**216**	**1**

otes: 1113 coastal landslides in the GSL databank are recorded as involving one or more bedrock lithologies. 1409 bedrock attributes are recorded

rather vague categories of 'heterogeneous' and 'head' (periglacial solifluction deposits) predominate. This is because of several reasons. First, interest in the recording and mapping of superficial deposits is relatively recent in terms of the history of geological mapping in this country, which means that the nature and extent of superficial deposits is frequently poorly known **at the detailed level**. Second, superficial deposits are, for the most part, the unsorted debris produced by geomorphological processes and therefore tend to be heterogeneous in nature. Third, solifluction deposits are ubiquitous and therefore it should come as no surprise that the category 'head' has the largest number of recordings. Finally, both clays and silt-sized materials usually form significant proportions of so-called heterogeneous deposits and head. Thus the association between landsliding and clay materials probably holds true.

Table 5.3 **The relationship between superficial deposits and landslide type in Great Britain, as revealed by the DOE survey.**

Superficial material	Number of recorded landslides	Unspecified	Rockfalls	Topples	Sagging failures	Single Rotational	Multiple Rotational	Successive Rotational	Compound	Translational	Flows	Complex	Cambered/foundered strata
unspecified material	531	309	11	–	–	83	1	1	4	60	19	42	1
clay	127	83	–	–	–	9	5	–	5	7	9	9	–
silt	5	1	–	–	–	1	–	–	1	–	1	1	–
sand	18	8	–	–	–	3	–	3	1	–	2	1	–
gravel	16	14	–	–	–	–	–	–	–	1	1	–	–
mixed argillaceous	271	26	–	–	–	13	5	–	5	56	49	117	–
mixed arenaceous	478	99	–	–	–	21	11	3	10	166	71	97	–
pedological soil	33	1	–	–	–	1	–	–	–	25	6	–	–
peat	27	18	–	–	–	1	–	–	–	–	8	–	–
heterogeneous	801	752	–	–	–	9	5	–	2	10	12	10	1
clay with flints	2	–	–	–	–	–	–	–	–	–	–	2	–
head	359	126	–	–	–	138	9	–	6	43	9	28	–
other	25	5	–	–	–	2	–	–	1	3	13	1	–
spoil	3	–	–	–	–	–	–	–	–	3	–	–	–
Total	**2696**	**1442**	**11**	**0**	**0**	**281**	**36**	**7**	**35**	**374**	**200**	**308**	**2**
Superficial origin													
unspecified origin	519	198	11	–	–	91	11	4	17	96	33	57	1
lacustrine	4	1	–	–	–	–	–	–	–	–	1	2	–
marine	9	3	–	–	–	4	1	1	–	–	–	–	–
fluvial	56	48	–	–	–	3	1	–	1	1	1	1	–
fluvio-glacial	37	21	–	–	–	4	1	2	1	2	3	3	–
glacial	1 613	1027	–	–	–	35	11	–	2	218	105	215	–
peri-glacial	401	126	–	–	–	141	10	–	9	51	36	27	1
aeolian	2	2	–	–	–	–	–	–	–	–	–	–	–
contemporary processes	55	16	–	–	–	3	1	–	5	6	21	3	–
Total	**2696**	**1442**	**11**	**0**	**0**	**281**	**36**	**7**	**35**	**374**	**200**	**308**	**2**

Having considered the weakly consolidated or unconsolidated deposits which underlie much of the ground surface, it is necessary to briefly return to the more durable materials. The change from normally consolidated to over-consolidated materials *(see Chapter 4 for definitions)* involves greater strength but increasing numbers of discontinuities. Such materials are usually referred to as 'soft rocks' and underlie most of Lowland Britain (*Figure 2.2*). The transition from 'soft rocks' to 'hard rocks' is arbitrarily defined but inevitably results in a group of materials with high friction angles and large to very large cohesion values (*Table 4.1*). If these materials were wholly intact they would be extremely durable. However, in reality they are not intact but flawed discontinuous materials and it is the nature and density of the discontinuities that determine the likelihood of failure. In these 'hard rocks' it is the presence of stress release jointing, faults, foliations, bedding planes

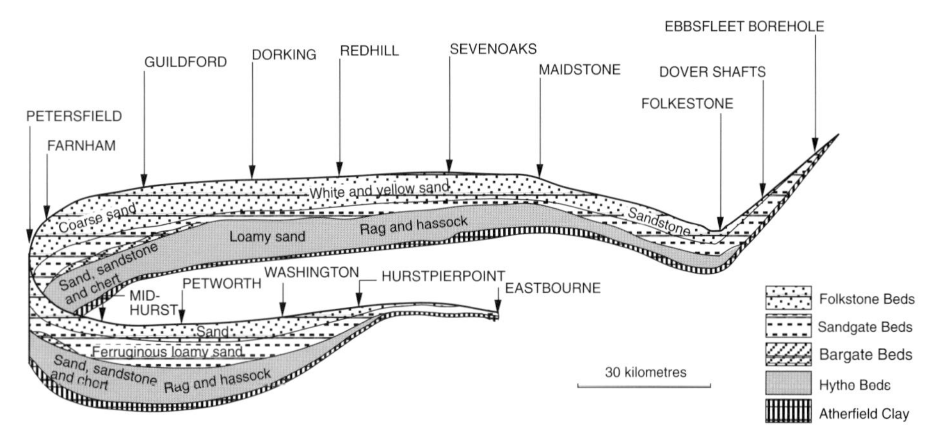

.2 Ribbon diagram showing he lateral variability of the ower Greensand beds across he Weald (after Gallios, 965). The beds of chert, andstone and ag-and-Hassock (thin Beds re relatively resistant to enudation and form the rominent escarpments of the estern and northern Weald.

and intra-formational shears that provide the lines of weakness, or flaws, which expose the rock to water penetration and weathering attack. The stability of slopes developed in these materials is therefore determined by the discontinuity framework or, to be more specific, the orientation, frequency, openness and smoothness of intersecting joint sets and faults within the rock mass[3].

Mention of weathering usefully serves as a reminder that the intact strength and mass strength of most materials are gradually reduced over time by the processes of physical and chemical weathering. As this was discussed in Chapter 4 it merely serves to quote the example of investigations at Mam Tor, Derbyshire (*Photo 2.10*), which have revealed that the Edale Shales involved in movement are sensitive to changes in pore-water chemistry[4]. As a consequence, the continued instability of this large ancient landslide is considered to be the result of repeated reductions in shear strength along the basal shear surface due to weathering of freshly exposed pyritic shales by naturally generated acidic groundwaters.

Classification of rocks or superficial deposits in a sequence is normally achieved by using either age or lithological characteristics. Age continues to be the most widely used method in stratigraphy and usually employs the remains of fauna and flora to indicate antiquity. This **biostratigraphical** approach has produced the divisions of time that make up the stratigraphic column (*Table 2.1*) which are capable of further sub-division. While the age of earth materials is clearly of considerable interest, their composition is of much greater importance in landslide studies. Therefore, the division of outcropping deposits based on lithological characteristics, or **lithostratigraphy,** is much to be preferred.

Lithological characteristics are to some extent incorporated into the classic stratigraphic divisions of British rocks, most obviously in the division of the Carboniferous succession into basal Carboniferous Limestone Series, Millstone Grit Series and overlying Coal Measure Series, or the equally well known division of Cretaceous strata into Weald Clay, Lower Greensand, Gault Clay, Upper Greensand and Chalk. But these units are by no means homogeneous and their names merely convenient labels which often can be very misleading to the non-specialist. In reality, they frequently display marked spatial variations in thickness and internal composition over their outcrops. For example, only part of the Carboniferous Limestone Series is actually composed of massive limestones and in some areas limestone is absent entirely. At a more detailed level, the landslide-prone Lower Greensand of the Weald (*Figure 5.2*) is composed of five groups of beds (formations)

– the Atherfield Clay, Hythe Beds, Bargate Beds, Sandgate Beds and Folkestone Beds – which differ markedly in character as well as displaying pronounced spatial variability in composition and thickness. Even the apparently uniform Chalk that forms the high perpendicular cliffs at Beachy Head, the Severn Sisters and along the Dover Straits, actually contains variations in composition (*Figure 5.3*), which become very important in the lowest levels where increased clay content in the Lower Chalk has led to inland failures (e.g. near Folkestone, Kent) of a deposit that is widely perceived to be very stable. Such variations were not emphasised in the traditional biostratigraphic division (*Figure 5.3*), but have recently been given prominence in new classifications based on sediment characteristics and dominant physical properties, which use lithostratigraphic, biostratigraphic and geophysical markers to define members[5].

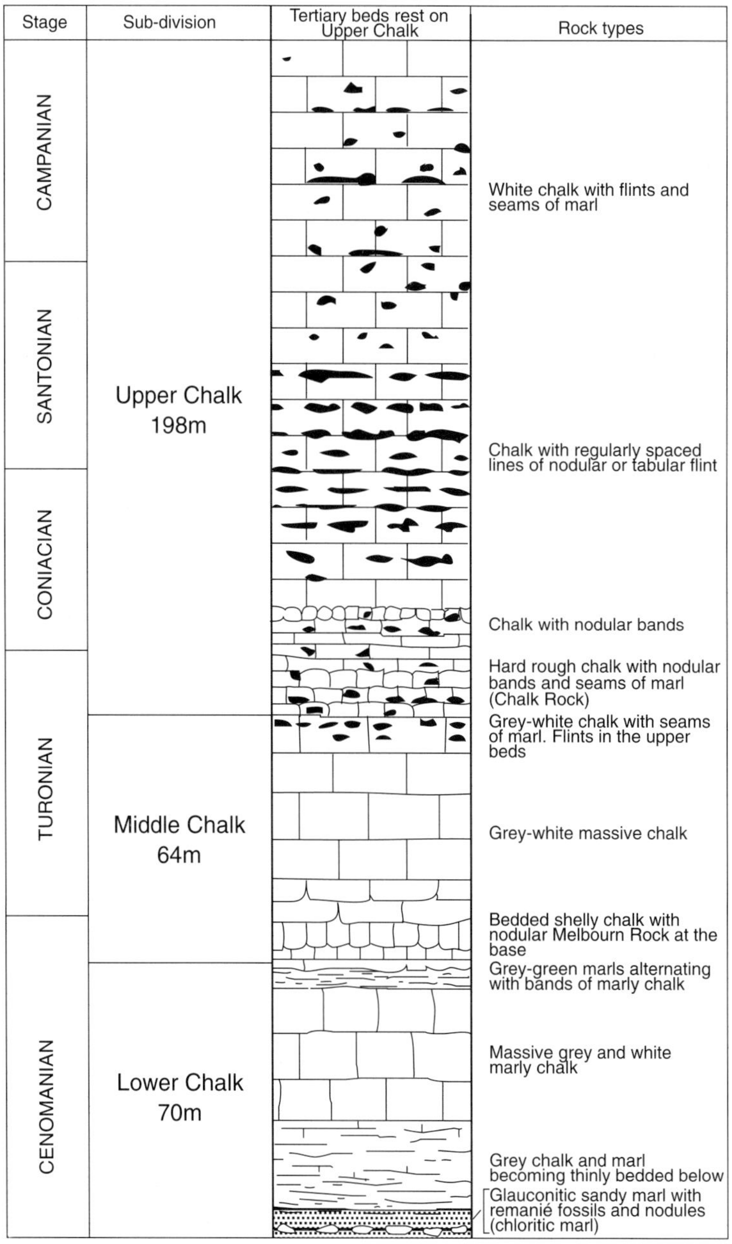

5.3 Lithological variation within the Chalk of Sussex.

It is the vertical variations in lithology between one bed of rock and another, or between bedrock and overlying superficial deposits, or even within superficial deposits themselves, that are of crucial importance in landslide studies because they produce the differences in water-bearing characteristics and strength. It is these factors that profoundly influence the siting of instability and largely determine the nature of the resultant landsliding. In alternating sequences of clays and sands or clays/shales and limestones, variations in strength will mean that fairly tough beds of rock will be involved in movements because of the failure of weaker underlying layers. This is well displayed at Stonebarrow, Dorset (*Photo 2.5*), where strong, rigid beds of Upper Greensand Chert have been involved in rotational landsliding due to failure in the underlying weaker Gault Clay, resulting in the production of coherent rotated blocks that give rise to well-defined relief features. The same is true of Folkestone Warren, Kent (*Photo 4.1, Figure 3.23*).

Emphasis on vertical associations of different lithologies must not be allowed to conceal the importance of horizontal variations. Indeed the occurrence of vertical associations conducive to slope instability is often the product of horizontal changes in thickness and composition, and this must be borne in mind when considering the relationship between landsliding and recognised geological formations (strata), as will be discussed in Chapter 6. For example, returning to the Lower Greensand of the Weald (*Figure 5.2*), it is important to recognise that the character, scale and intensity of landsliding varies considerably over the outcrop and is, for the most part, determined by the nature and thickness of the Hythe Beds and the association between the Hythe Beds lithology and the underlying Atherfield Clay and Weald Clay. The prominence of the landslide-prone Lower Greensand escarpment is a function of the strength of the Hythe Beds. Where resistant members such as rag-and-hassock or chert are present, the scarp is high and steep with numerous well-defined failures, some of which attain considerable size. However, where the Hythe Beds are thin and sandy, the scarp is more subdued and slope failure tends to be dominated by shallow translational movements. Where the Hythe Beds are too thin or weak to form an escarpment (eg in East Sussex), failures only occur where slopes are locally over-steepened by rivers or human activity.

The Cotswold escarpment provides a fitting final example of the relationship between rock type and landsliding. This prominent escarpment is widely known to be mantled by landslides and appears as a major concentration on the landslide distribution map (*Figure 2.14*). However, detailed investigations have revealed that the escarpment to the south of Stroud is relatively free of landslides (Figure 5.4), but that to the north is heavily affected[6]. The distribution and character of landsliding in these two zones is a reflection of two clay strata; the Lower Fuller's Earth Clay in the south and the Upper Lias Clay in the north.

To the south of Stroud, landsliding has occurred preferentially on the Lower Fuller's Earth Clay. This is a 10m thick calcareous, silty, overconsolidated mudstone with poorly developed, very thin bedding but with a close network of stress release fissures. Detailed mapping has revealed a marked concentration of failures on its outcrop (*Figure 5.5*), particularly mudslides which are especially concentrated to the east of Wooton-under-Edge, indicating that it is the least stable slope-forming

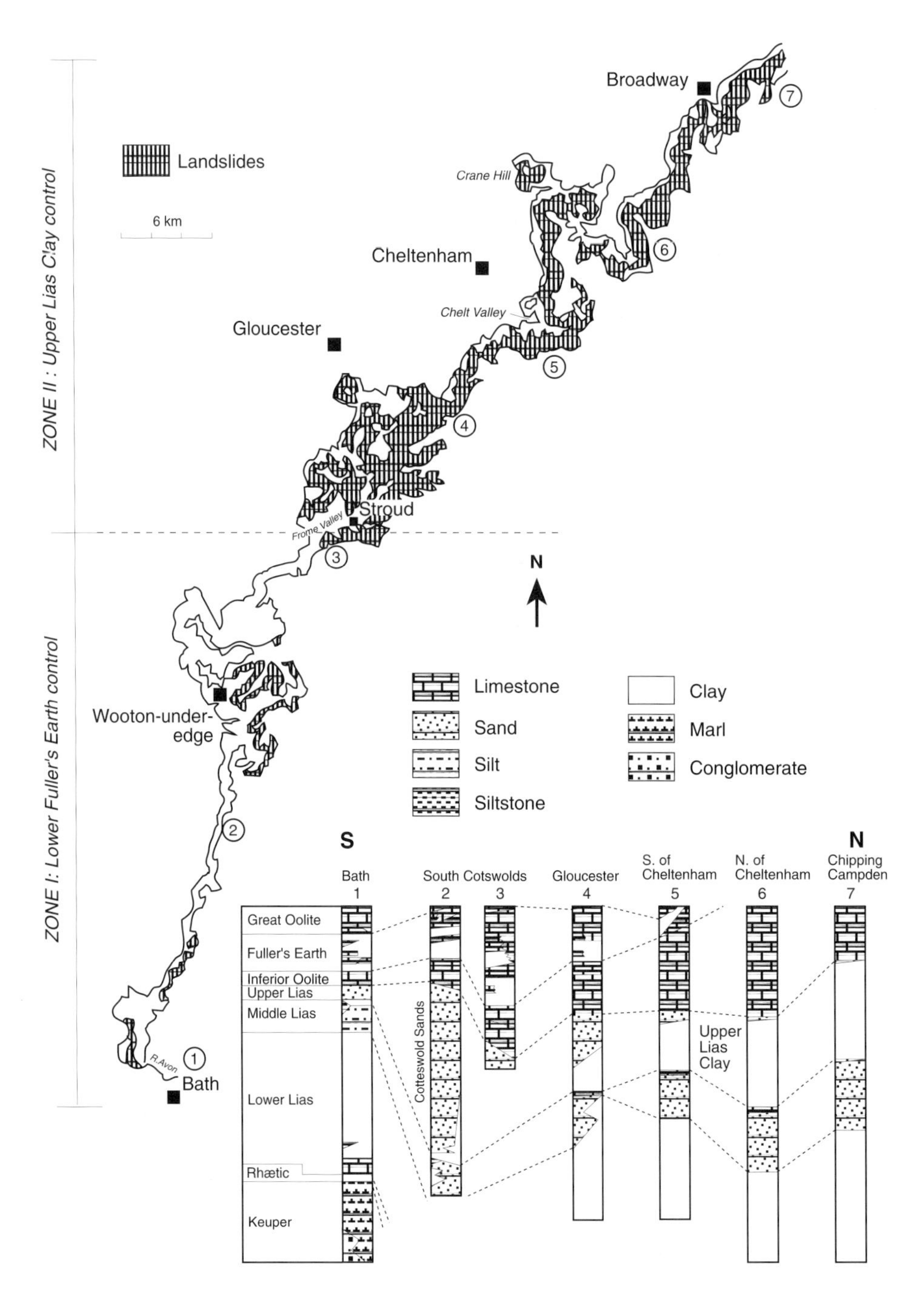

5.4 The importance of lithology in controlling the pattern of landsliding along the Cotswolds escarpment (after Butler, 1983).

material in the area. Indeed, the clay is likely to be unstable on slopes above 10° and has a long-term angle of stability of only 5.5°. However, over most of the escarpment between Bath and Stroud, the Fuller's Earth Clay is sandwiched between the stable Inferior Oolite and the overlying Great Oolite cap-rock, so that failures only occur where the cap-rock is thin or dislocated.

To the north of Stroud very different lithological conditions exist on the scarp. Both the Great Oolite and Fuller's Earth Clay are progressively eroded away, the underlying Inferior Oolite thickens northwards and becomes the scarp-former, and the Upper Lias Cotteswold Sands, which are prominent in the southern area, are progressively replaced by Upper Lias Clay (*Figure 5.4*). It is the existence of a

.5 Geological map of part of the southern Cotswolds howing the concentration of landslides on the Fuller's Earth Clay (after Butler, 983).

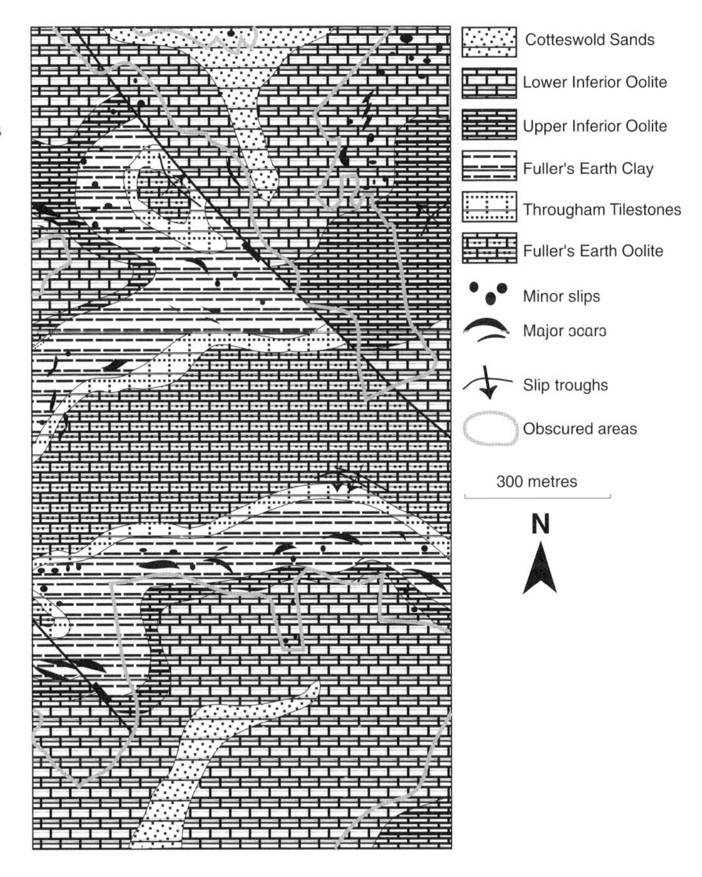

substantial thickness of Upper Lias Clay on the scarp face beneath the Inferior Oolite cap-rock that results in ubiquitous landsliding. Indeed, detailed survey of the Upper Lias outcrop in this area has revealed that 51% of the outcrop is mantled by landslides[6].

Hydrogeology

The real significance of variations in lithology lies in their effects on groundwater conditions, i.e. hydrogeology. Hydrogeology is a very important factor in determining the behaviour of slopes because of the influence of pore-water pressures on the effective strength of materials, as was described in Chapter 4. Pore-water pressures are generally very variable within the near surface layers of the ground, reflecting local topography and the different permeability of materials. However, it is common for a marked increase in density and, hence, a decrease of permeability to occur at roughly a constant depth, often corresponding to the interface between weathered and unweathered materials. This abrupt change results in water being forced to flow laterally, parallel to the slope. As was illustrated by the infinite slope model (*Figure 4.10*), in this instance, stability is dependent upon the water level within the material, with the worst-case scenario occurring when groundwater extends up to the surface. Although this condition may be very rare on some slopes, it has only to occur once to be the cause of instability.

Within deep-seated landslides the groundwater flow patterns and, hence, the pore-pressure distribution, is also dependent upon the relative permeabilities of the materials. This flow or **seepage** pattern can be graphically represented by a series of groundwater potential contours (**equipotential lines**; *Figure 5.6*). The direction of flow in a uniform soil is at right angles to these contours. However, in most slopes, variations in permeability due to bedding or changes in lithology, can give rise to more complex flow patterns. Changes in permeability at depth may also arise as a result of the increased overburden pressure, and hence effective stress, which will tend to reduce the porosity. In general, pore pressures tend to increase at a constant rate with depth (**hydrostatic** conditions), although the presence of an underlying more permeable material will give rise to a degree of **underdrainage** (*Figure 5.6*). Underdrainage will have the effect of reducing the pore-water pressures immediately above the permeable horizon, thereby improving the stability at this depth. Failure, of course, could still occur at a shallower depth where pore pressures are higher and available shear strength lower.

Broadening the perspective, it is necessary to recognise that the location of saturated zones within the ground is determined by the sequence of rocks. Groundwater is stored in permeable layers (aquifers), i.e. materials with relatively high proportions of voids, fissures and discontinuities such as sands and limestones. For these aquifers to operate they must be underlain by a confining bed of much lower permeability, often an impermeable layer (aquiclude) such as clay, and receive inputs of water from higher level by gravity-induced percolation or flow along joints, fissures and other discontinuities. The surface of the saturated zone is known as the water table (*Figure 5.7*). It is highly irregular reflecting the physical form of lithological units that can act as aquifers, geological structure, topography, location and scale of inputs and outputs, including groundwater extraction for human use. Water tables fluctuate, therefore, in response to climatic variability and human consumption. Several water tables may exist in a particular area, the term perched water table being applied to relatively localised zones of saturation.

Springlines or seepage zones often indicate the outcrop of impermeable materials on a hill-slope or along an escarpment. Springs represent areas of concentrated discharge from an aquifer and therefore a zone where high pore-water pressures may be reducing slope stability. If the aquifer is also overlain by an impermeable

5.6 Typical flow lines and pore pressure distribution within a clay slope (after Bromhead, 1986).

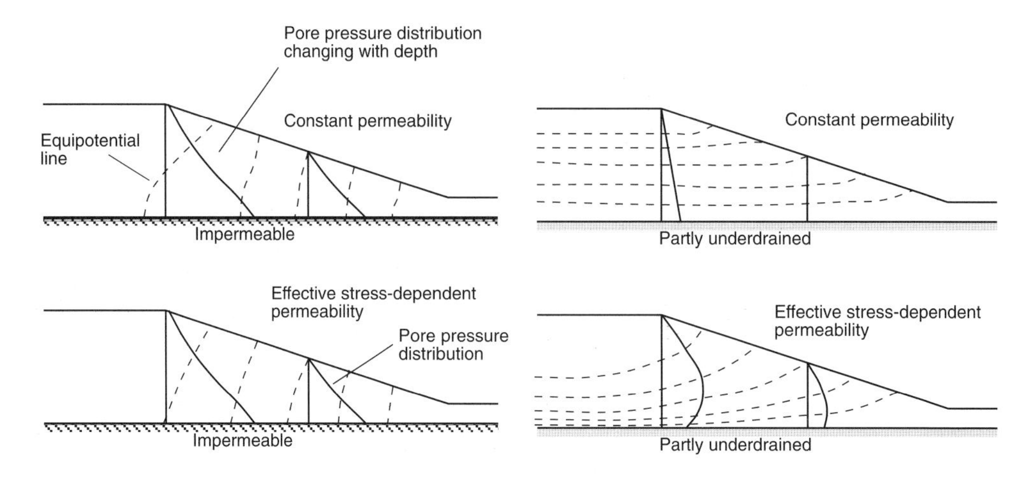

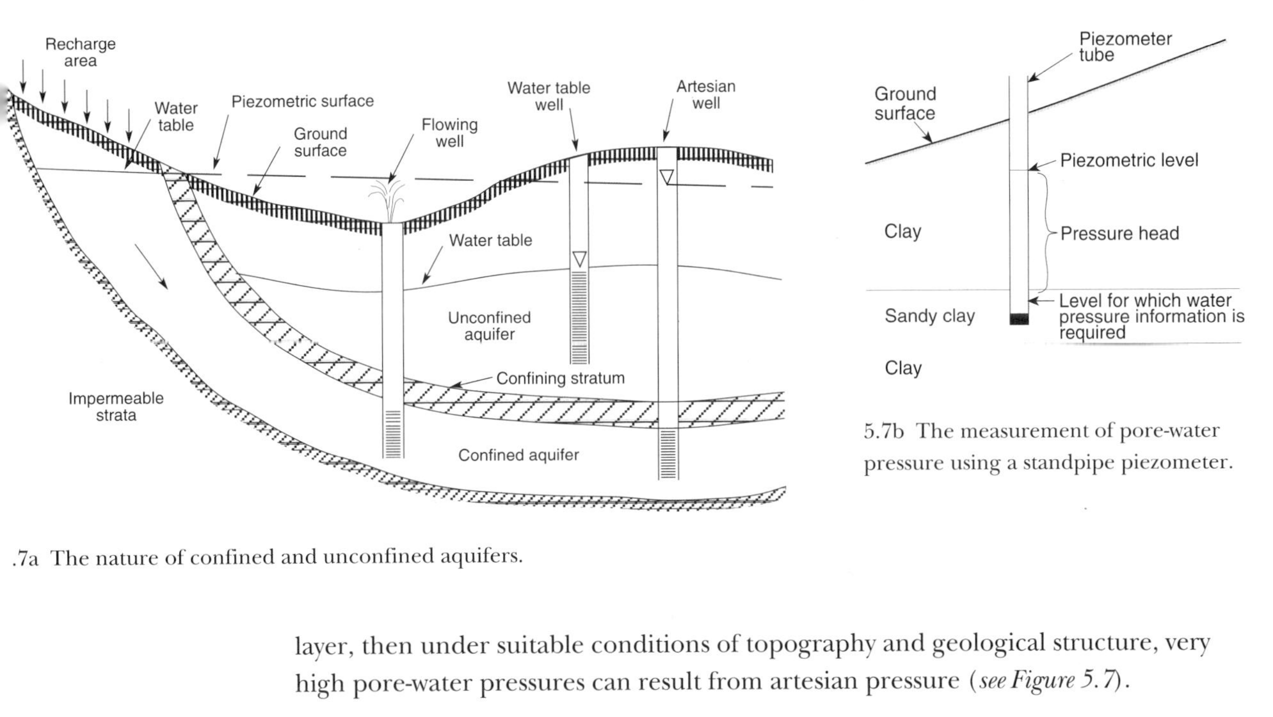

5.7b The measurement of pore-water pressure using a standpipe piezometer.

.7a The nature of confined and unconfined aquifers.

layer, then under suitable conditions of topography and geological structure, very high pore-water pressures can result from artesian pressure (*see Figure 5.7*).

The combined effects of lithological variation and associated groundwater conditions are well displayed in the coastal landslide complex at Fairlight Glen, Sussex[7] (*Photo 5.1; Figure 5.8*) where there has been a long history of mudsliding and complex failure in the lithologically variable Hastings Beds. Similarly, the presence of permeable gravels and sands overlying clays at Chewton Bunny, Hampshire (*Plate 29*) is also the cause of significant slope instability. Good examples of the influence of these factors on inland landsliding are revealed by the landslide distribution map (*Figure 2.14*) which shows marked concentrations on the variable Millstone Grit Series of the Southern Pennines, the Jurassic sandy-limestones and clays of the

.1 The Fairlight Glen andslide complex, East ussex (R Moore).

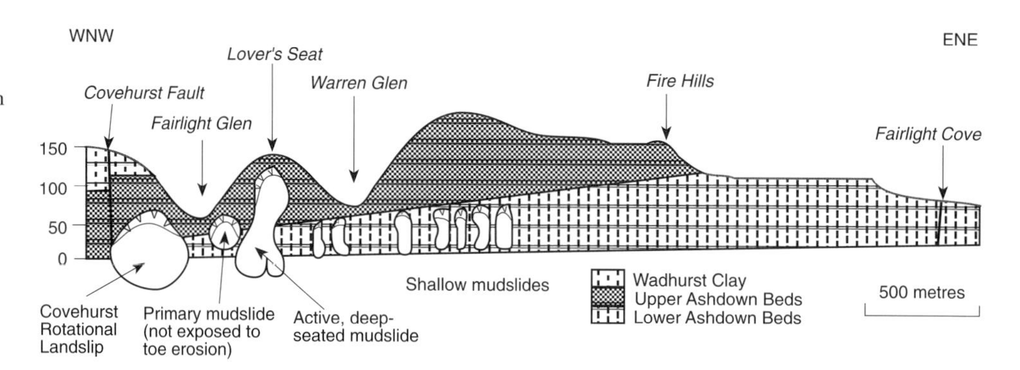

5.8 Section along the coast showing the pattern of landsliding in Fairlight Glen (after Moore, 1986).

Cotswold escarpment in Gloucestershire (*Figure 5.4*), and the variable deposits of the Lower Greensand in the Weald (*Figure 5.2*).

The influence of lithology and groundwater conditions has probably been most well established for the widespread slope failures of the South Wales valleys as a result of numerous recent studies[8]. The Upper Coal Measures of this area comprise a succession of massive well-jointed sandstones (Pennant Sandstones) separated by thin layers of mudstone or silty mudstone with coals underlain by seatearths (*Figure 5.9*). Water percolates down through the Pennant Sandstones until it reaches one of these impermeable layers where it forms a perched water-table. The aquifer is then forced to discharge onto the valley-side at the base of the sandstone as a series of seepage zones or springs. Softening, weathering and the removal of fines in these areas reduces the margin of stability to such a point that failures occur, often involving massive blocks of the overlying sandstones (*Figure 5.9*). This is best displayed in the large failure at Taren, in the Taff Valley immediately to the south of Aberfan (*Figure 2.10*), where major translational (rockslide) movements at various levels, involving 7-8 million m^3 of material, probably took place due to removal of support following the last withdrawal of glacier ice. But water issuing from these aquifers has also caused smaller, shallower failures, especially in the overlying mantles of glacial (till) and periglacial (head) deposits. This is well displayed in the Glynrhigos Farm failure described in Chapter 3.

The importance of hydrogeological conditions in explaining the susceptibility of particular geological units, or combinations of units, to failure has long been recognised[9]. This is clearly displayed in the combination of water-bearing sands overlying clays which is a controlling factor in the development of extensive and potentially damaging failures initiated by seepage erosion[10], e.g. Chale, Isle of Wight[11] (*Plate 16*). The presence of a permeable stratum overlying a soft impermeable stratum is a common feature of landsliding in many areas, especially

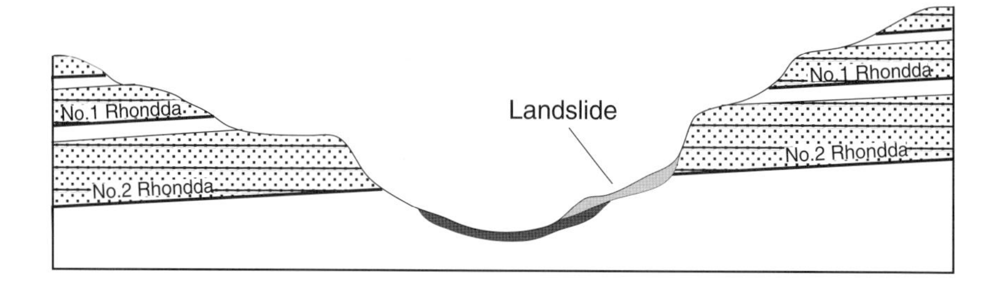

5.9 The relationship between landsliding and the dip of strata in South Wales (after Woodland, 1986).

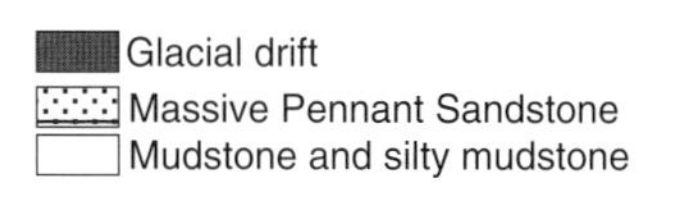

on the variable sequence of Mesozoic rocks in Lowland Britain (*see Table 2.1 and Figure 2.2*). Little wonder that groundwater investigations come high on the agenda in slope stability studies (*see Chapter 8*) and that slope drainage is seen to be the main tool in attempts to stabilise landslides in sedimentary rocks and glacial materials.

Structure

Structural factors may be considered to include both the internal partitioning of rock masses and the disposition of the rock masses themselves. These aspects only warrant brief consideration in the context of this book but further details are readily available elsewhere[12].

Discontinuities

Discontinuities, or partings, are breaks in the continuity of rocks and soils. They are the product of a wide range of processes that cause stress and include faults, joints, shears, foliations and cleavage planes. The importance of discontinuities in generating slope instability increases as rock strength increases. This is reflected in the often marked difference between intact strength and rock mass strength.

Joints are the commonest form of discontinuity and are ubiquitous in both rocks and soils. The significance of joints in partitioning rocks and facilitating the ingress of water has already been stressed on several occasions (*see especially rockfalls, topples and rockslides in Chapter 3*) and works irrespective of whether the jointing is the product of stress relief, tectonic stress, dehydration or cooling. What is important is the roughness of joint surfaces, their openness, density, orientation and interrelationships. These factors are graphically illustrated in Figure 5.10 which shows the affects of different relationships between joint sets and surface slopes in terms of possible modes of failure.

Although the significance of discontinuities will be apparent whenever rocks outcrop at the surface or are shallowly mantled by superficial deposits (*Figure 2.3*), their importance is most marked in the harder rock terrains of Highland Britain and Ancient Britain (*Figure 2.2*). For example, 65-70% of all documented landslides in Scotland are rock-slope failures involving sliding, toppling or sagging (*see Chapter 3*). These rock-slope failures display marked concentrations on the outcrops of Moinian and Dalradian metamorphic rocks (schists, slates, phyllites and granulites) and virtual absence on nearby gneiss, Torridon Sandstone and granite. Indeed, almost 80% of rock-slope failures occur on schist[13]. Although these metamorphic rocks have lower shear strengths than the other rock types in the area, the main reason for failures is the development of pronounced foliation planes. Where these foliations dip valleywards they inevitably form the basis for shear surface development.

.10 The effect of joint spacing and orientation on slope failure (after Selby, 1982).

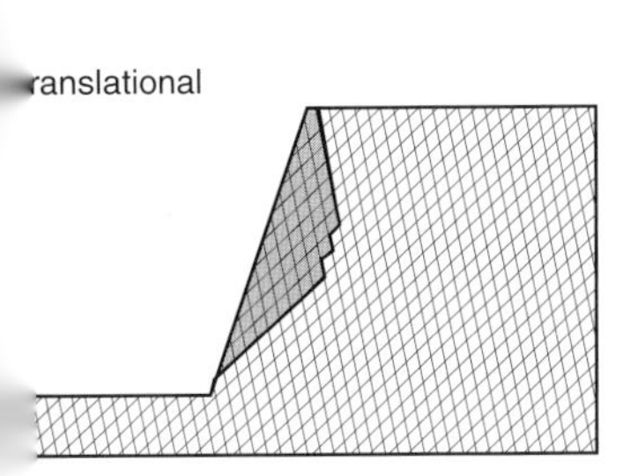

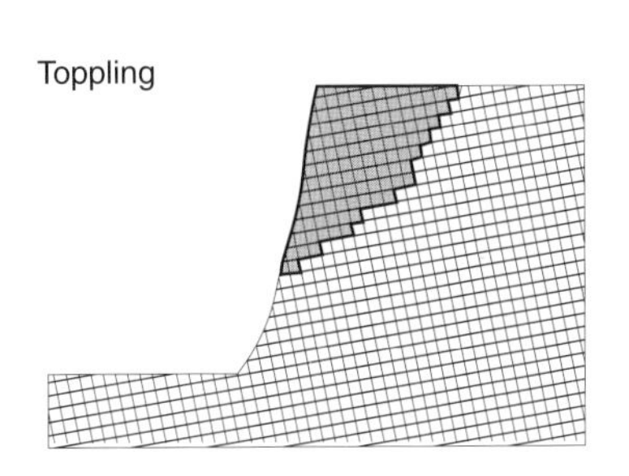

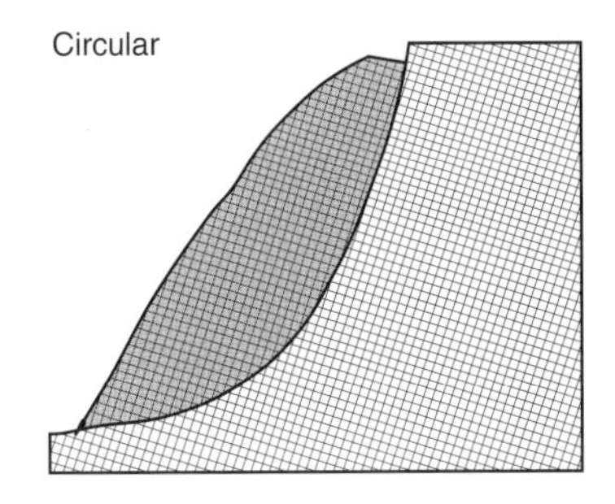

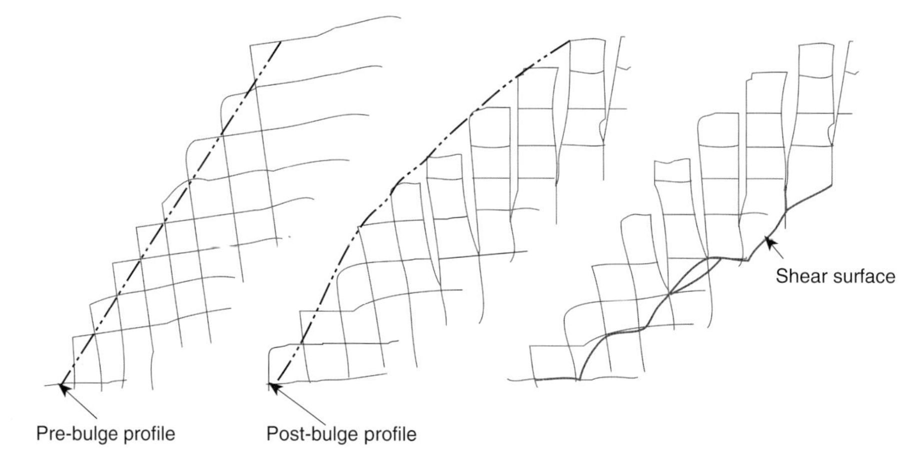

5.11 The opening and linking of joints in a cliff to form a shear surface.

Valleyward dipping joint systems have a similar effect, as has been illustrated with reference to the Carn Mor topple (*Figure 3.19*) and the hypothetical situation shown in Figure 5.11.

Faulting can also contribute to slope instability by compartmentalising rock masses and by allowing downward penetration of water. Faults become especially important in those locations where deep incision by rivers and ice results in steep slopes which intersect fault planes. Under these circumstances, the existence of a zone of weakness along the fault may be sufficient to initiate movement. This certainly appears to be the case in the Darren Dhu landslide, South Wales, where the backscar is coincident with a major fault in the Pennant Measures Sandstones. The backscar of the Taren landslide, South Wales (*Figure 2.10*), also appears to be partly controlled by the Kilkenny Fault.

As the frequency of faulting in Great Britain generally increases with the antiquity of surface materials, the influence of faults will also tend to be greatest in Upland, Highland and Ancient Britain (*Figure 2.2*).

Cambered/ Foundered Strata

There are many locations in Great Britain where beds of sedimentary rock can be seen to be dislocated and displaced, indicating significant disturbance following their creation. In many cases the dislocations are the result of faulting, for minor faults and master joints are widely developed in competent rocks. But there are numerous instances where displacements can be shown to be independent of faulting, indicating the existence of near-surface deforming processes. Subsidence is one cause of displaced strata, especially solution subsidence where the removal of mass from buried calcareous strata (especially limestones) as a consequence of subterranean water movements, can result in the development of underground caverns which can collapse, thereby removing support for overlying strata which also collapse. The result is known as **foundered strata** and often takes on the form of areas of competent strata which locally are arranged in a chaotic jumbled fashion with very variable directions of dip.

In many instances, however, the local disturbance to bedrocks cannot be explained by subterranean solution and in these instances the likely cause is ground movement due to the variable response of different superimposed rocks to intense freezing during the frequent arctic (periglacial) phases of the Pleistocene (*Figure 5.12*). The

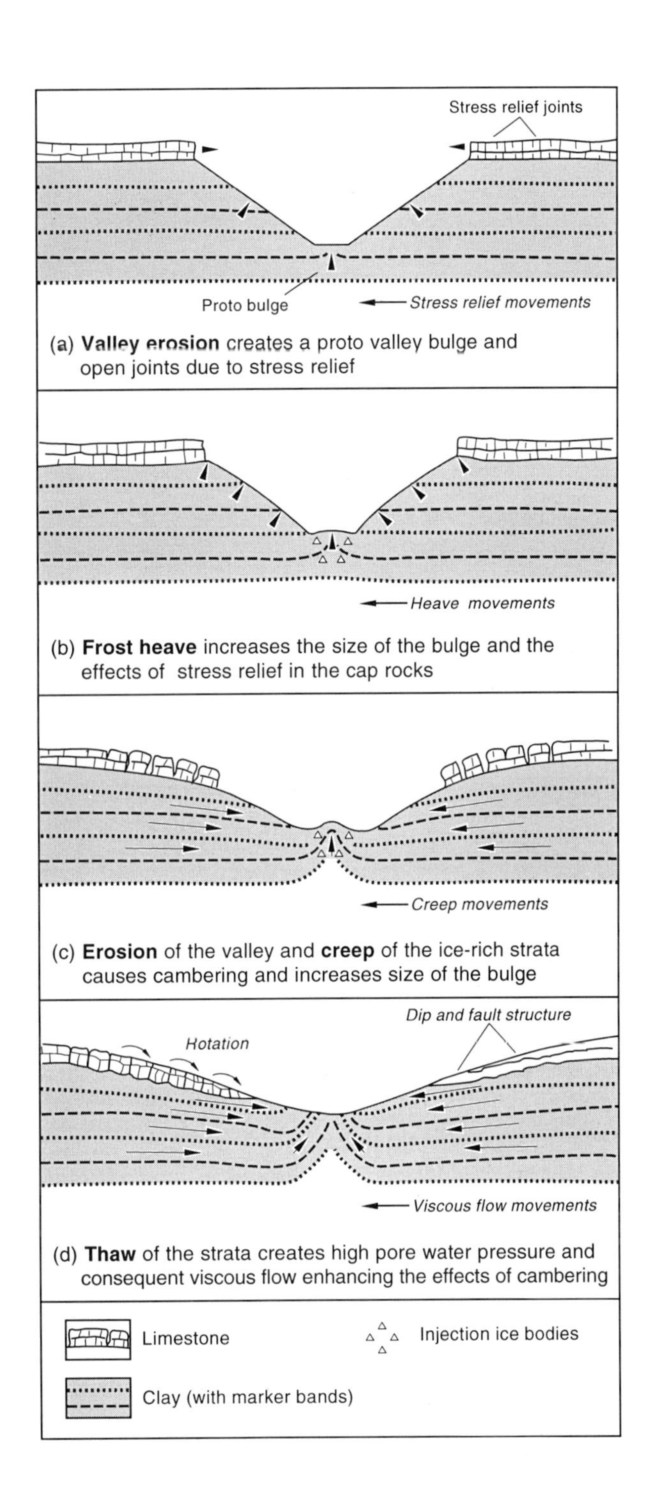

5.12 The development of cambered structures and valley bulges (after Parks, 1989).

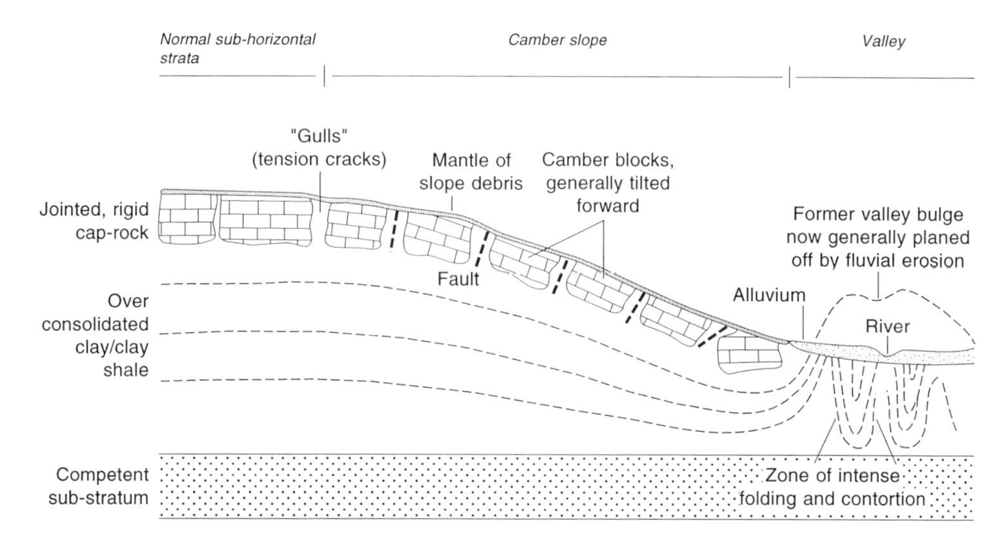

5.13 Typical feature associated with cambering (after Hutchinson, 1988).

term **cambering** has been applied to this process. Both cambering and foundering encompass large-scale, relatively deep-seated movements which are highly susceptible to, and often develop into, other forms of landsliding. Indeed, on morphological grounds they are often very difficult to distinguish from large landslides.

The umbrella term 'cambering' includes five main groups of features: the bending of competent strata (cambering properly defined); gulls; ridge and trough topography; dip and fault structures; and valley bulging.

Cambering is the downslope bending of a rigid surface layer of rock on the flanks of valleys (valley cambering) or on the crest of escarpments. It is best developed in those situations where valleys have been cut into sub-horizontal strata consisting of a rigid cap-rock, a thick layer of clay and an underlying rigid sub-stratum (*Figure 5.13*).

Gulls are essentially widened joints formed by the extensional movements of rocks affected by cambering. The tension created by the downward bending causes pre-existing major joints to open, thereby sub-dividing the cambered layer into individual blocks (*Figure 5.13*). As a consequence, it is common for gulls to be aligned parallel to the principal joint orientation, as was well described for Lincolnshire Limestone forming the Lincoln Scarp[14]. Gulls are normally filled with overlying superficial deposits which collapse into the voids as they open, and have been recorded up to 5m in width[15]. Detailed investigation of gulls in the Blue Lias limestone on a building site at Radstock, Avon, revealed that they were not continuous features but were often arranged en echelon and varied in character over short distances[15].

Ridge and trough topography describes the ground surface form that has been recognised on the interfluves between valleys at a number of locations in Great Britain. For example, complex patterns of ridges and troughs were described in the Windrush Valley in the Cotswolds[16]. At Upper Slaughter the troughs developed in the Inferior Oolite are 9m deep, up to 300m in length, and form a trellis-like pattern

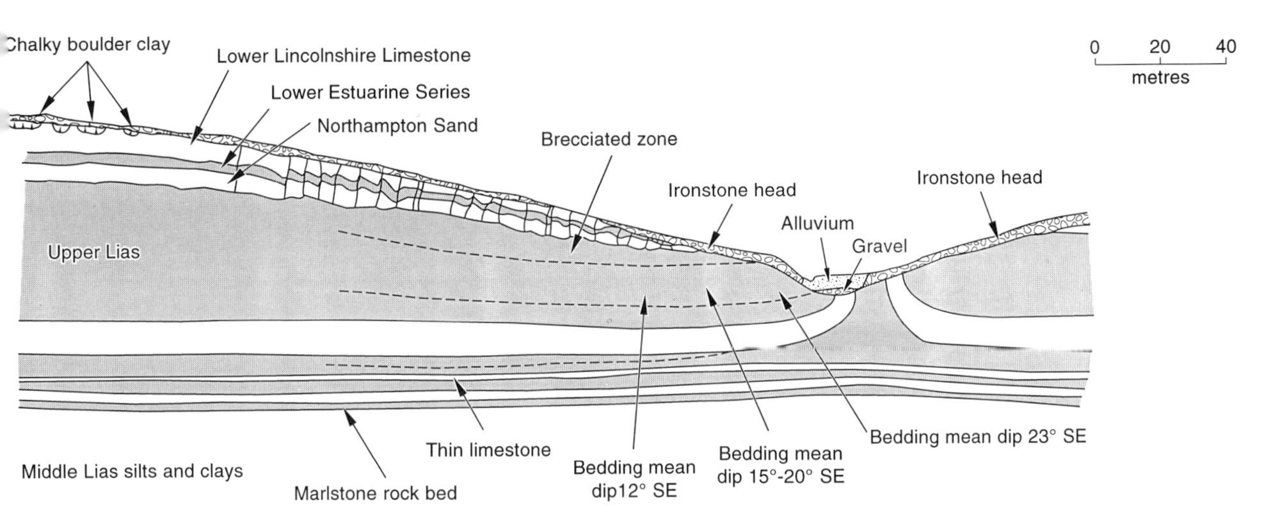

5.14 Cambered structures recorded at Empingham Dam, Leicestershire (after Horswill & Horton, 1976).

of shallow depressions which hang above the main Windrush Valley. Similar features have been described in parts of the Hambleton Hills in North Yorkshire[17], where Upper Jurassic Corallian sandstones and limestones have been widely affected by cambering. Ridge and trough topography is essentially a well developed network of collapsed gulls which have been only partially infilled by debris, so leaving a distinctive surface expression.

Dip and fault structures arise when there has been vertical displacement of strata across a gull, often with lowering on the uphill side. When this is repeated across a series of blocks, the overall result is of a downhill dip greater than the general dip. Such features were originally recognised in Northamptonshire, where large areas of the Northampton Ironstone have been affected by cambering[18]. During the investigation for the design of Empingham dam and Rutland Water a wide variety of cambered structures were recognised, including much dislocated blocks of Northampton Sand and Lincolnshire Limestone (*Figure 5.14*), often with up to 4m of vertical displacement between individual blocks[19].

Valley bulging results from the clay or shale strata underlying the cap rock, thrusting upwards in the floors of valleys. Often the tops of these bulges have been eroded away and they are normally part obscured by alluvium. The beds within a bulge structure are highly sheared and softened, with the base of the deformed zone marked by one or more planes of décollement[20]. Excellent examples of valley bulges have been described in the Frome Valley near Stroud[21] (*Figure 5.15*), consisting of

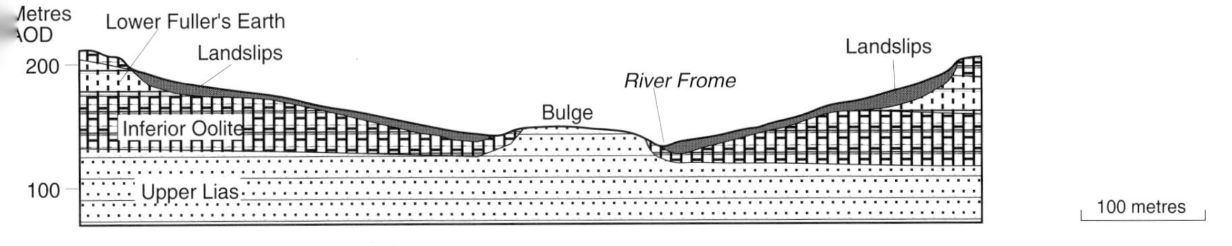

5.15 Valley bulge structures in the Frome Valley, near Stroud (after Ackerman & Cave, 1967).

plugs of Upper Lias sand and silt forced through the Inferior Oolite to form a ridge which is 7m above the valley floor at Pinbury Park, sufficient to block and then divert the course of the River Frome.

The features associated with cambering were first reported in the early years of this century following their identification in cut-off trenches associated with the development of reservoirs in the Pennine Valleys[22]. However it was not until 1944 that the cambering process was identified, following a detailed study of the structures exposed in the opencast ironstone mines of Northamptonshire[18]. Many processes have been proposed to explain the formation of cambering but there is now dominant support for the view that it developed during phases of intense freeze-thaw activity under a periglacial regime[23]. Providing the existence of suitable geological and topographical conditions, such as sub-horizontal sedimentary rocks forming a tableland dissected by valleys, then stress release would be sufficient to create the valley bulge structure and the basal shear surface below the valley floor[24] (*Figure 5.12*).

The DOE survey identified 158 instances of reported cambered/foundered strata (*Table 3.1*), the majority of which are located on the Jurassic rocks of the Midlands (Northamptonshire, Leicestershire, Oxfordshire, Gloucestershire), although similar features are also known to be widely developed on the Cretaceous Lower Greensand Hythe Beds escarpment of Kent and Surrey. Although a relict of the past, the resultant division of competent cap rocks into small blocks by the development of gulls, and their flexuring valleyward by cambering, is of great significance to more recent landsliding, as it prepares the rock for subsequent mass movement (*see Figures 3.27 and 5.13*).

Disposition

The term disposition is used here to cover the arrangement and attitude of rocks and soil. Geological forces not only lead to the rupturing of rocks to produce movement along faults, but also cause corrugation into folds as well as tilting. These larger-scale structural considerations are of importance to slope stability in several ways.

The bending of layers of rock into troughs (synclines) and arches (anticlines) results in the development of tensional forces on the stretched side of the layer (outside of the curve) and compressional forces on the inner margin. As a result, the crests of anticlinal structures tend to have well developed and relatively open joint systems which assist in weathering development and the initiation of mass movement. Such influences are best displaced in the more severely disturbed rocks of Highland and Ancient Britain (*Figure 2.2*), although they can also be of importance in Upland Britain where alternating sequences of sandstones and shales or limestones and shales have been deformed. Evidence from Lowland Britain is generally lacking because well developed minor folds are restricted to the extreme south (the Weald, Hampshire, Dorset and the Isle of Wight).

Of much more general importance is the effect of bedding inclination (dip) due to folding and faulting and the consequences in terms of outcrop elevation and groundwater characteristics. For example, one of the only substantiated examples of

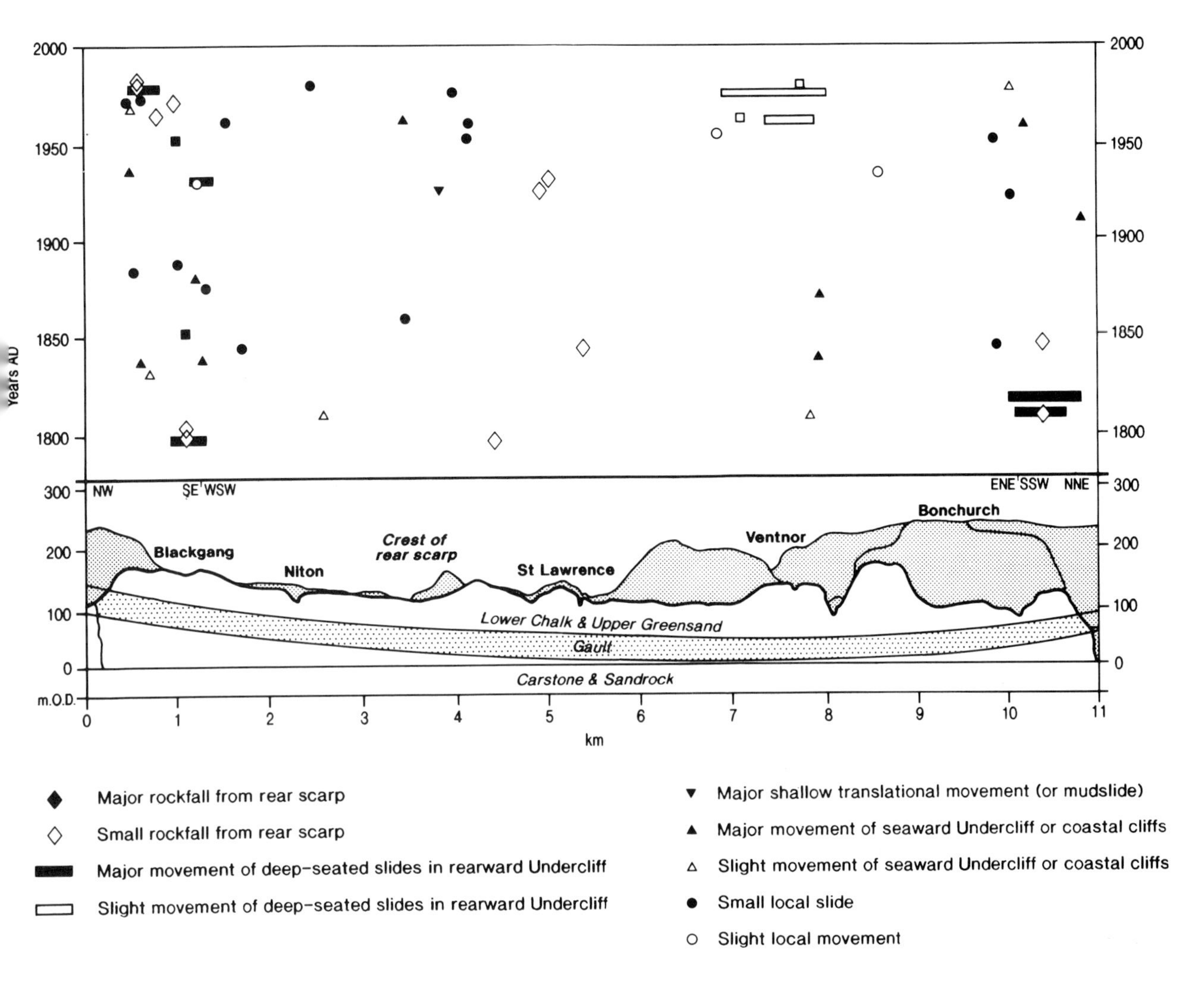

5.16 The pattern of recorded landslide activity along the Isle of Wight Undercliff showing the importance of the broad synclinal structure (after Hutchinson, 1965b).

the role of structure on landsliding in Southern England relates to the Undercliff, Isle of Wight. Here, analysis of landslide reports for the last 200 years has revealed a marked spatial variation in the frequency of occurrence, with greatest activity concentrated at the eastern and western ends of the 11km long landslide belt (*Figure 5.16*)[25]. This has been explained as a function of the elevation of the Gault Clay, itself determined by the existence of a broad syncline. The height of the clay and soft Lower Greensand cliffs increases at the eastern and western ends of the Undercliff, enabling rapid coastal erosion of around 0.4m per year and encouraging repeated landslide activity. Elsewhere along the Undercliff the displaced, massive, Upper Greensand blocks and debris aprons, which form the coastline, are much more resistant to erosion and the landslide complex is relatively inactive.

Inclined bedding is of crucial importance to slope stability, for the direction of dip defines the preferred direction of water movement in an alternating sequence of sediments as well as the direction of bedding plane slippage. If the direction of dip is into the valley-side or cliffline, then the slopes will be relatively stable despite the fact that the dip favours the ingress of water into the slopes. Shallow failures may occur, although these will almost certainly result from unfavourable jointing or the presence of faults. However, if the dip is towards the valley axis or the sea, then the slopes will be potentially unstable because slabs of rock can slide over bedding planes. The steeper the dips, the more extreme the potential stability or instability. On the other hand, relatively shallow dips facilitate both the movement of subsurface water and the generation of significant pore-water pressures. Thus along the Isle of Wight Undercliff (*Photos 2.4 and 3.10*), gentle seaward dip on the Upper Greensand beds above the Gault Clay aquiclude allows water to move towards the landslide complex, thereby assisting in the maintenance of instability and helping to create the characteristic stepped topographic profile. Without this water movement, there would be no massive failures and there would be high, relatively vertical clifflines, similar to those developed at both Compton Bay and Culver Cliff (also on the Isle of Wight) where the same beds dip steeply inland (*Plate 30*).

The influence of dip is also apparent in the case of valleys that trend with the strike of the geology (i.e. at right angles to the dip) so that opposite valley sides have markedly different stability conditions. On some occasions, usually when the dip is pronounced, river downcutting results in lateral migration of the valley down the dip (uniclinal shifting) so that it develops an asymmetrical form (*Figure 5.17a*). In this situation shallow instability will occur on the down-dip flank of the valley due to slope over-steeping. If the dip is shallower, however, asymmetry will be less marked or even absent, so that water movements become the most important factor in stability. In this case (*Figure 5.17b*), instability will occur on the up dip flank of the valley due to pore-water pressures. These two situations are readily apparent in many valleys in the Cotswolds where dips are generally shallow.

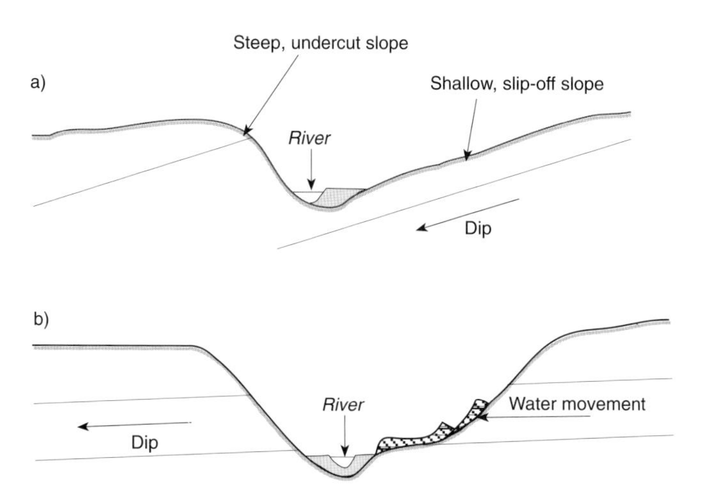

5.17 The importance of valleyward dip in promoting landsliding.

Finally, dip has a profound influence on topographical development because the outcropping end of a layer of resistant strata (limestone or sandstone) is usually attacked by denudation to yield a relatively steep, abrupt termination known as an escarpment. Escarpments are commonly developed on the Mesozoic and Tertiary strata of Lowland Britain (*Figure 2.2*), as well as on the rocks of Upland Britain. By definition, the dip must be into the escarpment thereby giving rise to stable conditions. However, escarpment slopes represent the major concentrations of recorded landslides in Great Britain (*Figure 2.14*), with extensive belts of landslipped ground associated with Millstone Grit Series escarpments in the Southern Pennines, the Cotswold escarpment in Gloucestershire, Bredon Hill (*Figure 2.12*), the Lower Greensand escarpment of Kent and Surrey and the Upper Greensand escarpment of Wiltshire. Instability in these situations is produced through the combination of erosional lowering of the adjacent clay vale, thereby leading to slope steepening, and water emerging on the scarp face from the contact between permeable cap-rock and impermeable lower horizons *(see earlier discussions of landsliding along the Cotswolds included under the section on 'Lithology')*. The gentler the dip of the strata the greater the proportion of groundwater that will emerge at the scarp face and therefore the more instability that will be generated.

Stratigraphy

Intuitively, it would appear both possible and desirable to relate the occurrence of landsliding to particular strata so that 'stable', 'intermediate' and 'unstable' outcrops could be identified for use in planning, engineering, the construction industry and other sectors concerned with land management. However, to achieve such a goal is immensely difficult. First, it has to be recognised that the factors that determine slope instability are many and varied (*Chapter 4*), and that their influence is not constant but extremely variable over space and through time. Second, the geological fabric of Great Britain is exceedingly complex and to varying degrees obscured by superficial deposits of widely ranging character and thickness (*Chapter 2*). Third, incomplete knowledge exists as to the actual distribution of landsliding on the ground *(see Chapter 2)* so any attempted correlations have to be recognised as preliminary estimations which are liable to error because of data bias, i.e. large numbers of records for small areas that have been searched in detail overwhelm limited data for extensive areas obtained through random reporting. Nevertheless, an attempt will be made in this section to relate existing levels of knowledge of landsliding to the various bedrock units that form the land area of Great Britain.

Landslide Prone Strata

The bedrock stratigraphy of Great Britain is known to varying degrees of detail. The basic stratigraphic division that has come to be used by the British Geological Survey on the 1:625,000 scale National Planning Series Geological Maps of Great Britain, consists of 115 separate units (formations) together with 18 amalgamations, partly based on chronostratigraphy (age) and partly based on nature in the case of ancient rocks and non-sedimentary materials (e.g. basalt, granite, etc.). These 133 "formations" are listed in Table 5.4, together with the extent of their outcrops as shown on the 1:625,000 scale geological maps, or at least their outcrops if all covering mantles of extensive and thick superficial deposits were removed. Also listed in Table 5.4 are the numbers of landslides reported to have involved each of these formations, (*but see discussion in Chapter 2 as to the problem regarding the great range in size of failures attributed the status of 'a landslide'*).

Table 5.4 **The frequency, density and type of landslides for British Geological Survey stratigraphic units.**

Stratigraphic Description *(Source: BGS 1:625,000 scale maps)*	**Number of recorded landslides**	**Total outcrop area (km²)**	**Density (total area) per 100 km²**	**Actual outcrop area (km²)**	**Density (actual area) per 100 km²**	**Unspecified landslides**	**Rockfalls**	**Topples**	**Sagging failures**	**Single rotational**	**Multiple rotational**	**Successive rotational**	**Compound**	**Translational**	**Flows**	**Complex**	**Cambered/foundered strata**
115 Norwich Crag, Red Crag & Chillesford Clay	26	2980	0.9	980	2.7	5	4	–	–	11	1	–	1	–	3	1	–
112–114 Neogene gravels (?Pliocene)	–	15	–	9	–	–	–	–	–	–	–	–	–	–	–	–	–
114 St Erth Beds	–	1	–	1	–	–	–	–	–	–	–	–	–	–	–	–	–
113 Coralline Crag	–	16	–	16	–	–	–	–	–	–	–	–	–	–	–	–	–
112 Lenham Beds	–	1	–	*	*	–	–	–	–	–	–	–	–	–	–	–	–
111 Hamstead Beds & Bembridge Marls	24	344	7.0	344	7.0	9	3	–	–	1	1	–	–	3	4	3	–
110 Bovey Formation, St Agnes Sands, Lough Neagh Clays etc	–	60	–	56	–	–	–	–	–	–	–	–	–	–	–	–	–
107–109 ?Eocene inter-lava beds	–	1	–	1	–	–	–	–	–	–	–	–	–	–	–	–	–
109 Barton, Bracklesham & Bagshot Beds	88	1682	5.2	1444	6.1	54	5	–	–	13	2	9	3	–	–	2	–
108 London Clay	356	6452	5.5	5095	7.0	145	5	–	–	95	9	17	32	8	19	26	–
107 Oldhaven, Blackheath, Woolwich & Reading & Thanet Beds	23	1089	2.1	857	2.7	6	2	–	–	5	–	–	–	–	3	1	6
106 Chalk	117	18178	0.6	1221	9.6	26	49	1	–	5	3	–	2	12	1	11	7
102–105 Undifferentiated Lower Cretaceous	–	301	–	203	–	–	–	–	–	–	–	–	–	–	–	–	–
105 Upper Greensand & Gault	273	2040	13.4	1865	14.6	159	5	1	–	10	26	3	7	11	4	46	1
104 Lower Greensand	166	1420	11.7	1395	11.9	78	7	1	–	20	29	6	1	2	8	13	1
103 Weald Clay	151	1722	8.8	1722	8.8	112	3	–	–	10	1	4	2	–	7	10	2
102 Hastings Beds	139	1873	7.4	1873	7.4	120	4	–	–	2	–	–	2	3	3	–	5
101 Purbeck Beds	21	59	35.6	51	41.2	8	2	–	–	–	–	–	–	2	–	9	–
100 Portland Beds	20	117	17.1	117	17.1	4	4	2	–	–	1	–	–	–	–	9	–
98–99 Ampthill & Kimmeridge Clays	11	2041	0.5	1111	1.0	7	–	–	–	–	–	–	1	1	–	2	–
99 Kimmeridge Clay	41	913	4.5	876	4.7	25	1	–	–	3	–	–	–	1	1	10	–
98 Corallian	57	897	6.4	897	6.4	26	–	–	–	2	5	–	18	1	–	3	2
97 Oxford Clay & Kellaways Beds	66	3461	1.9	2479	2.7	43	–	2	–	7	5	–	–	3	–	4	2
94–96 Undifferentiated Middle Jurassic	35	92	38.0	53	66.0	25	1	–	–	–	1	5	–	–	–	3	–

1 Terracettes produced as a result of soil creep (T Fell).

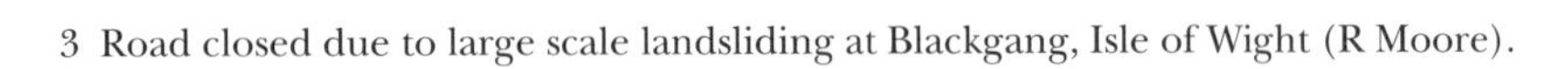

3 Road closed due to large scale landsliding at Blackgang, Isle of Wight (R Moore).

2 House abandoned as a result of coastal landsliding, Barton-on-Sea (E M Lee).

4 Severely damaged property, Holbeck Hall Hotel, Scarborough (E M Lee).

6 Degradation of coastal cliffs developed in Barton Beds and overlying Plateau Gravels, Naish Farm near Christchurch (E M Lee). (above)

5 Large rotational landslide in London Clay, Warden Point, Isle of Sheppy (R Moore). (above right)

7 The massive coastal landslide complex at Black Ven, Dorset (T Fell). (below right)

8 Large scale movements at The Spittles, east of Lyme Regis during the winter of 1986–1987 (E M Lee). (above)

9 The Storr, Isle of Skye: the largest and most dramatic landslide in Great Britain (T Fell). (above right)

10 Degradation of the abandoned sea cliff at Hadleigh has affected the castle remains (T M Dibb). (below right)

11 The abandoned sea cliff between Hythe and Lympne, overlooking Romney Marsh (E M Lee).

12 Degraded landslide features on the northern flank of Mam Tor, Derbyshire (T M Dibb).

13 An active mudslide system on the western flank of Bredon Hill, Hereford and Worcester (Geomorphological Services Ltd).

14 The spectacular scree slopes at Wastwater, Cumbria (T Fell).

15 The near vertical Chalk cliffs of Sussex which are retreating around 0.91m a year. Debris is rarely visible at the cliff foot because of the extremely rapid rates of removal (T Fell). (above left)

16 The Lower Greensand cliffs at Chale, Isle of Wight where rapid retreat is caused by a combination of marine and seepage erosion (R Moore). (below left)

17 The Blaencwn landslide, which involves an unusual combination of toppling and sliding (H Siddle). (below)

18 Shallow successive rotational landslides developed in London Clay, Hadleigh, Essex (T M Dibb). (above left)

19 A large rock slide in Glen Ogle, Central Region which affected the railway line to Killin (Reproduced by permission of the Director, British Geological Survey: NERC copyright reserved). (above)

20 A shallow debris slide developed in Wadhurst Clay, near Robertsbridge, East Sussex (D K C Jones). (below left)

21 The Fox Hill landslide on the A45 Daventry By-pass (S Biczysko).

22 The Newnham Hill landslide on the A45 Daventry By-pass (S Biczysko).

23 Coastal mudslide developed in Wealden Beds, Warbarrow Bay, Dorset (R Moore).

24 Peat slide in Upper Teesdale, July 1983 (P Carling).

25 The major landslide which occurred in 1971 during construction of the Stromeferry By-pass in the Scottish Highlands (Reproduced by permission of the Director, British Geological Survey: NERC copyright reserved). (above left)

26 The 1962 Cliff House landslide at Lyme Regis, which occurred during slope regrading (J N Hutchinson). (below left)

27 The 1975 landslide during construction of the Portavadie Dry Dock in western Scotland (A R Clark). (above)

28 House destroyed by landsliding at Blackgang, Isle of Wight, in January 1994 (R Moore). (above left)

29 The rapidly degrading cliffs at Chewton Bunny, Hampshire (E M Lee). (below left)

30 Eroding Lower Greensand and Gault Clay cliffs of Compton Bay, Isle of Wight; contrast with the massive landslides of the Undercliff developed in similar strata shown in Photo 2.4 (E M Lee). (above)

31 Debris slide developed in Hastings Beds, Fairlight Glen (R Moore).

32 Major rock slide developed in Devonian limestone and shales, Torbay (E M Lee).

33 Gormire Lake, formed as a result of a major multiple rotational landslide blocking an ice-margin drainage channel between 15,000–10,000 years ago (E M Lee).

35 Construction of a 550m long, 30m wide concrete armoured berm to protect the cliffs at Fairlight Village, East Sussex (R Moore).

34 Building close to the cliff edge, Fairlight Village, East Sussex (R Moore).

36 Concern for the fate of Luccombe Village led to considerable publicity and the formation of the Luccombe Residents Association (Photo courtesy of Luccombe Residents Association).

37 Ground movement causes repeated road maintenance problems in parts of Ventnor (E M Lee). (above left)

38 Subsidence of a graben-like feature has caused considerable damage to this playground in Ventnor (E M Lee). (below left)

39 Despite repeated efforts to repair the road the A625 at Mam Tor finally had to be abandoned in 1979 (D Anderson). (above)

40 Subdued rotational landslide features, Dovers Hill, Gloucestershire (E M Lee). (above left)

41 The major Taren landslide in the Taff Vale, South Wales (Rendel Palmer and Tritton). (below left)

42 Construction of shallow trench drains, Whitby, North Yorkshire (E M Lee). (above)

43 Cliff face treatment works in progress, Shanklin, Isle of Wight (Rendel Geotechnics).

44 Coastal erosion at Whitby has led to a number of cliff top properties being threatened (Rendel Geotechnics).

45 Slope reprofiling and emplacement of rock armour at Whitby (E M Lee).

46 The Castle Hill landslide which marks the point where the Channel Tunnel begins (G Birch).

47 Landslip Potential Map from the Rhondda Valleys, South Wales (Sir William Halcrow and Partners).

48 The massive Holbeck Hall landslide of June 1993, Scarborough (Photoair).

Table 5.4 **The frequency, density and type of landslides for British Geological Survey stratigraphic units** *(continued)*.

Stratigraphic Description *Source: BGS 1:625,000 scale maps)*	Number of recorded landslides	Total outcrop area (km²)	Density (total area) per 100 km²	Actual outcrop area (km²)	Density (actual area) per 100 km²	Unspecified landslides	Rockfalls	Topples	Sagging failures	Single rotational	Multiple rotational	Successive rotational	Compound	Translational	Flows	Complex	Cambered/foundered strata
6 Cornbrash	8	688	1.2	634	1.3	4	–	–	–	1	–	1	–	–	–	–	2
4–95 Great & Inferior Oolite including Great Estuarine Series	68	1360	5.0	932	7.3	45	2	–	–	1	–	2	–	1	1	12	4
5 Great Oolite	176	2139	8.2	1767	10.0	128	2	–	–	1	15	–	–	3	11	9	7
4 Inferior Oolite	397	1819	21.8	1598	24.8	111	–	–	–	42	17	95	4	14	7	23	84
1–93 Undifferentiated Lower Jurassic	20	95	21.1	86	23.3	15	1	–	–	1	–	–	–	–	–	1	2
3 Upper Lias	662	1920	34.5	1591	41.6	316	–	1	–	50	14	96	10	14	50	29	82
2 Middle Lias	344	1417	24.3	1228	28.0	272	1	–	–	9	3	11	–	7	7	21	13
1 Lower Lias	310	5573	5.6	3235	9.6	191	9	–	–	14	15	2	4	15	17	39	4
0 Triassic mudstones (inc "Keuper Marl" & Rhaetic)	212	11557	1.8	5368	3.9	181	6	–	–	6	2	–	3	4	3	5	2
9 Permo–Triassic sandstones (inc "Bunter" & "Keuper")	44	8755	0.5	5172	0.9	36	1	–	–	3	1	–	–	2	–	1	–
8 Budleigh Salterton Pebble Beds	8	53	15.1	53	15.1	4	2	–	–	–	–	–	–	–	–	2	–
7 Permian mudstones (inc Mid/Upper Marls, Eden & St Bees shales)	22	658	3.3	296	7.4	21	–	–	–	–	–	–	1	–	–	–	–
6 Magnesian Limestone (Permian)	83	1347	6.2	591	14.0	71	5	–	–	–	–	–	4	–	–	1	2
5 Permian basal breccias, sandstones & mudstones	136	1725	7.9	688	19.8	31	53	2	–	3	–	–	7	34	2	4	–
4 Westphalian & ?Stephanian, undivided of "Barren Red" lithology	6	1600	0.4	960	0.6	1	–	–	–	–	1	–	–	1	2	1	–
2–83 Undifferentiated Westphalian	40	1474	2.7	215	18.6	35	1	–	–	–	–	–	–	3	–	1	–
1–83 Undifferentiated Upper Carboniferous (SW England)	56	3075	1.8	3075	1.8	37	–	1	–	3	–	–	–	9	–	6	–
3 Upper Westphalian (C+D) inc "Pennant Measures"	618	3313	18.7	1989	31.1	116	3	2	–	27	4	1	–	153	70	241	1
2 Lower Westphalian (A+B) mainly "Productive Coal Measures"	285	5567	5.1	2982	9.6	170	3	–	–	26	2	1	8	21	7	47	–

Table 5.4 **The frequency, density and type of landslides for British Geological Survey stratigraphic units** *(continued).*

Stratigraphic Description *(Source: BGS 1:625,000 scale maps)*	Number of recorded landslides	Total outcrop area (km^2)	Density (total area) per 100 km^2	Actual outcrop area (km^2)	Density (actual area) per 100 km^2	Unspecified landslides	Rockfalls	Topples	Sagging failures	Single rotational	Multiple rotational	Successive rotational	Compound	Translational	Flows	Complex	Cambered/foundered strata
81 Namurian ("Millstone Grit Series")	1254	10008	12.5	4717	26.6	936	8	1	-	184	47	-	12	22	2	36	6
78–80 Devonian Carboniferous transition group	2	255	0.8	255	0.8	2	-	-	-	-	-	-	-	-	-	-	-
80 Tournaisian & Visean (Carb Limestone Series, L Carb in SW Eng.)	737	10636	6.9	3732	19.7	668	13	4	-	11	1	-	3	17	3	9	8
79 Carboniferous basal conglomerate	2	152	1.3	78	2.6	2	-	-	-	-	-	-	-	-	-	-	-
77–78 Middle & Upper Devonian & Upper Old Red Sandstone	6	1029	0.6	1029	0.6	1	-	1	-	-	-	-	-	-	1	3	-
75–78 Undifferentiated Devonian	11	855	1.3	833	1.3	7	-	-	1	-	-	-	-	3	-	-	-
78 Upper Devonian & Upper Old Red Sandstone	66	3211	2.1	1585	4.2	36	14	1	-	-	-	-	3	9	-	3	-
77 Middle Devonian (England), Middle Old Red Sandstone	155	4166	3.7	2102	7.4	23	81	3	-	-	1	-	8	36	-	3	-
76 Lower Devonian (England & Wales)	63	1548	4.1	1520	4.1	19	13	2	-	-	-	-	1	25	-	3	-
75 Lower Old Red Sandstone including Downtonian	74	9307	0.8	4569	1.6	44	2	5	-	5	-	-	1	6	1	10	-
73–74 Undifferentiated Wenlock & Ludlow	9	*	*	*	*	-	-	-	-	-	-	-	-	5	1	3	-
72–74 Undifferentiated Silurian	-	175	-	18	-	-	-	-	-	-	-	-	-	-	-	-	-
74 Ludlow	175	3477	5.0	2061	8.5	174	1	-	-	-	-	-	-	-	-	-	-
73 Wenlock	54	1797	3.0	972	5.6	53	-	-	-	-	1	-	-	-	-	-	-
72 Llandovery	10	8435	0.1	4321	0.2	8	-	-	-	1	-	-	-	1	-	-	-
70–71 Upper Ordovician (Caradoc & Ashgill)	39	4376	0.9	2377	1.6	38	-	-	-	-	-	-	-	1	-	-	-
71 Ashgill	-	499	-	391	-	-	-	-	-	-	-	-	-	-	-	-	-
70 Caradoc	17	1213	1.4	2377	0.7	7	6	-	-	2	-	-	-	-	-	2	-
68–69 Lower Ordovician (Arenig, Llanvirn & Llandeilo)	4	563	0.7	203	2.0	-	-	-	-	-	-	-	-	3	-	1	-

Table 5.4 **The frequency, density and type of landslides for British Geological Survey stratigraphic units** *(continued).*

Stratigraphic Description *Source: BGS 1:625,000 scale maps)*	**Number of recorded landslides**	**Total outcrop area (km²)**	**Density (total area) per 100 km²**	**Actual outcrop area (km²)**	**Density (actual area) per 100 km²**	**Unspecified landslides**	**Rockfalls**	**Topples**	**Sagging failures**	**Single rotational**	**Multiple rotational**	**Successive rotational**	**Compound**	**Translational**	**Flows**	**Complex**	**Cambered/foundered strata**
9 Llandeilo	–	86	–	69	–	–	–	–	–	–	–	–	–	–	–	–	–
8 Llanvirn & Arenig	13	1210	1.1	808	1.6	10	1	–	–	1	–	–	–	–	1	–	–
7 Durness Limestone (Ordovician & Cambrian)	–	103	–	72	–	–	–	–	–	–	–	–	–	–	–	–	–
4–66 Undifferentiated Cambrian	3	75	4.0	75	4.0	–	3	–	–	–	–	–	–	–	–	–	–
4–65 Undifferentiated Lower & Middle Cambrian	–	*	*	*	*	–	–	–	–	–	–	–	–	–	–	–	–
6 Upper Cambrian including Tremadoc	10	499	2.0	401	2.5	1	2	–	–	1	–	–	–	4	–	2	–
5 Middle Cambrian	–	101	–	73	–	–	–	–	–	–	–	–	–	–	–	–	–
4 Lower Cambrian	–	231	–	153	–	–	–	–	–	–	–	–	–	–	–	–	–
3 Serpulite Grit & Fucoid Beds (Cambrian)	–	56	–	37	–	–	–	–	–	–	–	–	–	–	–	–	–
2 Cambrian pipe–rock & basal quartzite	–	368	–	252	–	–	–	–	–	–	–	–	–	–	–	–	–
1 Torridonian sandstone & grit	11	1808	0.6	1328	0.8	5	3	–	–	–	–	–	–	1	–	2	–
0 Precambrian rocks of Anglesey, Lleyn, Charnwood, Longmynd etc.	3	482	0.6	126	2.4	2	–	–	–	–	–	–	–	1	–	–	–
9 Tertiary undifferentiated tuff	–	17	–	17	–	–	–	–	–	–	–	–	–	–	–	–	–
8 Tertiary rhyolite, trachyte & allied type	–	8	–	8	–	–	–	–	–	–	–	–	–	–	–	–	–
7 Tertiary basalt & spilite	123	1477	8.3	1076	11.4	99	2	–	–	4	5	9	–	–	–	4	–
6 Permian basalt, splilite & rhyolite	–	60	–	22	–	–	–	–	–	–	–	–	–	–	–	–	–
5 Carboniferous undifferentiated tuff & agglomerate	11	76	14.5	35	31.4	11	–	–	–	–	–	–	–	–	–	–	–
4 Carboniferous rhyolite, trachyte & allied types	1	0.7	31	3.2	1	–	–	–	–	–	–	–	–	–	–	–	–
3 Carboniferous basalt & spilite	47	1470	3.2	386	12.2	40	1	–	–	1	–	–	–	–	–	5	–
2 Devonian & Old Red Sandstone tuff, mainly andesitic	4	92	4.3	62	6.5	4	–	–	–	–	–	–	–	–	–	–	–
1 Devonian & Old Red Sandstone rhyolite, trachyte & allied types	3	66	4.5	35	8.6	2	–	–	–	–	–	–	–	–	–	1	–

Table 5.4 **The frequency, density and type of landslides for British Geological Survey stratigraphic units** *(continued).*

Stratigraphic Description *(Source: BGS 1:625,000 scale maps)*	**Number of recorded landslides**	**Total outcrop area (km^2)**	**Density (total area) per 100 km^2**	**Actual outcrop area (km^2)**	**Density (actual area) per 100 km^2**	**Unspecified landslides**	**Rockfalls**	**Topples**	**Sagging failures**	**Single rotational**	**Multiple rotational**	**Successive rotational**	**Compound**	**Translational**	**Flows**	**Complex**	**Cambered/foundered strata**
50 Devonian & ORS andesitic & basaltic lava & tuff, undifferentiated	15	1890	0.8	1106	1.4	12	3	–	–	–	–	–	–	–	–	–	–
49 Devonian & Old Red Sandstone basalt & spilite	1	9	11.1	2	50.0	1	–	–	–	–	–	–	–	–	–	–	–
48 Lower Palaeozoic undifferentiated tuff, mainly andesitic	6	534	1.1	515	1.2	1	3	–	–	–	–	–	–	–	1	1	–
47 Lower Palaeozoic rhyolitic tuff, including i gnimbrite	11	231	4.8	231	4.8	8	2	–	–	–	–	–	–	–	1	–	–
46 Lower Palaeozoic undifferentiated rhyolitic lava & tuff	5	388	1.3	84	6.0	3	–	–	–	1	–	–	–	–	1	–	–
45 Silurian & Ordovician andesitic tuff	–	1	–	1	–	–	–	–	–	–	–	–	–	–	–	–	–
44 Lower Palaeozoic undifferentiated andesitic lava & tuff	2	370	0.5	267	0.7	1	1	–	–	–	–	–	–	–	–	–	–
43 Silurian & Ordovician basaltic tuff	–	10	–	3	–	–	–	–	–	–	–	–	–	–	–	–	–
42 Lower Palaeozoic basalt, spilite, hyaloclastic & related tuffs	–	87	–	2	–	–	–	–	–	–	–	–	–	–	–	–	–
41 Precambrian undifferentiated rhyolitic & trachytic lavas & tuffs	–	225	–	81	–	–	–	–	–	–	–	–	–	–	–	–	–
40 Precambrian basalt, spilite & related tuff	–	9	–	9	–	–	–	–	–	–	–	–	–	–	–	–	–
39 Precambrian andesitic lava & tuff	–	7	–	7	–	–	–	–	–	–	–	–	–	–	–	–	–
38 Intrusive neck agglomerate	3	80	3.8	30	10.0	3	–	–	–	–	–	–	–	–	–	–	–
7 Intrusive rhyolite, trachyte, felsite, elvans & allies	–	524	–	242	–	–	–	–	–	–	–	–	–	–	–	–	–
36 Intrusive porphyrite, lamprophyre & allied types	2	536	0.4	334	0.6	2	–	–	–	–	–	–	–	–	–	–	–

Table 5.4 **The frequency, density and type of landslides for British Geological Survey stratigraphic units** *(continued).*

Stratigraphic Description *(Source: BGS 1:625,000 scale maps)*	**Number of recorded landslides**	**Total outcrop area (km²)**	**Density (total area) per 100 km²**	**Actual outcrop area (km²)**	**Density (actual area) per 100 km²**	**Unspecified landslides**	**Rockfalls**	**Topples**	**Sagging failures**	**Single rotational**	**Multiple rotational**	**Successive rotational**	**Compound**	**Translational**	**Flows**	**Complex**	**Cambered/foundered strata**
35 Intrusive basalt, dolerite, camptonite & allied types	58	2937	2.0	1473	3.9	48	5	–	–	1	1	–	–	1	–	2	–
34 Intrusive granite, syenite, granophyre & allied types	32	7646	0.4	3442	0.9	24	3	–	1	1	–	–	–	–	1	2	–
33 Intrusive diorite & allied intermediate types	14	425	3.3	197	7.1	12	–	–	–	–	–	–	–	–	–	2	–
32 Intrusive gabbro & allied types	3	644	0.5	173	1.7	2	1	–	–	–	–	–	–	–	–	–	–
31 Intrusive ultrabasic rock	3	68	4.4	15	20.0	2	–	–	–	1	–	–	–	–	–	–	–
30 Precambrian gneiss, mica schist	6	183	3.3	55	10.9	3	1	–	–	1	1	–	–	–	–	–	–
29 Precambrian Hornblende schists	3	75	4.0	54	5.6	3	–	–	–	–	–	–	–	–	–	–	–
28 Moinian & Dalradian foliated granite, syenite & allies	2	172	1.2	130	1.5	1	–	–	–	–	–	–	–	–	–	1	–
27 Moinian & Dalradian epidiorite, hornblende schist & allies	16	939	1.7	567	2.8	11	–	–	1	–	–	–	1	1	–	2	–
26 Moinian and Dalradian serpentine	–	115	–	83	–	–	–	–	–	–	–	–	–	–	–	–	–
25 Upper Dalradian limestone	–	1	–	1	–	–	–	–	–	–	–	–	–	–	–	–	–
24 Dalradian limestone	12	743	1.6	235	5.1	8	–	1	–	–	–	–	1	1	–	1	–
23 Dalradian graphitic schist & slate	11	585	1.9	241	4.6	6	1	–	4	–	–	–	–	–	–	–	–
22 Upper Dalradian black shale with chert	–	14	–	14	–	–	–	–	–	–	–	–	–	–	–	–	–
21 Upper Dalradian slate, phyllite & mica-schist	5	175	2.9	124	4.0	3	–	–	–	–	–	–	–	–	–	2	–
20 Dalradian slate, phyllite & mica-schist	31	1762	1.8	992	3.1	24	–	–	2	–	–	–	–	4	–	1	–
19 Upper Dalradian quartz-mica schist, grit, slate & phyllite	169	4710	3.6	1801	9.4	122	–	5	4	–	–	–	3	18	–	17	–
18 Dalradian quartzose-mica schist	19	1895	1.0	331	5.7	9	1	1	3	–	–	–	1	2	–	2	–

Table 5.4 **The frequency, density and type of landslides for British Geological Survey stratigraphic units** *(continued).*

Stratigraphic Description *(Source: BGS 1:625,000 scale maps)*	**Number of recorded landslides**	**Total outcrop area (km^2)**	**Density (total area) per 100 km^2**	**Actual outcrop area (km^2)**	**Density (actual area) per 100 km^2**	**Unspecified landslides**	**Rockfalls**	**Topples**	**Sagging failures**	**Single rotational**	**Multiple rotational**	**Successive rotational**	**Compound**	**Translational**	**Flows**	**Complex**	**Cambered/foundered strata**
17 Dalradian quartzite, grit, quartzose–mica schist	26	2008	1.3	1230	2.1	17	–	–	1	–	–	–	–	6	–	2	–
16 Dalradian boulder bed & conglomerate	–	41	–	17	–	–	–	–	–	–	–	–	–	–	–	–	–
5 Upper Dalradian epidote–chlorite schist, hornblendic–Green Beds	7	261	2.7	106	6.6	5	–	–	–	–	–	–	–	1	–	1	–
14 Dalradian epidote–chlorite schist, hornblendic–Green Beds	–	13	–	13	–	–	–	–	–	–	–	–	–	–	–	–	–
13 Undiff. Dalradian schist & gneiss of Shetland & Co. Tyrone	14	438	3.2	438	3.2	14	–	–	–	–	–	–	–	–	–	–	–
12 Moinian granitic gneiss	–	175	–	108	–	–	–	–	–	–	–	–	–	–	–	–	–
11 Moinian mica–schist, semi–pelitic schist & mixed schists	109	2146	5.1	1102	9.9	74	–	5	8	–	–	–	1	13	–	8	–
110 Moinian quartz–feldspar granulite	117	11810	1.0	4069	2.9	88	1	6	8	–	–	–	–	8	–	6	–
9 Moinian quartzite	–	37	–	16	–	–	–	–	–	–	–	–	–	–	–	–	–
8 Undifferentiated Moinian rocks	6	656	0.9	344	1.7	4	–	1	–	–	–	–	–	1	–	–	–
7 Lewisian gneissose granite, granite & pegmatite	–	46	–	46	–	–	–	–	–	–	–	–	–	–	–	–	–
6 Lewisian intermediate & basic rock	–	336	–	304	–	–	–	–	–	–	–	–	–	–	–	–	–
5 Lewisian ultrabasic rock	–	33	–	29	–	–	–	–	–	–	–	–	–	–	–	–	–
4 Lewisian anorthosite	–	14	–	14	–	–	–	–	–	–	–	–	–	–	–	–	–
3 Lewisian marble	–	3	–	3	–	–	–	–	–	–	–	–	–	–	–	–	–
2 Lewisian metasediments	–	103	–	103	–	–	–	–	–	–	–	–	–	–	–	–	–
1 Lewisian undifferentiated gneiss	17	4486	0.4	3843	0.4	7	1	–	4	–	–	–	–	3	–	2	–
TOTAL	**8741**	**220620**	**4.0**	**117959**	**7.4**	**5348**	**357**	**49**	**37**	**586**	**215**	**262**	**145**	**521**	**243**	**735**	**243**

Notes: 7524 landslides in the GSL databank are recorded as involving one or more stratigraphic units.
8741 stratigraphic attributes are recorded.
Total area refers to rockhead outcrop.
Actual area refers to surface outcrop.
* indicates unknown area and density.

This preliminary stage of analysis reveals that 62% of recorded landslides are associated with the following 11 formations:

- the Carboniferous Namurian formation (the Millstone Grit Series) with 1254 recorded landslides;
- the Carboniferous Tournaisian and Visean Formation (the Carboniferous Limestone Series) with 737 recorded landslides;
- the Upper Lias, with 662 recorded landslides;
- the Carboniferous Upper Westphalian, including the Pennant Measures of South Wales, with 618 recorded landslides;
- the Middle Jurassic Inferior Oolite, with 397 recorded landslides;
- the Eocene London Clay, with 356 recorded landslides;
- the Middle Lias, with 344 recorded landslides;
- the Lower Lias, with 310 recorded landslides;
- the Carboniferous Lower Westphalian, mainly Productive Coal Measures, with 285 recorded landslides;
- the Cretaceous Upper Greensand and Gault, with 273 recorded landslides;
- the Triassic mudstones, with 212 recorded landslides.

Absolute numbers of recordings are misleading, however, because of the markedly differing extents of formation outcrops. Density of reported landsliding per unit area (i.e. per $100km^2$) is clearly a much better basis for measuring susceptibility to failure. Outcrop areas were therefore calculated from the 1:625,000 scale geology maps to yield the values and densities shown in Table 5.4. These estimates of densities of recorded landslides includes the occurrence of coastal slides, as these features provide a valuable indication as to how susceptible particular rocks are to continued basal erosion.

The totals can be misleading, however, because significant portions of many outcrops are concealed beneath extensive and thick mantles of superficial deposits (*Figure 2.3*). It is therefore essential that reported occurrences of landsliding are related to the actual outcrop area of each formation, by excluding the portions of outcrop that are concealed beneath thick superficial deposits.

The resultant outcrop areas are shown in Table 5.4 (actual outcrop), together with the density values, and reveal that the most landslide prone strata on the basis of currently available information appear to be:

- the Undifferentiated Middle Jurassic rocks of the islands of Skye, Raasay and Eigg, with a density of 66 landslides/$100km^2$;
- the Upper Lias, with a density of 42 landslides/$100km^2$;
- the Upper Jurassic Purbeck Beds with a density of 41 landslides/$100km^2$;

- the Undifferentiated Carboniferous tuffs and agglomerates of Derbyshire, with a density of 31 landslides/100km^2;
- the Carboniferous Upper Westphalian, including the Pennant Measures, with a density of 31 landslides/100km^2;
- the Middle Lias, with a density of 28 landslides/100km^2;
- the Carboniferous Namurian Formation (the Millstone Grit Series), with a density of 27 landslides/100km^2;
- the Middle Jurassic Inferior Oolite, with a density of 25 landslide/100km^2;
- the Undifferentiated Lower Jurassic rocks of Mull and Raasay with a density of 23 landslides/100km^2;
- the Permian basal breccias, with a density of 20 landslides/100km^2;
- the Carboniferous Tournaisian and Visean Formation (the Carboniferous Limestone Series), with a density of 20 landslides/100km^2;
- the Undifferentiated Westphalian Formations of North East England and the Central Lowlands of Scotland, with a density of 19 landslides/100km^2;
- the Upper Jurassic Portland Beds with a density of 17 landslides/100km^2;
- the Permian Budleigh Salterton Pebble Beds, with a density of 15 landslides/100km^2;
- the Cretaceous Upper Greensand and Gault, with a density of 15 landslides/100km^2;
- the Permian Magnesian limestone, with a density of 14 landslides / 100km^2.

These figures relate to all recorded landslides held in the database. However, the results may be considered to be biased by the inclusion of data on coastal landsliding, for the influences of coastal failures will vary with the length of coastal outcrop, intensity of coastal erosion and the varying degree to which coastal landsliding has been investigated and documented (strongly biased to South-eastern England). The resultant figures for inland landsliding are therefore as follows:

- the Upper Lias, with a density of 41 landslides/100km^2;
- the Undifferentiated Carboniferous tuffs and agglomerates of Derbyshire with a density of 31 landslides/100km^2;
- the Carboniferous Upper Westphalian, including the Pennant Measures, with a density of 30 landslides/ 100km^2;
- the Carboniferous Namurian Formation (the Millstone Grit Series), with a density of 27 landslides/100km^2;
- the Middle Lias, with a density of 26 landslides/ 100km^2;
- the Middle Jurassic Inferior Oolite, with a density of 25 landslides/100km^2;
- the Undifferentiated Middle Jurassic rocks of the Islands of Skye, Raasay and Eigg, with a density of 25 landslides/100km^2;
- the Upper Jurassic Purbeck Beds with a density of 20 landslides/100km^2;

- the Carboniferous Tournaisian and Visean Formation (the Carboniferous Limestone Series), with a density of 19 landslides/100km^2;
- the Undifferentiated lower Jurassic rocks of Mull and Raasay, with a density of 19 landslides/100km^2;
- the Undifferentiated Westphalian Formations of North East England and the Central Lowlands of Scotland, with a density of 16 landslides/100km^2.

If attention is restricted to inland landsliding in England and Wales, then the resultant distribution is as shown in Figure 5.18. These formations clearly may be

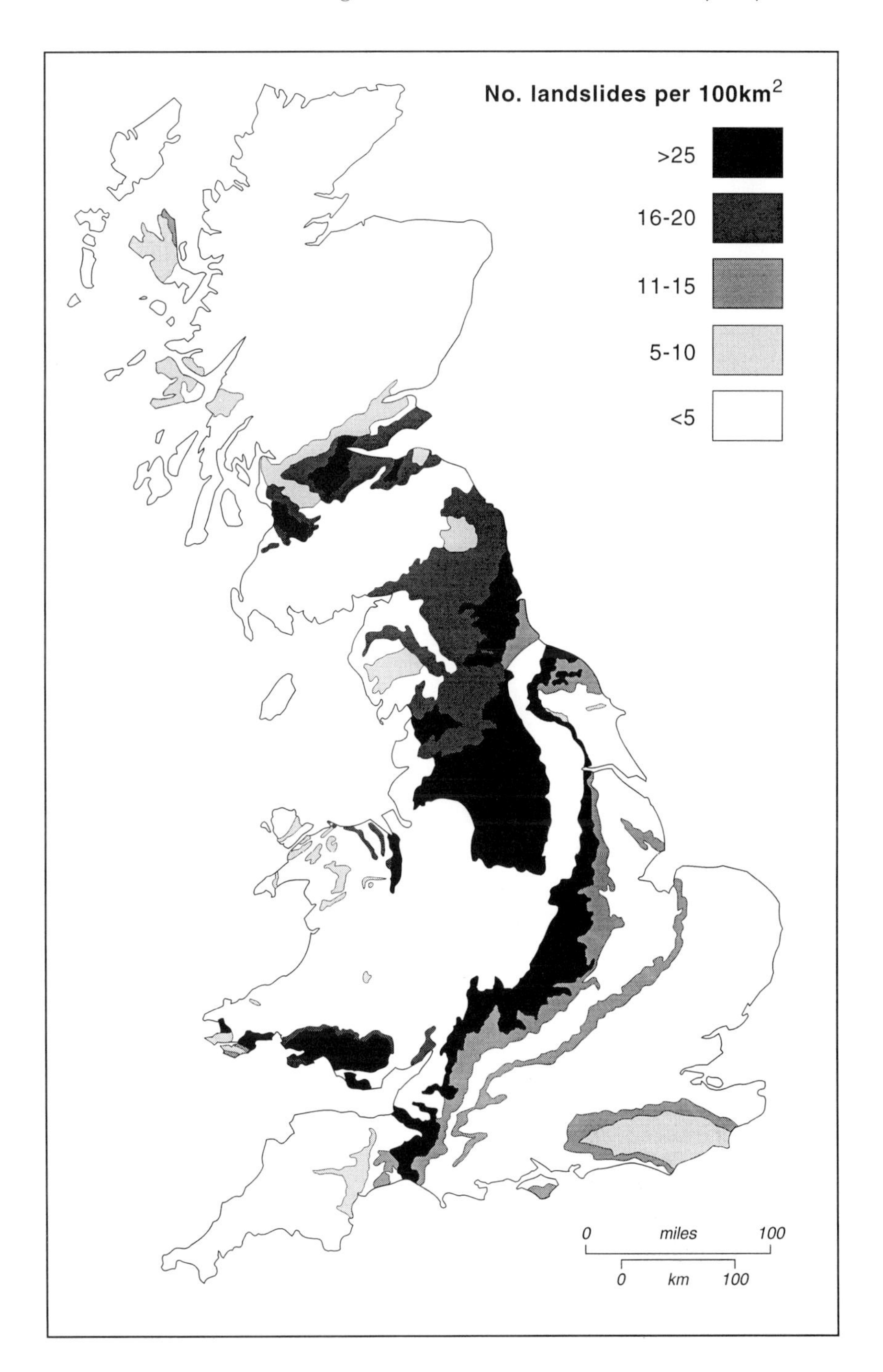

.18 The density of inland ndsliding across the major eological formations. NB 'his map shows the **total** **utcrop** area of the various ndslide prone strata and ot the **actual** areas of xposure where not oncealed by extensive and ick mantles of superficial eposits.

considered to be **landslide prone**. However, the reverse is not true, that no reports of failures or small numbers of recorded instances of landsliding implies stability. This is because the pattern of past research into landsliding has resulted in variable levels of knowledge regarding the occurrence of slides on different strata. For example, much of the research has been concentrated on the behaviour of stiff, fissured clays (e.g. London Clay, Lias Clay, Barton Clay) and mudrocks (e.g. the Carboniferous Coal Measures of South Wales and Carboniferous Millstone Grit Series of the Southern Pennines) rather than other outcrops such as Devonian shales and slates (e.g. in North Devon, and the Black Mountains in South Wales), or the Hastings Beds of the Central Weald.

The assessment presented here must be recognised as preliminary, and the eventual figures of landslide densities or ranking of landslide-prone formations may become significantly different to those presented here.

There are, however, additional considerations that have to be noted. Conditions with regard to landslide potential are far from constant or uniform over the outcrop of each stratigraphic unit. Topography is the most obvious variable. The relief developed on an outcrop can reflect a number of variables: lithological variability within the stratigraphic unit, relative durability of materials compared with those of adjacent outcrops (i.e. high ground can only be formed if adjacent materials are more readily eroded to form low ground), geomorphological history (glaciated portions of an outcrop will tend to have lower relative relief than unglaciated parts, as is clearly shown with respect to the Chalk outcrop to the east and west of Hitchin) and distance from the sea. Groundwater conditions may also be extremely variable reflecting differences in topographic conditions and lithology (see earlier). But it is lithology that is probably the most important variable.

It has been stressed already that formations are not homogeneous units in terms of lithology (*see previous section on lithology*) but this point has to be emphasised again here. Some of the 133 formations used in this analysis are, in reality, convenient groupings of generally similar rocks that have limited outcrops (e.g. Unit 21 - Dalradian slate, phyllite and mica-schist). Others are generalisations covering larger areas which have yet to be mapped in sufficient detail to facilitate internal division or are not deemed sufficiently important through location to warrant division for portrayal on maps at 1:625,000 scale (e.g. Unit 20). But most important, the formational names used to describe the very extensive occurrence of varied sedimentary strata in Great Britain (*see Figure 2.2*) must, through necessity, be gross generalisations. The lithology of sedimentary strata is extremely variable over an outcrop and vertically within a sequence, as has already been discussed earlier in this chapter with reference to the Chalk and Lower Greensand. Such variations also occur in formations whose names imply uniformity (e.g. Weald Clay) as well as in those whose names indicate variety through use of plurals (e.g. Hastings Beds), generalisation names (e.g. Cornbrash) or the term "series" (e.g. Carboniferous Limestone Series). In view of the large numbers of reported landslides in South Wales, this point is probably best amplified with respect to the Upper Carboniferous Coal Measure Series (Units 82 and 83).

The Coal Measure Series consist of sequences of sandstones, shales, thin coal seams and seatearths, all of which have different geotechnical properties. Indeed, the presence of rock units of contrasting strengths and permeabilities within a formation will often have a far greater effect on the stability of the slope than the dominant material as indicated by the formation name. Sand or rock horizons within clays are particularly important as they give rise to preferential flow paths of water within a slope which promote weathering and softening adjacent to the permeable horizons, thereby creating zones of weakness. Detailed knowledge is therefore an important prerequisite in understanding the causes of landsliding, as exemplified by a recently described example[26] of a failure passing through a 50mm thick softened horizon of Lias Clay above a 250mm limestone band at Beechen Cliff in Bath, which had clearly been promoted by the presence of this permeable band within the clay.

It follows for the above that extrapolation of local results across outcrops is potentially very misleading unless supported by detailed consideration of ground form and ground materials. The figures and maps presented in this book must therefore be considered as first approximations that may change significantly as new data becomes available. The same is true also of the relationships between landslide types and geological formations which are presented in Table 5.4 but will not be discussed in detail.

Characteristic settings for landslides

On the basis of the results presented above it appears that landsliding in Great Britain is most frequently associated with a number of different geological settings:

Group A; stiff fissured clays and mudrocks where low shear strengths and a high susceptibility to weathering and softening has led to large numbers of failures on oversteepened slopes. The most common forms of landsliding on these materials include single and successive rotational slides, debris slides and mudslides (*Plate 31 and Photo 5.2*). The most widely investigated materials in this category are the London Clay and Lias Clays of Southern and Eastern England;

Group B; well-jointed, faulted, cleaved and foliated hard rocks in which the pattern of discontinuities provides potential failure surfaces or weak zones within the rock mass. Rockfalls, topples, sagging failures and rock slides are the dominant modes of failure of these rocks, which include Upper Dalradian quartz-mica-schist and Moinean mica-schist in the Scottish Highlands, Permian Basal Breccias and Devonian limestones in Torbay, the Carboniferous limestones of Wales and Northern England, and the Chalk along the south coast of England, (*Plate 32*).

Group C; the occurrence of sequences of lithologically variable rock types which create potentially unstable conditions. For example, many areas of known landsliding are associated with the presence of thick horizons of impermeable fissured clays or mudrocks overlain by a massive, but well-jointed, permeable caprock of sandstone, limestone or volcanic rock. Cambering and valley bulging, multiple rotational slides and compound failures are the dominant forms of

5.2 Coastal mudslides developed in Jurrasic clays, Osmington, Dorset (Crown copyright/MOD).

5.3 Major multiple rotational slide at Beinn a' Cheorcaill, Isle of Skye (Cambridge University Collection: copyright reserved).

.4 Compound landslide
eveloped in Chalk in Gault
lay, Beer, Devon
Cambridge University
ollection: copyright
eserved).

landsliding associated with this setting (*Photos 5.3 and 5.4*). Classic examples of areas with landsliding promoted by these unstable combinations of rocks include: the variable sequences of sandstones and mudrocks in the Millstone Grit Series of the Pennines; the Carboniferous Coal Measures of South Wales; the Inferior Oolite overlying the Lias Clay in the Cotswolds and East Midlands (*Figure 5.4*); the Lower Greensand within the Weald of Surrey and Kent, and the Upper Greensand overlying the Gault Clay in Wiltshire and along the Isle of Wight Undercliff.

Group D; rocks which weather to produce sandy regoliths with high infiltration rates, which are susceptible to debris flow or debris slide activity, especially during intense rainstorms (*Photo 5.5*). Indeed, observations of debris flows in Scotland have revealed that they are more frequent on rocks which weathered to a coarse grained matrix, such as granites and sandstones, rather than those, such as schist, which yield cohesive clay-rich soils[27].

Superficial Deposits

It is necessary to briefly consider the significance of superficial deposits which mantle an estimated 41% of the land area (*Figure 2.3*). The presence of superficial deposits on many valley sides has a significant influence on slope stability. Of the 8835 landslides recorded in the database, 3042 (34%) were reported as either involving, or being wholly developed in, superficial deposits (2696 in inland situations and 346 on the coast). In South Wales, for example, it has been estimated that over 75% of the c. 1000 landslides recognised on the valley slopes were shallow failures involving superficial materials[28].

5.5 Debris slides in weathered basalts, the Gargunnock Hills, Scotland (Cambridge University Collection: copyright reserved).

5.6 Rotational landslides developed in weak superficial deposits, Walton-on-the-Naze, Essex (Cambridge University Collection: copyright reserved).

Superficial deposits are exceedingly diverse in character. The most susceptible to slope instability appear to be glacial tills (1743 landslides, 20% of all recorded landslides in Great Britain). A total of 1613 inland landslides are reported as involving till, making it the most commonly recorded superficial material associated with slope failure (60%, *see Table 5.4*), with marked concentrations in South Wales, the Conway valley of North Wales, the Pennines and the Vale of Eden. Where glacial tills outcrop along the coast they are particularly prone to rapid erosion, as along the Holderness and Norfolk coastlines of Eastern England (*Photo 5.6*) and the Clwyd coast of North Wales.

Tills are generally heterogeneous deposits, often rich in clay and silt but with layers and lenses of sand and gravel and, occasionally, boulders and cobbles. A distinction has to be drawn between **lodgements tills** which were laid down beneath ice-sheets and are usually overconsolidated, and **ablation tills** which accumulated by melt-out from decaying ice-masses and are normally consolidated (as are associated fluvio-glacial sediments). Stress relief fissuring is common in lodgement tills and have been recorded as influencing the stability of till slopes. For example, cutslope failures during the construction of the Hurlford Bypass, Ayrshire, were found to be controlled by softening along fissures within the stiff tills[29]. Slope failures occurred where fissure orientation was locally unfavourable, but otherwise the slopes remained stable. Where discontinuities are absent, or in non-critical orientations, slopes of considerable height can develop in tills. In Lower Teesdale and Peterlee[30], for example, river cliffs up to 50m high occur in the low plasticity tills. Measurements of slope angles indicate that there is no decrease in stable slope angle with increasing height above 10m. The higher slopes appear to behave as if they were purely frictional materials *(see Chapter 4)*. However, if this were true, an apparent angle of friction ($Ø'$) of 50° would be necessary to explain the stability. Such a value would be unusually high and is certainly unrealistic. What actually happens is that the materials rely on a cohesion component for their strength, but only on the higher slopes. This conclusion is supported by the fact that deep-seated slides occur on the higher slopes, since only shallow slides would be anticipated with purely frictional behaviour.

The great diversity of tills reflects both the differing ages of the deposits *(see Chapter 6)* and the variations in materials eroded by the ice-sheets that contributed to the eventual composition of till. Clearly the nature of a till is largely a reflection of the materials overridden by the ice, so that a route predominantly over sandstone bedrock would give rise to a coarse-grained sandy till while another part of the same ice-sheet that crossed tracts of mudstone or pre-existing fine-grained till, would produce a silt-clay rich deposit. Clay-rich tills mainly occur in Lowland Britain. However, over much of Highland and Upland Britain, glacial erosion resulted in the formation of arenaceous or silty tills, often with large volumes of coarse gravels or cobbles. Clearly a wide range of shear strength values can be anticipated depending on the nature of the material. However, there are marked differences between low plasticity sandy clays and plastic clays. The former generally show only limited brittleness (*Chapter 4*) and, hence, are less susceptible to progressive failure or softening. In the South Wales Coalfield, for example, natural failures in the coarse tills are generally shallow and have moved only short distances because of the small drop from peak to residual strength[31].

Periglacial solifluction (head) deposits are also well represented in the data-set (*Table 5.3*). This is not surprising for two reasons. First, head deposits are ubiquitous. Indeed they are far more widespread than is generally appreciated and so are, quite naturally, involved in numerous instances of slope failure. Second, many clayey head deposits are also particularly prone to failure, because of the presence of relict shear surfaces at or close to residual strength within the deposit. Such materials are widespread on inland slopes developed in clays and mudrocks, and renewed movements have often caused problems during engineering projects, as occurred during the construction of the Sevenoaks Bypass (*Figure 3.13*). Landslides involving head deposits are commonly single rotational failures, shallow translational slides and flows, and have been widely reported in South Wales, the Pennines, the Cotswolds and the Weald.

It is quite impossible to describe the diversity of head deposits that exist in Great Britain within the space available. It merely serves to note that the thawing of previously permanently frozen ice-rich ground (soil and rock) resulted in the release of larger volumes of meltwater than could be accommodated in the available pore spaces. With drainage impeded by permanently frozen ground below, the water was frequently unable to drain away fast enough, with the result that excess pore-water pressures (excess of hydrostatic) were generated[32]. When the pore-water pressures equalled normal stress, the coefficient of friction was zero and the available strength was equivalent to the undrained shear strength (Cu; *see Chapter 4*). Instability was therefore possible on very low angled slopes and a very wide range of materials were involved in movements, with considerable intermixing of different parent materials. Thus head deposits range in character from soliflucted calcareous Chalk (distinctive pale grey coloured coombe rock) through a range of fine-grained deposits to fairly rubble-like deposits of markedly varied colour and character. Three major groups of head deposits can be recognised on the basis of their parent materials[33]:

(i) head deposits derived from non-argillaceous bedrock, which tend to be poorly sorted with a silty or sandy matrix;

(ii) head derived from till, which tends to be crudely bedded with a silty, sandy matrix, although high clay contents do occur, e.g. head derived from the chalky tills of East Anglia;

(iii) head derived from argillaceous bedrock; periglacial weathering of the clays of Southern England and the East Midlands resulted in brecciation and softening. Thawing of the permafrost resulted in extensive shallow translational sliding, often on angles as low as 3° for Gault Clay (7° is necessary for landsliding in present day conditions)[34]. At Sevenoaks Weald it was estimated that sliding could have occurred on slopes as low as 2°.

The nature of a head deposit is important in determining its mode of origin and post-glacial stability. **Sand-silt** heads have low plasticity and were derived from non-argillaceous bedrocks. They tend to be particularly sensitive to changes in water content, susceptible to loss of strength and **flowage** when water content is high (i.e.

gelefluction) and are characterised by a gradual layer by layer accumulation. Today these deposits are highly permeable, stable, with no relict shear surfaces.

Silt-clay heads (e.g. derived from the weathered Culm Measures in Central Devon) generally have medium plasticity, and were probably formed by plastic or ductile flow rather than sliding[35]. However, these soils are now liable to shallow translational sliding induced by high groundwater levels.

Clay rich heads have high plasticity and the mode of failure was likely to have been by sliding. Depth of permafrost limited slide depth to the thawed 'active layer' and produced the laterally extensive shear surfaces commonly reported within and beneath these materials. These deposits have to be treated with caution as there is a considerable hazard of reactivation of relict shear surfaces.

Topography

It has already been stressed that suitable topographic conditions are a necessary prerequisite for mass movement because of the need for gravitational stress to overcome material strength. As with water, material must have the capability of change in elevation if potential energy (the energy due to position) is to be converted into kinetic energy (the energy of motion). However, the slope angle required for landsliding to occur varies considerably with material characteristics and pore-water pressures *(see Chapter 4)*.

The slope angle at which there is no possibility of further movement is known as the **'angle of ultimate stability against landsliding'**. It is clear that in the case of dry non-cohesive soils and rock rubble the maximum angle that a slope can remain stable is equal to the angle of internal friction of the soil, and representative values are given in Table 4.1. However, soils are rarely completely dry in Great Britain. In the case of saturated cohesive soils with the water-table at the ground surface and seepage downslope, the value declines to roughly half the angle of friction with respect to effective stresses (normally referred to as $\emptyset'$). As a consequence, the threshold slope angles are extremely varied ranging from 6°–14° in the case of weathered clays and shales (8° for London Clay), to 19°–33° for semi-frictional sandy soils in upland areas. Thus it is not necessary to have steep slopes in order to generate mass movement. Indeed, there are records of solifluction sheets which moved over slopes of less than 3°.

Few parts of Great Britain are so flat that 'natural' landsliding is confined to the banks of rivers and drainage ditches. Such areas do exist, however, in the major tracts of recently deposited alluvium (the Fens, the Humberhead Lowlands, Romney Marsh - *see Figure 2.3*), the spread of fluvio-glacial deposition in the Vale of York, the relatively flat, low-lying glacial deposits of Holderness and the Lincolnshire Marsh and the elevated marine deposits of the Lancashire Plain and the Sussex Coastal Plain. Over the remainder of Great Britain, landscapes developed on both thick sheets of superficial deposits (*Figure 2.3*) and bedrocks contain relief which increases in scale on the tougher, older rocks in the west and north (*Figure 2.2*). Within these terrains the topographic conditions most conducive to slope failure (*Figure 5.19*) occur quite widely.

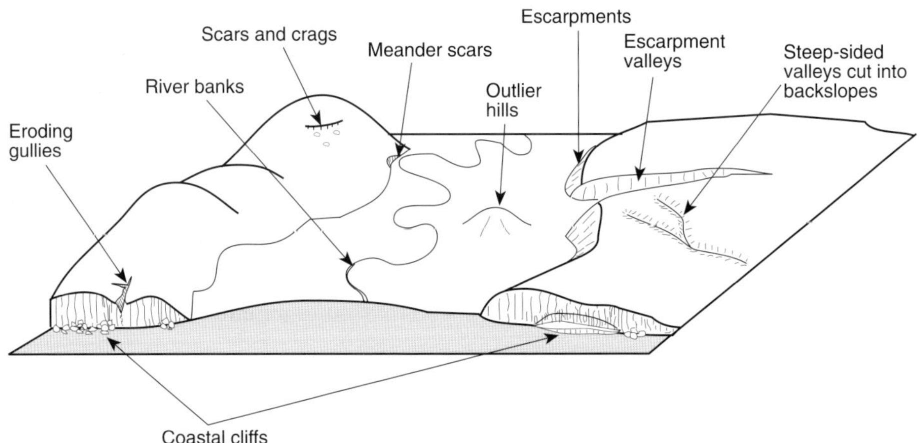

5.19 Topographic positions susceptible to landsliding

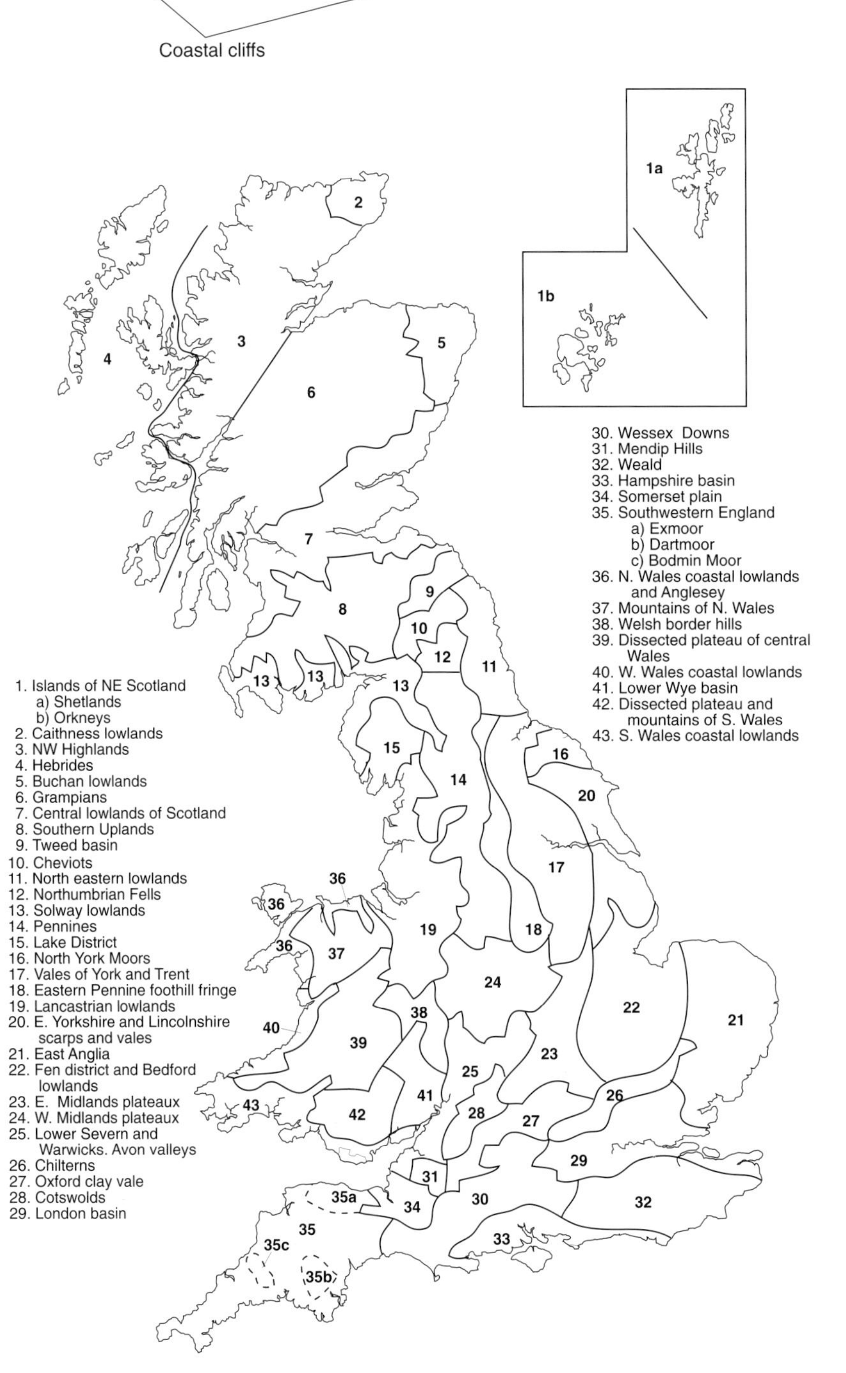

5.20 Relief regions of Great Britain (after Warwick, 1964).

It must be recognised, however, that landslide potential is not related to elevation but to relative relief, i.e. the differences in height between adjacent topographic 'highs' and "lows", and the consequences in terms of slope gradient. It is this, in combination with the local factors of geology, hydrogeology, etc., discussed in this chapter, that determines the likelihood of slope failure. Thus any version of the relief map of Great Britain (*see Figure 2.1*) will provide a poor indication of landslide occurrence because it lacks information on the geology. For example, there are few landslides on the granite uplands of Dartmoor or on the prominent Chalk downlands of South east England. What is actually required is a map combining slope steepness categories with susceptible lithologies and hydrogeology. Unfortunately, such a map does not yet exist.

Some ideas of the potential complexity of such a map can be gained from the broadscale division of Great Britain into relief regions, as shown in Figure 5.20. Comparison with the landslide distribution map (*Figure 2.14*) indicates that the most conspicuous groupings are located in South Wales and the Southern Pennines, both well researched, and the Cotswold escarpment, which has also been well mapped (*see Figure 5.4*). These dense concentrations provide some indication of the possible extent of landsliding in other areas, not merely the steep uplands of Wales, the Lake District, the Pennines, the Southern Uplands and the Highlands of Scotland, but also within Lowland Britain.

.21 The relationship between ecorded landslides and the division f England and Wales into Lowland nd Upland/Highland Britain.

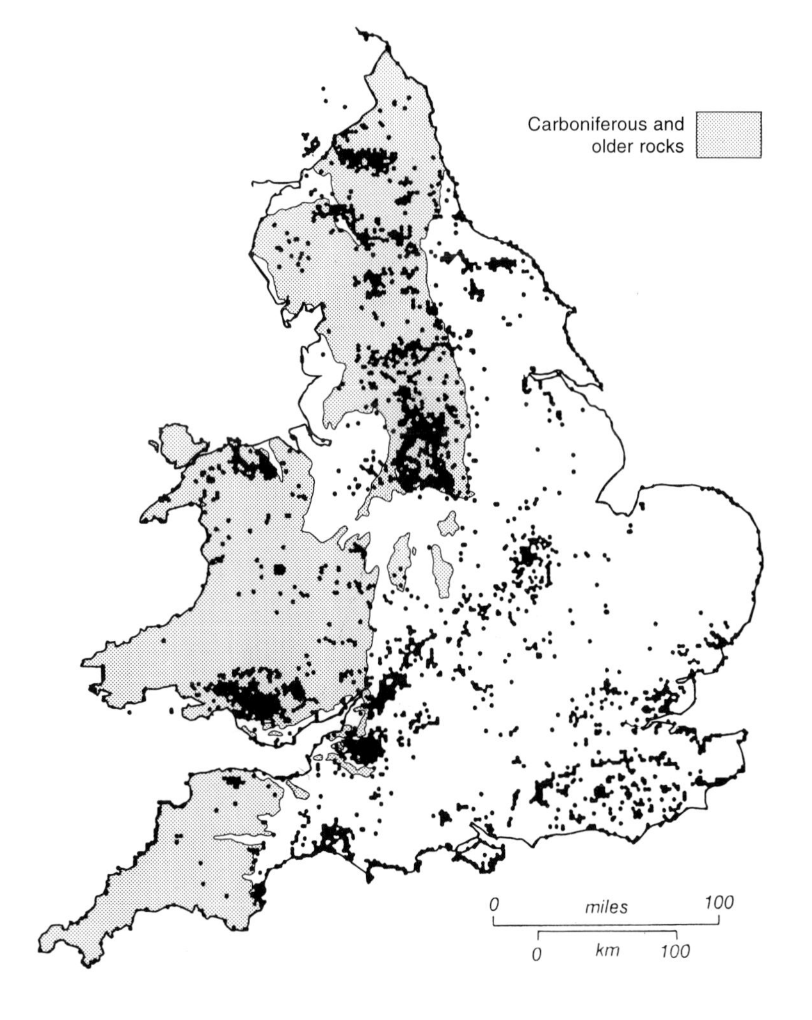

The distribution of landslides in Lowland Britain is shown on Figure 5.21. Three belts of reported landsliding are prominent: the Cotswold escarpment in Gloucestershire, the Upper Greensand escarpment in Wiltshire and the Lower Greensand (Hythe Beds) escarpment in the Weald. In addition, clusters of recorded landslides also occur in West Dorset, Avon and Leicestershire which, if taken together with the Cotswold concentration, indicate that landsliding may be widespread on steep slopes (escarpments and valley flanks) developed on strata of Lower and Middle Jurassic age. There are signs of a similar concentration on the lithologically varied strata of the Central Weald (the Hastings Beds) where the flanks of deeply incised valleys have suffered failure. Thus it must be concluded that landsliding is widely developed in those areas where locally steep terrain is produced by the unstable combination of limestone/sandstone overlying clay/shale. These conditions will be especially pronounced in extra-glacial Lowland Britain (*see Figure 2.4*) where topography has been able to evolve over lengthy time spans without the influence of glacial erosion or deposition (*see Chapter 6*).

Climate

The links between climate and mass movement are most apparent through the facts that weathering and groundwater both act to reduce material strength. On a global scale the linkage is very clear, for variations in temperature and rainfall have a profound effect on weathering types and weathering rates, with strong chemical rotting in the Humid Tropics and strong physical disintegration in the more humid portions of the Arctic and Antarctic.

In the case of Great Britain the variations are much less obvious. The latitudinal extent of 9° (960km for the main island) does provide for variation of the temperate climate and this is exacerbated by the effects of relief (1343m) which reinforces the increasing harshness of climate towards the north. Mean annual temperatures decrease northwards as average rainfall values increase (400mm in parts of the South east England to in excess of 2500mm in parts of Wales, the Lake District and the Scottish Highlands). Consideration of temperatures and rainfall totals suggests that most of Upland, Highland and Ancient Britain are characterised by moderate rates of chemical weathering and weak rates of frost weathering, while Lowland Britain has moderate chemical weathering but insignificant frost weathering. Certainly, the higher summits over 800m display signs of active frost attack which become increasingly conspicuous with increasing altitude and more northerly latitude.

There are so many factors that influence mass movement and so many attributes of climate, that it remains extremely difficult to identify significant variations. The higher rainfall totals of the uplands undoubtedly contributes to instability. But it is not merely the extreme wetness. The intensity of rainfall events, the strong winds and the fluctuations of ground surface temperatures in winter, all contribute to a harshness of climate that results in a range of active processes similar to those active in periglacial environments. For example, operating solifluction features are widespread on many upland slopes above 550m[36], although measured rates of movement are very low in comparison with those recorded in other contemporary periglacial environments. In the Drumochter Hills, Central Grampians, for example, surface displacements of only 50mm over a 3 year period were recorded[37].

Associated with these active solifluction features are ploughing boulders located at the downslope end of vegetation covered furrows. Such features have been reported throughout Upland Britain and isolated examples recorded at elevations as low as 450m[38].

By contrast, the greater variability of climate in Lowland Britain may also be a significant factor in promoting slope failure. Here the increased occurrence, intensity and duration of drought facilitates dessication; rainfall totals are more variable, thunderstorm rainfall is more frequent and phases of intense cold are not unknown.

There are, of course, significant differences in climate between regions. Nevertheless, the majority of landslide events appear to be related to heavy rainfall or saturated soils (*see Chapter 4*). Conditions which initiate failure generally arise when the groundwater table is highest and corresponds with periods when the difference between evapotranspiration and rainfall is greatest. A useful general measure of soil wetness is the average period per annum during which rainfall exceeds evapotranspiration and there is zero soil moisture deficit (**the field capacity period**). During this period the actual field conditions vary according to soil properties and site hydrology. In winter, the water content of well drained soils fluctuates with each rainfall event, with waterlogging only occurring for a few hours during, or immediately after, heavy rain. Clayey soils, on the other hand, are commonly waterlogged throughout the field capacity period. The variability of the field capacity period across the country (*Figure 5.22*) is a valuable indicator of the

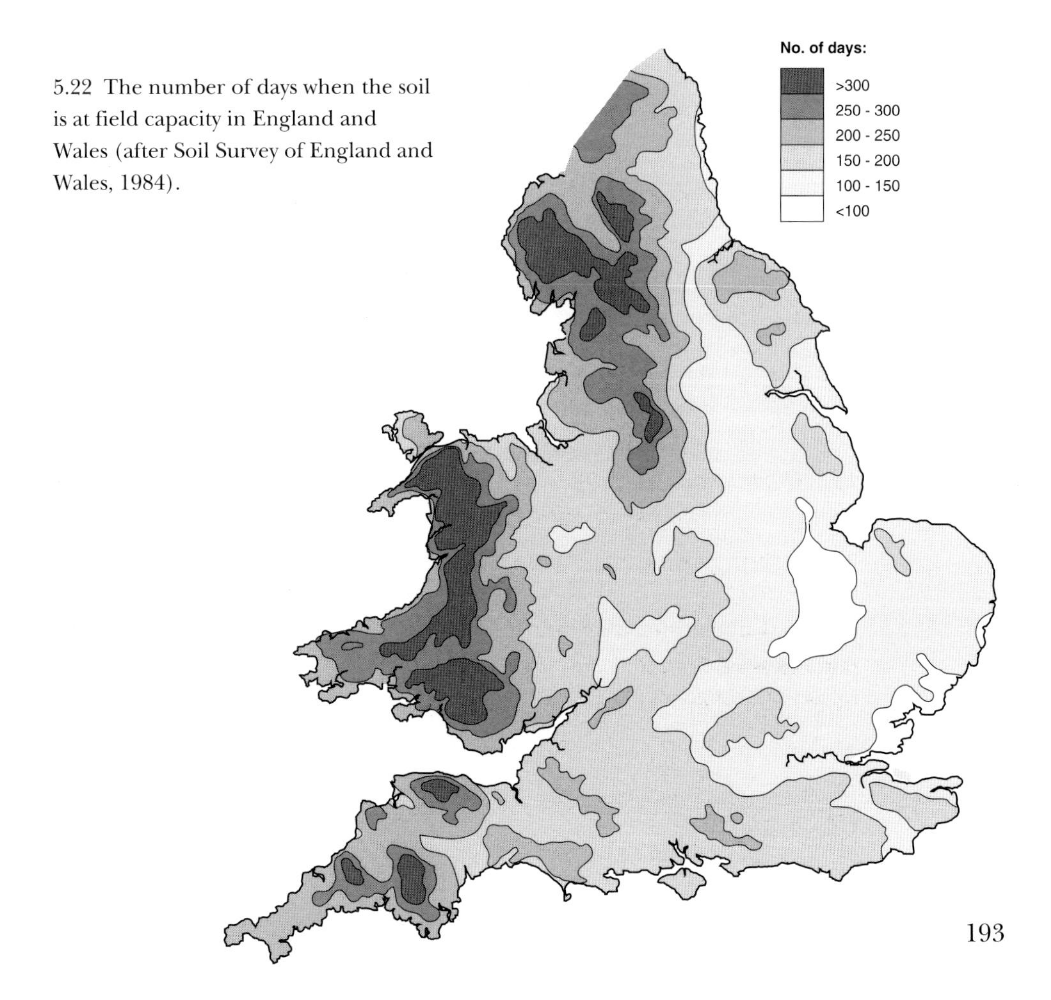

5.22 The number of days when the soil is at field capacity in England and Wales (after Soil Survey of England and Wales, 1984).

relative frequency that groundwater-triggered landslide events could occur in different areas. Figure 5.22 highlights a general trend for decreased soil wetness and, hence, depressed groundwater levels towards the south east. For many years, therefore, the water levels in slopes developed on the sedimentary strata of Lowland Britain, such as London Clay, Gault Clay, Weald Clay, Wadhurst Clay. etc., may be significantly depressed, suggesting that landslide triggering climatic events may be relatively infrequent. By contrast, over the upland areas of Northern England, South-west England and Wales, where the soil may be saturated for over 250 days a year, the potential for groundwater-triggered landslide events is much greater.

Uncertainty as to the significance of contemporary variations in climatic parameters is replaced by certainty as to their significance over time. The legacy of climatic change is so important that it will be considered separately in the next chapter.

NOTES

1. Innes, 1982
   Innes, 1983
2. Crozier, 1986
3. Selby, 1982
4. Steward and Cripps, 1983
5. Mortimore, 1986
   Mortimore, 1990
6. Butler, 1983
7. Moore, 1986
8. Morgan, 1986
   Kelly and Martin, 1986
   Halcrow, 1988
9. Terzaghi and Peck, 1948
   Denness, 1972
   Hutchinson, 1982
   Bromhead, 1987
10. Hutchinson, 1982
11. Hutchinson et al., 1981a
12. Jaeger, 1972
    Hoek and Bray, 1977
13. Holmes, 1984
14. Penn et al., 1983
15. Hawkins, A.B. and Privett, 1981
16. Briggs and Courtney, 1972
17. Cooper, 1979
    Cooper, 1980
18. Hollingworth et al., 1944
19. Horswill and Horton, 1976
20. Hutchinson, 1988
21. Ackerman and Cave, 1967
22. Watts, 1905
    Boyd-Dawkins, 1904
    Lapworth, 1911
23. Parks, 1991
24. Vaughan, 1976
25. Hutchinson, 1965b
    Hutchinson et al., 1985a
    Geomorphological Services Ltd, 1991
26. Hawkins, A.B., 1988
27. Ballantyne, 1981
    Innes, 1982
    Innes, 1983
28. Forster and Northmore, 1986
29. McGown et al., 1974
30. Vaughan and Walbancke, 1975
31. Conway et al., 1980
32. Hutchinson, 1974
33. Harris, 1987
34. Weeks, 1969
35. Grainger and Harris, 1986
36. Ballantyne, 1987
37. Chattopadhay, 1983
38. Tufnell, 1972

CHAPTER 6

Time and the Development of Unstable Terrain

Age of Landsliding

Reference to the existence of landslides at specified locations automatically focuses attention on the identified existence of morphological evidence for movement at some time in the past. The evidence may be scars, displaced blocks and tension cracks that are sharply defined and fresh-looking – sometimes called broken ground – or subdued undulations indistinguishable from other undulations to the untutored eye. Thus the distribution of the 8835 reported landslides *(Figure 2.14)* must not be interpreted as representing a rash of failures in recent decades, but the sum total of presently available data on the existence of landslides. Some undoubtedly have developed in recent years, others may be of considerable antiquity. For example, the reactivated failure that delayed construction of the Sevenoaks Bypass in the early 1960s had moved little over the last 10,000 years and there is no reason to believe that this translational slide is the oldest mass movement feature to have survived, even at this site. Indeed, its relatively pronounced morphology indicates quite the opposite.

One important aim of the recently completed DOE survey was to provide an appreciation of the pattern of known landsliding through time. This was achieved by recording information on antiquity, but only where it had been established and reported. Survey personnel were not empowered to make judgments on age based on other knowledge or experience. A five category age classification was employed:

(i) **'active':** indicating a currently unstable site, such as a marine cliff, or sites which display a cyclical pattern of movement with a periodicity of up to 5 years, or first-time failures and reactivated failures that had occurred within 5 years of the start of the survey (i.e. since 1980);

(ii) **'recent':** indicating slope instability sites known to have suffered movements within the last 100 years, and where the sites remain in a state of marginal stability with well defined morphological expression;

(iii) **'relict':** indicating mass movement within the historic time-scale (100–1,000 years), where the landforms are still clearly recognisable but show no obvious signs of contemporary movement;

(iv) **'fossil':** indicating instability features that developed in early historic and prehistoric times and normally associated with rather different environmental conditions (e.g. periglacial); and

(v) **'unknown':** indicating that no statement on age was provided in the source material.

The numbers of recorded landslides in each of these age groups is shown in Table 6.1 and, not unexpectedly, reveals a preponderance in the 'unknown age' category. The division into 'inland' and 'coastal' landslides is provided for the main reason that evidence for older episodes of mass movement is likely to survive longer on inland slopes than along the coast, because wave attack tends to quickly remove evidence of failure.

Table 6.1

The relative frequency of landslides of different ages in Great Britain

	Inland		Coastal		Total	
	No.	**%**	**No.**	**%**	**No.**	**%**
Active	296	3.9	483	37.1	779	8.8
Recent	927	12.3	423	32.5	1350	15.3
Relict	970	12.9	66	5.1	1036	11.7
Fossil	669	8.9	45	3.5	714	8.1
Unknown age	4671	62.0	285	21.8	4956	56.1
Total	**7533**		**1302**		**8835**	

It is readily apparent that the age classification adopted was intended to focus attention on landsliding due to human activity, both directly and indirectly, and on the location of sites with contemporary instability. This was for two reasons. First, knowledge of landsliding in the recent past can be extremely useful in reducing the costs of landsliding in the future. Second, there has been growing awareness of the large numbers of landslides caused by human activity since the time that population numbers grew to a sufficient level to cause significant environmental change. Commercial forestry practices (e.g. the Scottish Highlands[1]), drainage alteration (e.g. the Hedgemead landslide in Bath[2]), urbanisation (e.g. the abandoned housing development at Ewood Bridge in the Irwell Valley[3]), coastal modification (e.g. Folkestone Warren, Kent[4]), the effects of mining (e.g. the South Wales Coalfield[5]) and the development of transport infrastructure (e.g. the Cullompton Bypass on the M5 in Devon, 1968–9[6]) have all been cited as prominent causes of slope instability. However, more recent studies[7] have indicated that human populations may have had a significant effect on forest cover, and thereby slope instability, as early as 5,000 years ago (some put the figure at 8,000 years). Thus the distinction between 'relict' and 'fossil' is somewhat artificial, and the two were subsequently merged together and labelled 'ancient'. Problems were also encountered in differentiating between 'active' and 'recent', with the result that the two were merged to form a 'youthful' group.

The recorded distributions of landslides reported to be of 'active', 'youthful' and 'ancient' ages are shown in Figures 6.1–6.3 and those of 'unknown age' in Figure 6.4. It is fair to assume that a very large proportion of the landslides recorded as of 'unknown age', are, in reality, more than 100 years old and therefore should be added to the 'ancient' grouping.

What do these distributions reveal? The pattern of 'active' failures *(Figure 6.1)* emphasises coastal instability, as is to be expected, and most especially the crumbling clifflines of Lowland Britain where weak superficial deposits and 'soft' bedrocks are undermined by wave attack *(see Chapter 2)*. By contrast, inland failures are more scattered, although major concentrations are apparent in those areas where detailed investigations have been undertaken. It is this factor that accounts for the dense concentrations in the South Wales Coalfield[8], where significant numbers of active landslides have been reported in the counties of Mid Glamorgan (73), Dyfed (48) and Gwent (41). Other occurrences are associated with the London Clay of Essex, the Lower Greensand of Kent, and the Ironbridge Gorge of

6.1 The distribution of recorded active landslides in Great Britain

6.2 The distribution of recorded youthful landslides in Great Britain

Shropshire *(see Photo 2.12)*. It must be stressed, however, that the 779 recorded failures in this group do not represent all the movements that occurred in the period 1980-85 and subsequently updated to 1990, for there must have been innumerable examples of minor failures on coastal cliffs, river banks, cuttings and fills which did not warrant recording, as well as numerous small slides, falls and debris flows in the uplands of Wales and the Highlands of Scotland *(see Chapter 3, especially Figure 3.20 and Photo 3.6)*. Nevertheless, the majority of significant failures are likely to have been recorded, although the densities revealed in those areas subjected to detailed investigations indicate that signs of contemporary slope instability are probably far more widespread than is suggested by the recorded data.

This conclusion is supported by the data on 'youthful' landsliding *(Figure 6.2)*, which continues to emphasise coastal instability but also reveals the much more widespread occurrence of inland slope instability. Although many of the 2129 failures are natural, the distribution also contains numerous cases of instability due to human intervention, many of which have been described in this book.

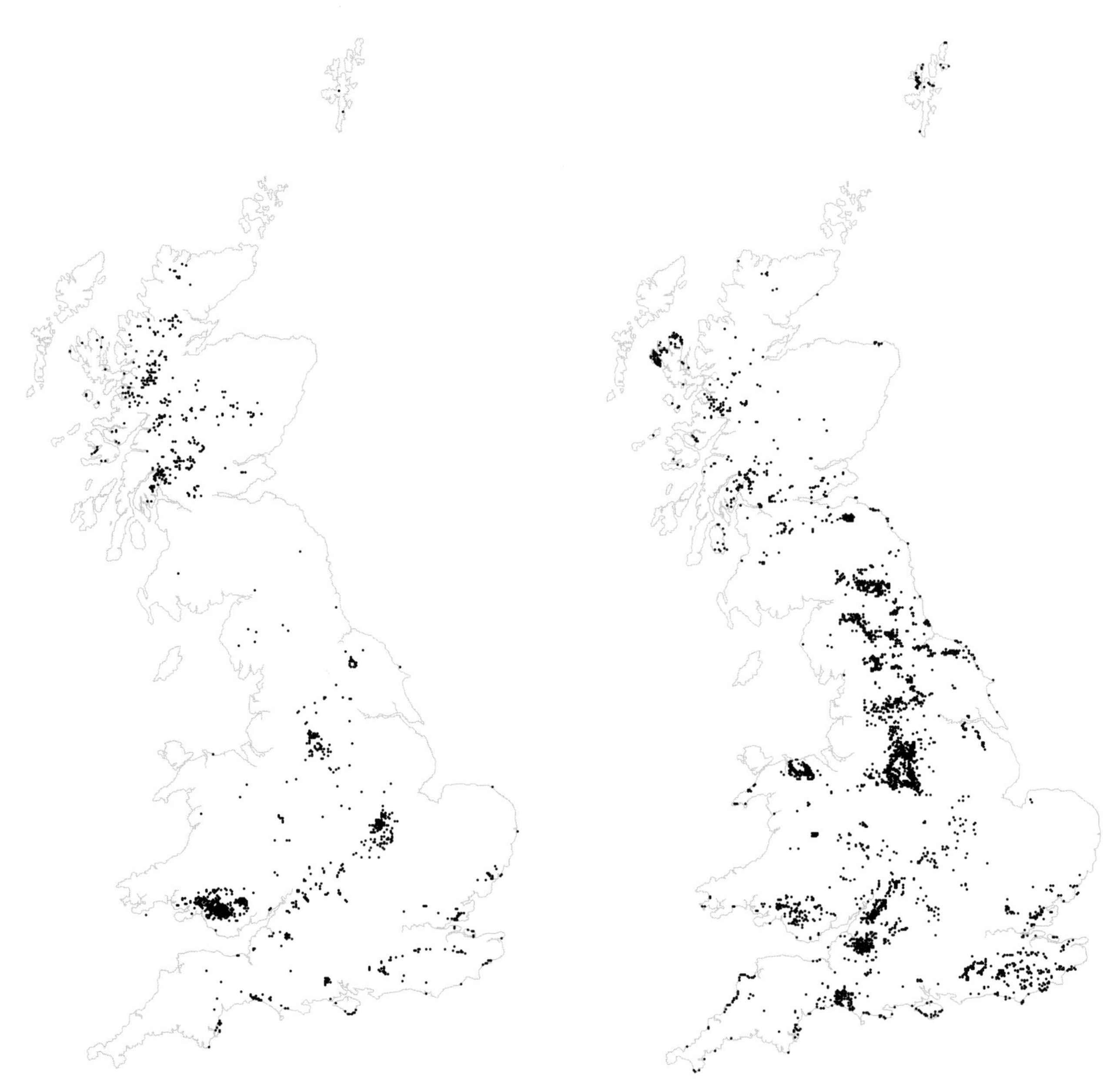

6.3 The distribution of recorded ancient landslides in Great Britain

6.4 The distribution of recorded landslides of unknown age in Great Britain

Ancient landslides account for almost 20% of the database (1750 recordings). The distribution *(Figure 6.3)* focuses on inland landsliding for the reasons already discussed, and reveals strong concentrations on the Lower Greensand (Hythe Beds) escarpment of the Weald; the Jurassic rocks of the Devon-Dorset-Somerset borders, Avon, the Cotswolds, the Northamptonshire-Leicestershire borders and the western margins of the North Yorkshire Moors; the Carboniferous strata of the South Wales Coalfield and the Derbyshire Pennines; and the ancient rocks of the Scottish Highlands.

The distribution of the 4956 reported landslides for which no age information is recorded *(Figure 6.4)*, appears, at first sight, similar to the pattern for 'ancient' failures *(Figure 6.3)*, but there are subtle differences. The Pennines display both increased numbers and a wider distribution, as is the case with the North Yorkshire Moors. New concentrations appear on the Hastings Beds of the Central Weald, in north Clwyd and within the Southern Uplands of Scotland. Conversely, the previous focus on the South Wales Coalfield is greatly diminished, which shows the value of detailed investigation.

Combining the distributions for 'ancient' and 'unknown' age landslides will give a reasonable approximation of the distribution of old instability features, except that certain major gaps in knowledge become apparent. For example, why are the overwhelming majority of recorded landslides in Wales confined to the South Wales Coalfield and Clwyd, or why are there separate clusters of reported landslides distributed over the outcrop of Jurassic strata in England, scattered along the length of the Pennines and apparently randomly distributed within the Southern Uplands of Scotland? This clustering may be real and these areas actually contain the greatest numbers of failures, or the most conspicuous failures, or even the most significant failures with regard to impacts on society. On the other hand, it seems equally probable that they reflect the patchy nature of knowledge and merely serve as indicators to the existence of a much larger and more widely distributed population.

If these arguments are correct, then inland ancient mass movement features (i.e. more than 100 years old) outnumber contemporary failures (ie. within the last century) by at least 4:1 and probably very much more. Even in the case of landslides older than 1,000 years, the data suggest a ratio of at least 2:1. The existence of such a rich legacy of ancient landslides requires explanation. In particular, reasons have to be found to explain why the patterns of contemporary and ancient landsliding often differ and why ancient landsliding appears to have been so widespread and intense in certain locations. Thus, to focus on the example of Stonebarrow, Dorset, described earlier *(see Chapter 2)*, the main question is not so much why the coastal landslide complex has continued to operate over the last 1,000 years *(see Photo 2.5 and Figures 3.25 & 3.39)* but why the extensive inland landsliding on adjacent valley slopes ceased *(see Figure 2.13)*. There are several possible explanatory factors that can account for the changing pattern and intensity of landsliding over time, including topographic development, climatic change, sea-level change and human interference. These will be examined in the following sections.

Topographic Development

The inland topographic landscape is not immobile but is evolving through the action of geomorphological processes, albeit at an exceedingly slow rate from a human perspective. The existence of topographic relief is both testimony to the efficacy of erosional processes and indicative of the continuing potential for erosion; the sediment load of rivers confirms that denudation is occurring.

Rates of topographic lowering vary with respect to geology, tectonic activity, climate and location. Mean rates for temperate latitudes vary between 10 mm/1,000 years and 50 mm/1,000 years, with much lower rates on very resistant rocks, such as granite (<5mm/1,000 years), but significantly higher rates on easily eroded terrains (up to 100mm/1,000 years or more) and mountains such as the Alps (400–1,000 mm/1,000 years). These values appear minute, but convert directly into metres per million years. Thus a rate of 50 mm/1,000 years represents 120m of overall lowering during the Quaternary *(see Table 2.1)*.

But topographic landscapes rarely suffer uniform lowering over significant lengths of geological time and the British landscape certainly has not. For the most part landscape evolution appears to have occurred in bursts of activity, or pulses, stimulated by major changes to prevailing conditions caused by tectonic uplift, warping, climatic change, etc. If it is assumed, for the sake of simplicity, that these changing circumstances affect a landscape that is initially fairly level and uniform, then during the early stages of subsequent landscape evolution, valley downcutting and the creation of relative relief will account for most of the overall lowering value, which tends to be fairly high. Later, when valley deepening has all but ceased, the reduction in elevation of the intervening ridges accounts for most of the overall lowering, which will now be fairly low. However, considerable time is required for such a cycle to run its course and it is rare for stability of environmental conditions to persist for sufficiently long. This is especially true of the Quaternary which is known to have been characterised by frequent and dramatic fluctuations of climate *(see later)*. Commonly, therefore, evolution is punctuated by changes in environmental conditions which accelerate, decelerate or change the character of landscape development so that polycyclic (stepped) landscapes emerge which display signs of several phases of geomorphological activity.

Applying these notions to Great Britain is difficult because of the very variable nature of the bedrock fabric *(Figure 2.2)*, the dramatically differing effects of glaciation *(Figures 2.4 and 2.5)*, variations in crustal movements in terms of direction, duration and scale, and the varying stages of topographic development[9]. For example, some parts of Great Britain have experienced intense glacial erosion *(Figure 2.5)* which has resulted in the development of over-deepened valleys (glacial troughs) with over-steepened walls that have proved a focus for instability *(Figure 6.5)*. By contrast, parts of glaciated Lowland Britain have experienced repeated phases of erosion and deposition (Figure 2.4), sometimes with little net overall effect.

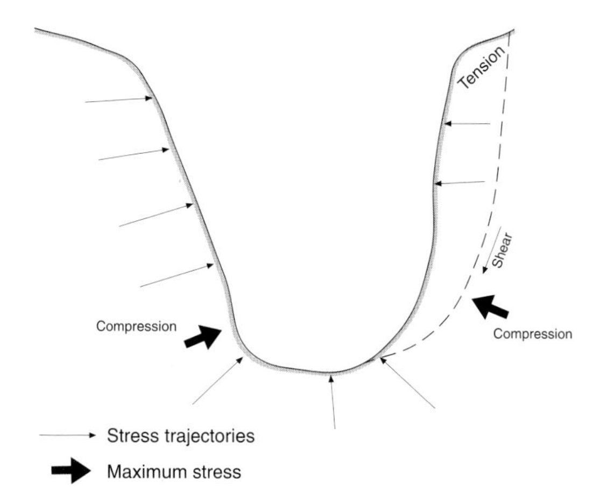

6.5 The distribution of stresses in a recently deglaciated valley (after Selby, 1982)

The simplest place to start considering this topic is in extra-glacial Southern Britain *(Figure 2.4)*. The presently accepted model of landscape development envisages a phase of marked uplift in the early Pleistocene *(Table 2.1)*, about 2 million years ago. The total uplift varied over the region, but values of 100–150 metres are often quoted and some consider that it may locally have exceeded 200m. This uplift provided the basis, in terms of potential energy, for river incision into the uplifted rocks. Valley deepening (lowering) over time resulted in the progressive development of topographic relief, with scarps and vales being excavated from rocks of differing resistance to erosion[9]. This growth in relative relief saw the creation of relatively long, steep slopes. Once either the slope angle had reached a limiting value, or slope height had reached a critical value, or dissection had exposed a landslide-prone stratum (usually a clay layer), then landsliding would have become a major mechanism of slope remodelling. Thus the progressive development of topography during the Quaternary must be seen as increasingly conducive to mass movement over time. This basic model of increasing severity of landslide behaviour in response to continued river erosion was first outlined with regard to the evolution of boulder clay slopes at Shotton, County Durham *(Figure 4.13)*[10], but is now seen to have more general applicability.

In reality, the evolutionary sequence was very much more complex than the simple model outlined above, and also showed variations from area to area. The repeated growth and decay of the huge Pleistocene ice-sheets *(see next section on climatic change)* resulted in great fluctuations of sea-level from up to 30m above present level during warm inter-glacials to –121m at the height of the last glacial phase (18,000 years ago), and probably to as low as –150m in a previous phase about 450,000 years ago when ice extended as far south as the Scilly Islands. Thus interglacial coastlines with cliffs and landslides were repeatedly left 'high and dry' as shorelines receded away towards the edge of the Continental Shelf, only to become reoccupied and reactivated when the ice sheets melted and sea-levels returned to their former levels.

The glaciations were also the cause of major shifts in climate, so that the oscillations of sea-level were accompanied by dramatic changes in temperature and rainfall. Harsh periglacial conditions with deeply frozen ground and limited vegetation predominated, interspersed with warmer, temperate episodes when the ground thawed, the snow was replaced by rainfall, and dense tree vegetation recolonised the land surface.

Thus, landscape development in extra-glacial Britain should be envisaged as having been progressive but not continuous, with episodes of rapid evolution separated by phases of quiescence. Landsliding may also be considered to have been episodic for reasons that will be explained in the next section. For the moment it merely suffices to note that as the relief amplitude developed over the Pleistocene, so each successive period which saw conditions suitable for mass movement resulted in increased scales of activity. This point can be illustrated by reference to the landsliding that extensively mantles the Lower Greensand (Hythe Beds) escarpment in the Weald *(Figure 5.2)*. Most of the evidence indicates three pulses of movement over the last 300,000 years, with evidence for the youngest phases most widely preserved[11]. But evidence for slope failure can be traced back to at least 500,000 years ago and almost certainly began even earlier in the Pleistocene.

The evolutionary sequence in glaciated parts of Great Britain is considerably complicated by the effects of glacial erosion, glacial deposition, drainage diversions and the consequences of vertical movements of the crust. Patterns of sea-level change were considerably modified by repeated crustal depression due to the weight of thick ice-sheets and crustal rebound upon deglaciation. These isostatic effects may have resulted in apparent sea-levels in Scotland differing from those of Southern England by up to 150m. Nevertheless, increasing relative relief and progressive valley deepening are considered characteristic of most areas, so a modified model of increasing propensity for landsliding through time appears acceptable.

There are, however, clear indications that the dramatic nature of certain landscape changes were of such great magnitude as to result in significant alterations to the pattern and scale of landsliding. Over-steepened rock slopes would have failed once the supporting or binding effects of ice had been removed. The reestablishment of river networks in areas covered with thick and extensive sheets of glacial till resulted in the establishment of linear belts of landslide activity as the rivers recreated valleys and the same would have happened in coastal areas as inter-glacial rising sea-levels reoccupied coastlines abandoned during previous cold "glacial" phases. But it is drainage diversion by glacial ice that has resulted in some of the most impressive landslide developments. The case of the Ironbridge Gorge is one good example *(see next section)*, the Cotswold escarpment is another.

The prominence of the Cotswold escarpment in Gloucestershire, the scattering of outlier hills (e.g. Robin's Wood Hill, Churchdown Hill, Bredon Hill, etc.) and the widespread occurrence of landslides *(Figure 2.14)* are all symptomatic of recent, intense geomorphological sculpturing. It has recently become apparent that the evolution of the Lower Severn Valley is dominated by three relatively recent events: glacial erosion during the Anglian phase (480,000–428,000 years ago)[12]; the formation of a huge pro-glacial meltwater lake between Bredon Hill and Leicestershire *(Figure 6.6)*, also in the Anglian, which eventually drained away along the line of the Lower Severn, resulting in the diversion of formerly eastward flowing drainage into the Severn (e.g. the Warwickshire Avon); and thirdly, the diversion of the present Upper Severn drainage southwards through the Ironbridge Gorge as a consequence of the development of a further pro-glacial meltwater lake in the Cheshire lowlands. Thus it is now believed that 'the period of the Anglian Glaciation also included the erosion of the low land of the Severn Basin, and hill ranges like the Cotswolds became prominent features'[12]. No doubt prior to the Anglian, the Cotswold escarpment had been a minor feature with limited landslide activity. After

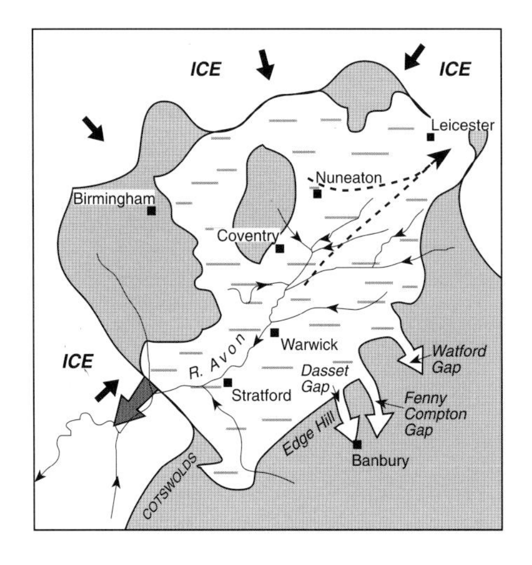

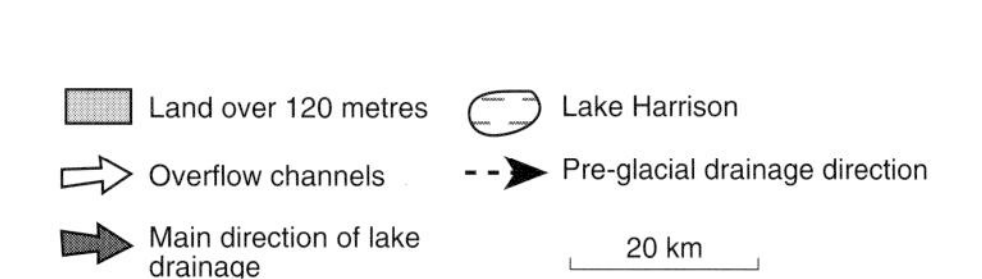

6.6 The extent of the pro-glacial Lake Harrison in the Midlands

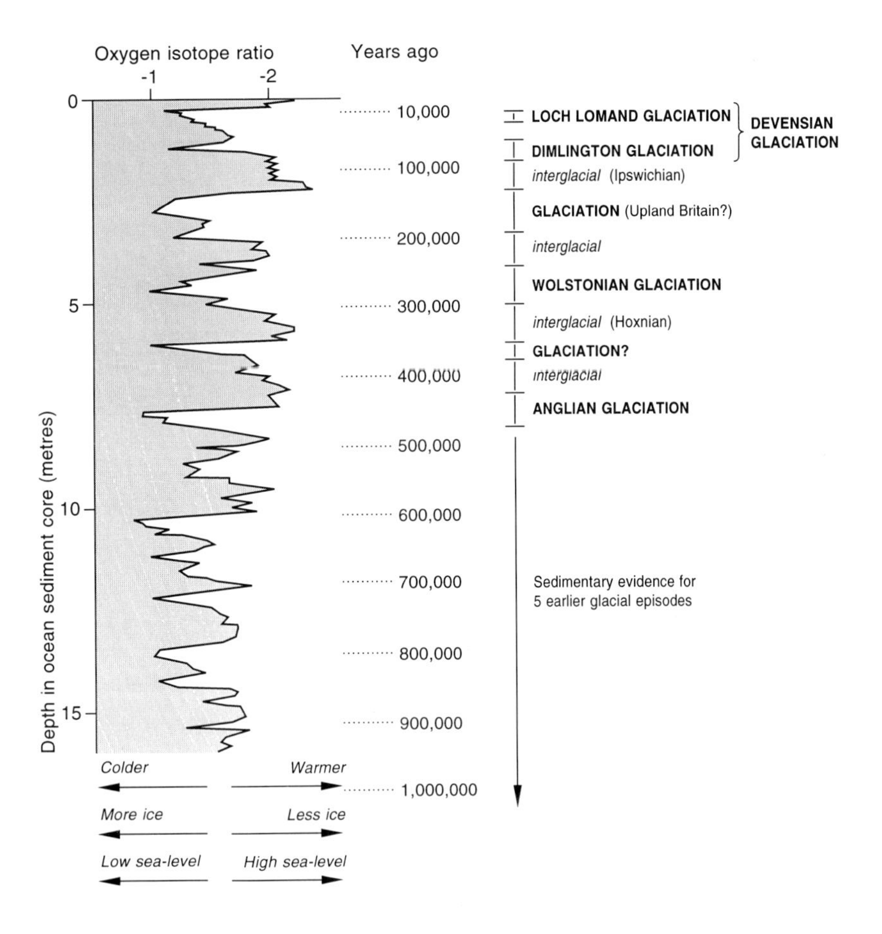

.7 An example of an oxygen-isotope curve from a deep-sea sediment core, showing the variability of climate and sea level over the last 1 Million years (based, in part, on Bowen et al, 1986).

this glaciation, the effects of ice erosion and pronounced river incision by the newly enlarged and much invigorated River Severn, resulted in rapid landscape evolution, thereby stimulating intense and widespread slope instability. This accounts for the prominence of the Cotswold escarpment on the landslide distribution map *(Figure 2.14)*.

Climatic Change

Over the last three decades analysis of sediment cores from the deeper parts of the Atlantic and Pacific oceans have provided much valuable information on climatic change during the Quaternary. Whereas the record on land is fragmentary and of limited duration because geomorphological processes, including glacial ice, keep disturbing, destroying or concealing evidence of earlier phases of landscape evolution, the quiet conditions of the deep ocean floor provide ideal environments for continuous sedimentation. Analysis of the contents of these cores has revealed variations in chemistry which indicate repeated oscillations in climate with a periodicity of a little over 100,000 years *(Figure 6.7)*. As a consequence, the climatic record of the Pleistocene is now considered to consist of about 20 cool/cold phases, of which 11–16 were of sufficient severity to cause the build-up of ice sheets and the development of permafrost far to the south of its present limits[13].

Exactly how many of these climatic deteriorations resulted in the development of ice sheets in Great Britain remains unknown. At the present time, three main ice-advances are generally recognised[12] *(Figure 2.4)*: Anglian, c.450,000 years ago; the Paviland/Welton ice advance, formerly known as the Wolstonian, c.270,000 years ago; and the Dimlington/Late Devensian ice-sheet with a maximum extent 18,000 years ago, together with fragmentary evidence of the possible existence of up to two earlier phases. In addition, there was a final minor readvance between 11,500 and 10,800 years ago, known as the Loch Lomond Stadial or Loch Lomond Readvance, which was mainly confined to the Scottish Highlands *(Figure 2.4)*, although small accumulations of ice (corrie glaciers or hanging glaciers) did develop in Snowdonia and the Lake District.

The huge physical size of the major ice sheets and their enormous impact on landforms and landform shaping processes is often difficult to comprehend. Each build-up of glacial ice would have resulted in the creation of huge ice-domes that submerged even the highest summits beneath hundreds of metres of ice, and powerful ice streams would have flowed across the pre-existing topographic grain to create diffluent cols. For example, one reconstruction of the last, Devensian, ice-sheet as it may have existed a mere 18,000 years ago *(Figure 6.8)*, reveals that the ice-dome surface could have lain at least 400m above the summits of Ben Nevis and the highest peaks in the English Lake District. During the advance and retreat stages of such ice-sheets, ice would have been restricted to valleys as valley glaciers, and sub-glacial erosion would have resulted in valley deepening. These growth and decay stages were often remarkably rapid, as is testified by the complete melting of the Devensian ice-sheet by 12,700 years ago and the creation of the Loch Lomond Readvance in about 500 years[14]. Thus the majority of glacial erosion and deposition were achieved in relatively short periods of time, especially during deglaciation when huge volumes of water would have been released.

The significance of glacial remodelling of topography through erosion and deposition has already been described *(see Figures 2.3–2.5)*, as has the influence of glacial erosion in the creation of steep, unstable rock slopes *(see Chapter 4 and Figure 6.5)*. Detailed investigations in the Scottish Highlands have revealed that the

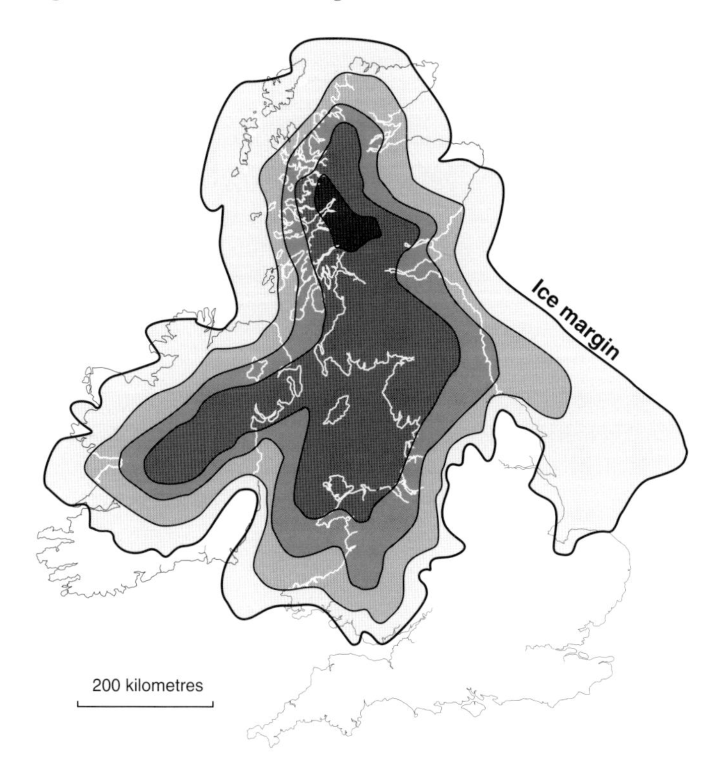

6.8 The late Devensian ice-sheet in the British Isles: one of several reconstructions advanced in recent years. It is possible that such reconstructions still over-estimate the surface elevations of the ice-dome surface above certain areas of the Uplands and Highlands.

.9 The relationship between major rock slope ailures in Scotland and the limits of the Loch .omond stadial (after Smith, D.I., 1984).

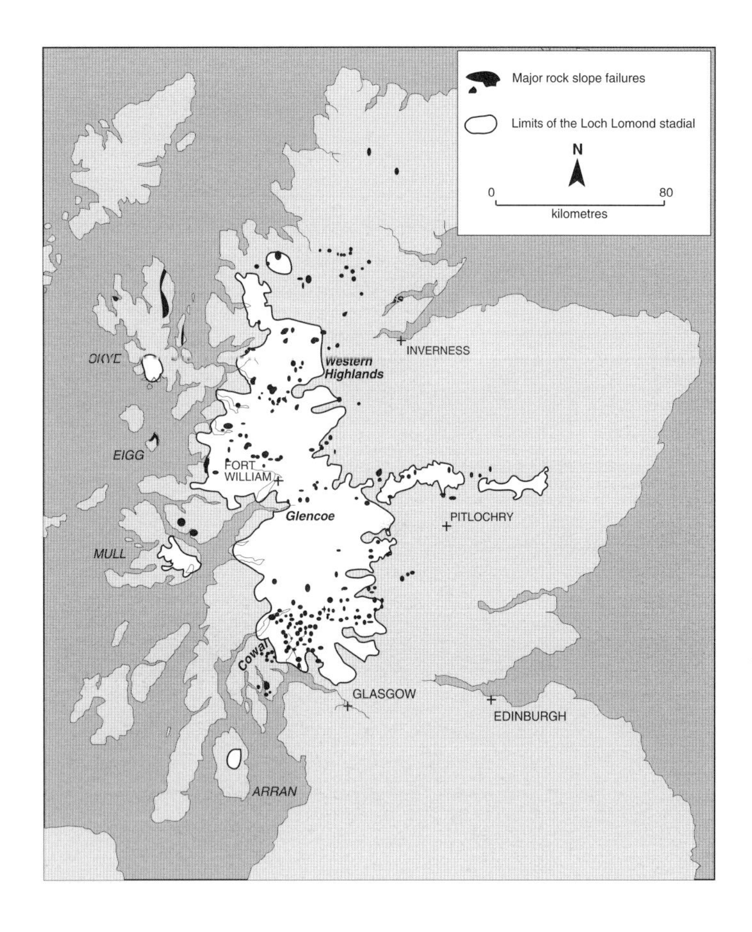

majority of presently visible rock slope failures are associated with the most recent Loch Lomond Stadial (11,500–10,800 years before present) but occur within the defined ice limit *(Figure 6.9)*. This indicates that failures occurred following ice decay, probably over the period 10,000–5,000 years ago, and were the result of a delayed action response associated with stress relief which involved fissuring and shearing[15]. This long-term decline in rock slope stability suggests that many recent failures, such as those at Stromeferry *(Plate 25)* and Farigaig, may be a legacy of glacial over-steepening[16].

The emplacement of internally variable tills and other superficial deposits had a very variable influence on stability. In some areas thick sequences of deposition completely negated the influence of bedrocks *(Figure 2.3)*, although internal variability and the superimposition of one depositional sequence upon the remains of an earlier sequence often resulted in instability. In other areas, deposition of superficial sediments on steep rock slopes has led to failure because of undercutting by streams *(Photo 6.1)*, changing groundwater conditions or human modification. This has been most clearly identified in South Wales where the majority of valley slopes have shallow landslides[8], and the intense pressure on land has meant that urban growth has often taken place on potentially unstable ground[17].

The Pleistocene glaciations had a further direct effect on landform development and slope stability. On many occasions during deglaciation, large lakes formed between masses of dying ice and topographic features, draining away via overflow

6.1 Small debris slides in oversteepened till slopes, South West Scotland (E M Lee).

channels that had long-term effects on the drainage pattern. The best example of such a feature was Lake Lapworth, which developed in the Cheshire Basin during the Wolstonian and Devensian glaciations. This overspilled southwards along the line of the Ironbridge Gorge, thereby considerably enlarging the present lower portion of the River Severn. The intense erosion that accompanied this drainage diversion has been the main reason for the concentration of failures in the Ironbridge area[18] *(see Photo 2.12).*

However, it is often the indirect effects of glaciation that are of the greatest significance to slope stability studies. The repeated deteriorations in climate saw increased frost weathering in uplands, most especially during deglaciation where previously stressed rock was exposed to freezing. Elsewhere, reductions in temperature witnessed the accumulation of snowfields underlain by ground materials that were permanently frozen to a depth of several tens of metres. Ice development within the ground caused rupturing, deformation and heave, all of which played a role in subsequent slope instability. These effects are most dramatically displayed in the cambered strata and valley bulging phenomena described in Chapter 5 *(Figures 5.12 and 5.13).*

These episodes of harsh, tundra conditions often lasted for tens of thousands of years and appear to have been characteristic of much of the Pleistocene climatic record. Levels of harshness varied over time and there is evidence that the intensity of the periglacial regime diminished westwards because of the moderating influence of the sea *(Figure 6.10).* However, during the summer months, thawing of the surface layer and melting of accumulated snow resulted in the production of a saturated zone or 'active layer' above the impermeable frozen subsoil. This results in the surface material moving downslope by sliding and flowing in a process that is widely known as solifluction, although some authors prefer gelefluction. Rates of movement range from 0.1 to 0.5m/pa and can occur on slopes with very low inclinations *(see below)*.

Solifluction deposits (head) are ubiquitous in Great Britain and occur as extensive, dissected sheets or spreads, sometimes several metres thick, often associated with clearly defined mudslides and rotational failures. This is most clearly displayed at Sevenoaks, Kent, where the prominent Lower Greensand Hythe Beds escarpment is blanketed by failures which diminish downslope to lobes and sheets of brecciated Weald Clay mixed with sand and chert, which had moved up to 500m away from the scarp on slopes as low as 2°[11] *(Figure 3.13)*. Somewhat surprisingly, solifluction was frequently excluded from discussions of landsliding, until sub-surface investigations, often in cases of mass movement reactivation, revealed that solifluction sheets are often internally divided by shears and overlie basal shear surfaces.

The significance of the periglacial regime in creating slope instability has been well established for many areas. One of the best documented examples is the Barnsdale slope in the Gwash Valley, Leicestershire[19]. This valley-side slope in Upper Lias Clay was formed as a result of erosion by the River Gwash, and originally stood at an angle of around 30° *(Figure 6.11)*. Degradation initially involved solifluction and shallow landslide activity under periglacial conditions in the early Devensian (around 111,000 – 60,000 years ago). However, the presence of displaced, but

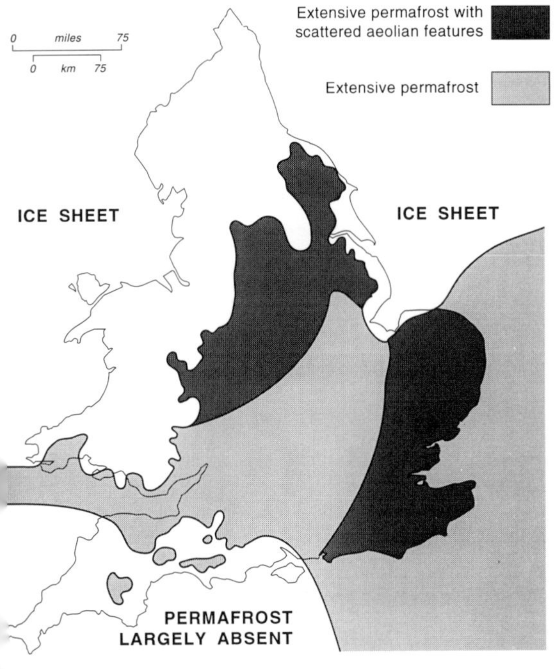

10 The extent of periglacial conditions in England and ales (after Hutchinson & Thomas-Betts, 1990).

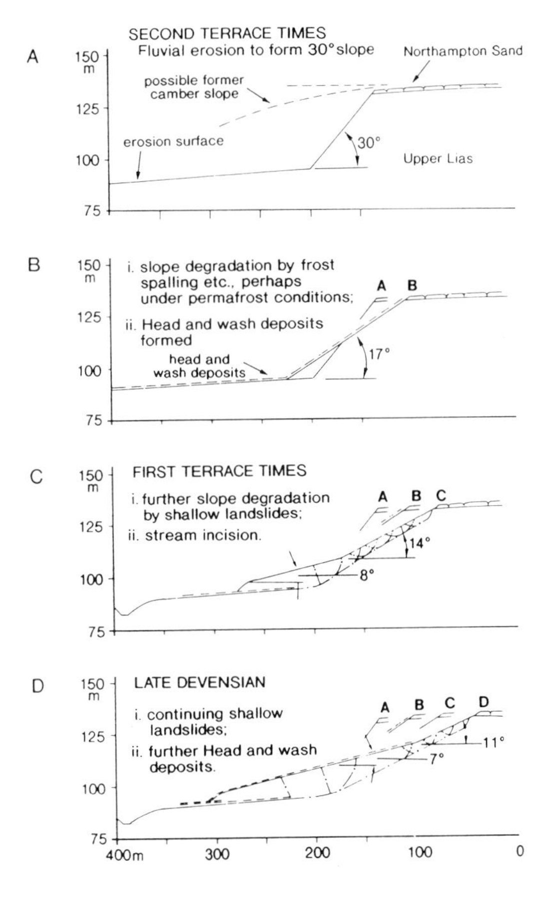

11 The degradation of the Barnsdale slope, icestershire (after Chandler, R.J., 1976).

unweathered, Lias clay above these early Devensian solifluction deposits indicates that there has been a later phase of deep-seated landsliding, most probably during the Middle Devensian (First Terrace times, *Figure 6.11*), and was probably associated with the warmer climates of the Chelford Interstadial (c. 60,000 BP) and the Upton Warren Interstadial (c. 43,000 BP). A further phase of landslide activity is believed to have occurred around 13,000 years ago, coincident with the warm Windermere Interstadial, i.e. the warm phase between the Devensian Glaciation and the Loch Lomond Readvance.

The stimulation of slope instability due to a shift in climate from harsh arctic conditions to warm temperate environment has also been established for Whitestone Cliff landslide in the Hambleton Hills, Yorkshire[20]. This large multiple rotational slide is developed in Jurassic strata (Inferior Oolite limestone and calcareous grits overlying Upper Lias Clays) and is believed to have blocked an ice-marginal drainage channel adjacent to the Devensian ice-sheet, thereby resulting in the formation of Gormire Lake, *(Plate 33)*. Analysis of pollen taken from cores in the lake bed sediments indicate that movements occurred between 15,000 and 10,000 years ago, coincident with the Windermere Interstadial.

The reasons for accelerated slope instability during phases of climatic change from arctic to temperate conditions are well known and have already been introduced in Chapters 3, 4 and 5. The progressive thawing of permafrost resulted in saturated surface layers and high pore-water pressures. Precipitation changed from snowfall to rain and the water penetrated into the ground easily because of the churning effects of solifluction. Also there was little vegetation to intercept and use water and limited stabilisation by root systems. Thus deglaciation probably witnessed phases of widespread slope instability, initially by shallow solifluction, but later by deeper-seated failures as the ground thawed to greater depths. These phases drew to a close, either because the slopes had degraded to stable angles or because of the stabilising effects of forest colonisation, or through a combination of the two.

The Pleistocene is therefore seen as a period characterised by numerous, relatively short-lived, pulses of landsliding, *(Appendix B)*. This climatically induced pattern has to be superimposed on the model of landscape development outlined in the previous section, which envisages increasing development of relative relief over time. Combining the two results in a model whereby the intensity of landsliding associated with each climatic oscillation will increase through time as conditions become more conducive to mass movement. It is wholly to be expected, therefore, that evidence for the most recent phases of instability (late Devensian, Loch Lomond Readvance) should be widely preserved in the landscape, for not only were they possibly amongst the most intense phases of landsliding but also they are the youngest, having occurred within the last 18,000 years. Evidence of earlier episodes will prove more difficult to find because of the sculpturing effects of later glaciations and subsequent landslide movements. Despite this, the reasons for a rich legacy of ancient landslides is clear and the potential problems of reactivation obvious.

The harsh, cold climate conditions of the Pleistocene ceased about 10,000 years ago, since when there have been generally warmer and more humid conditions, albeit with some variability. Variations in rainfall totals and vegetation patterns may have resulted in further phases of slope instability. Certainly the detailed investigations at Hadleigh Castle, Essex, indicate pulses of instability[21], although the true nature of the linkage between climatic variations and slope instability has yet to be established in detail. The evidence in the available literature has been gathered together and is presented as Appendix B. The main phases that have been identified are as follows.

Between 9,500 and 7,500 years before present (BP) (the pre-Boreal and Boreal periods), the climate was warm and dry and coincided with the expansion of deciduous woodland over much of Great Britain. At Hadleigh Castle in Essex *(Plate 10)*, the Boreal period appears to have been a phase of relative slope stability[21] and this may well be so elsewhere. However, the end of the Boreal was marked by an increase in rainfall characteristic of the following Atlantic period (7,500–5,000 BP). The first half of the Atlantic period has been identified as a phase of landslide activity in Europe[22], although there is only limited dated evidence in Great Britain to support this. At Hadleigh Castle, the early Atlantic period was found to be characterised by mudslide activity[21]. In the High Peak of Derbyshire, pollen analysis at Bradwell Sitch has indicated movements between 7,000 and 6,500 BP[23], whilst a date of around 7,500 BP was suggested for the initiation of the Buckstones Moss landslide in the Southern Pennines[24].

The succeeding sub-Boreal period (c.5,000–2,500 BP) was warm and dry, and is likely to have been a period of relative slope stability. The climate deteriorated, however, around 2,500 BP, becoming wetter and cooler. This sub-Atlantic period coincided with a steady increase in forest clearance for agricultural land, and is probably characterised by widespread landslide activity. This was demonstrated on the Inferior Oolite and Upper Lias Clay slopes near Rockingham, Northamptonshire, as indicated by the identification of Iron Age charcoal fragments at the base of the slope debris[25]. The landslide activity at this site was considered to be related to a period of erosion following deforestation, together with the increased rainfall. Radiocarbon dates from birch wood beneath mudslide debris at the Upper Bradwell Sitch landslide in the High Peak of Derbyshire, indicate that slope failures occurred around 2,0000 BP[23]. Also in the High Peak, the Mam Tor landslide probably became active around 2,500–2,000 BP, as suggested by analysis of peat deposits in surface depressions within the slide[26]. This slide also partly destroyed an Iron Age hill fort which had been first occupied around 3,000 BP[27], again suggesting landslide activity in the sub-Atlantic period.

Although there is evidence of landslide activity at Lympne, Kent, between the 6th and 7th centuries AD[28], the next significant phase of inland landslide activity appears to coincide with the 'Little Ice Age' of AD 1550–1850 (400–100 BP). During this period Great Britain suffered an appreciable climatic deterioration, with cold winters and wet summers. A review of historical evidence[29] suggests that there was a marked increase in inland landslide activity, notably in 1550–1600 and 1700–1850, which may have been the result of these climatic changes. During this period a significant number of large landslides were reported to have occurred, including the 'Wonder at Marcle' in Hereford and Worcester, which destroyed a chapel and a number of houses in 1575.

"The Wonder" was described by Camden:

> "A hill, which they call Mardley Hill, in the year 1575 roused itself up, as it were, out of steep, and for three days together shoving its prodigious body forwards, with a horrible roaring noise, and over-turning all that stood in its way, advanced itself..." (Camden, 1722).

The preacher John Wesley gives a fascinating description of a major failure at Whitestone Cliff, in the Hambleton Hills, in his journal for 1755:

> "On Tuesday March 25 last, being the week before Easter, many persons observed a great noise near a ridge of mountains in Yorkshire, called Black Hamilton. It was observed chiefly in the south-west side of the mountain, about a mile from the course where the Hamilton races are run; near a ridge of rocks, commonly called Whiston Cliffs, or Whiston-White-mare; two miles from Sutton, about five from Thirsk.
>
> The same noise was heard on Wednesday by all who went that way. On Thursday, about seven in the morning, Edward Abbott, weaver, and Adam Bosomworth, bleacher, both of Sutton, riding under Whiston Cliffs, heard a roaring (so they termed it), like many cannons, or loud and rolling thunder. It seemed to come from the cliffs, looking up to which they saw a large body of stone, four or five yards broad, split and fly off from the very top of the rocks. They thought it strange, but rode on. Between ten and eleven a larger piece of the rock, about fifteen yards thick, thirty high, and between sixty and seventy broad, was torn off and thrown into the valley.
>
> About seven in the evening, one who was riding by observed the ground to shake exceedingly, and soon after several large stones or rocks, of some tons weight each, rose out of the ground. Others were thrown on one side, others turned upside down, and many rolled over and over. Being a little surprised, and not very curious, he hasted on his way.
>
> Of Friday and Saturday morning the ground continued to shake and the rocks to roll over one another. The earth also clave asunder in very many places, and continued so to do till Sunday morning."

Further examples of inland landslides reported during the 'Little Ice Age' include the Birches landslide in Ironbridge Gorge during 1773, the Hawkley slip near Selborne in 1774[30], the Beacon Hill landslide in Bath during 1790[31] and on the north cast slopes of the Isle of Portland in 1792.

A colourful description of the Hawkley slip has been provided by Gilbert White in **The Natural History of Selborne:**

> "The months of January and February, in the year 1774, were remarkable for great melting snows and vast gluts of rain. The beginning of March also went on in the same tenor; when, in the night between the 8th and 9th of that month, a considerable part of the great woody hanger at Hawkley was torn from its place. It appears that this huge fragment, being perhaps sapped and undermined by waters, foundered, and was engulfed, going down in a perpendicular direction.

These people (local farmers) in the evening observed that the brick floors of their kitchens began to heave and part; and that walls seemed to open, and roofs to crack. The miserable inhabitants, not daring to go to bed, remained in the utmost solicitude and confusion, expecting every moment to be buried under the ruins of their shattered edifices. When day-light came they were at leisure to contemplate the devastations of the night; they found that a deep rift, or chasm, had opened under their houses, and torn them, as it were in two.

About fifty acres of land suffered from this violent convulsion; two houses were entirely destroyed; one end of a new barn was left in ruins, the walls cracked through the very stones that composed them; a hanging coppice was changed to a naked rock; and some grass grounds and an arable field so broken and rifted by the chasms as to be rendered, for a time, neither fit for the plough or safe for pasturage, till considerable labour and expense had been bestowed in levelling the surface and filling in the gaping fissures".

The Isle of Portland has a well documented history of landsliding throughout the 17th and 18th centries, with particularly valuable descriptions provided by Hutchins in "The History and Antiquities of the County of Dorset" published in 1803. One failure, on February 13th 1792, involved more than a mile of hillside on the north east side of the Island, and is believed to have been one of the largest landslides to have occurred in historical times. Hutchins described the event as follows:

"the season had been very wet, which made the blue marle or clay very slippery; the week preceding there had been some strong gales of wind, and remarkable high tides, that had drawn off a great deal of the rocks and beach, that served as a natural buttress to the land above. Early in the morning, the road was observed to crack and rent, which soon increased."

"The late Francis Steward, esq. receiver-general of the land tax for the county, was in the island before two o'clock at which time the ground had sunk several feet, and was one continued motion, attended with no noise but what was occasioned by the separation of the roots and brambles, and now and then a falling rock. At night it seemed to stop a little, but soon moved again; and before morning the ground from the top of the cliff to the water side had sunk in some places 50 feet perpendicular. The extent of the ground that moved was a mile and a quarter from North to South, and 600 yards East to West. At the ends little alteration is to be seen; but in the centre, between the two piers, it is thrust into the sea about 30 yards. The ruins of the King's Pier is so much further into the sea than it was; and the other pier, which is quite overset, seemed moved as much to the Southward, or nearer the Kings's Pier. Rocks that were some depth under the water were raised above water."

"The main road that the stone was hauled on the water side, and had been in use above half a century, is so torn and ruined, that in many places it cannot be distinguished from the general rubbish. These piers were rendered entirely useless; indeed, great part of the King's Pier is washed into deep water, with a great quantity of stone, and much more that was hauled near the water ready for shipping is either washed into the sea, or so buried as to be irrecoverably lost.' (Hutchins, 1803).

There is wide ranging evidence for episodes of inland failure preceding the Industrial Revolution and consequent large-scale human intervention. As a result, there are very good reasons why extensive tracts of ancient mass movement features exist in the landscape, often in areas of contemporary relative stability and devoid of large-scale human activity. Thus the variations in patterns and densities between the distributions of youthful and ancient/unknown age landslides *(Figure 6.2–6.4)* should not cause surprise. The main problem concerns identifying the true nature, distribution and origins of these relatively old instability features.

Sea Level

One of the most dramatic consequences of Quaternary climatic fluctuations was sea level change. The repeated advance and retreat of the ice-sheets was synchronised with oscillations in sea level that frequently had amplitudes of 120–150m, so that glacially lowered sea surfaces were often over 120m lower than those of today. These oscillations of sea surface elevation, in combination with tectonic and isostatic crustal movements, led to major fluctuations in shoreline position, so that during the most significant glacial phases much of the shallow sea floor was exposed as dry land. Certainly the southern part of the North Sea and the English Channel floor were exposed on many occasions[32]. As a consequence, the river networks repeatedly extended over the exposed sea floor and incision took place within the former land area. Thus sea level change helped facilitate the landscape evolution described in the previous section on 'topographic development'.

Sea level change has clearly contributed significantly to the development of inland landsliding in Great Britain but has rarely been the dominant factor. Episodes of lowered sea level stimulated erosion, while the succeeding high sea levels led to the flooding of the lower reaches of valleys to form estuaries where wave attack could undermine slopes, thereby creating instability until valley sedimentation removed the marine influence. The degraded cliffline around the inland margin of Romney Marsh, Kent, is a good example of such a feature *(Figure 6.12)*. But such features are of limited occurrence. Good examples of youthful shoreline features that have been raised above contemporary sea level are mainly confined to Scotland, where

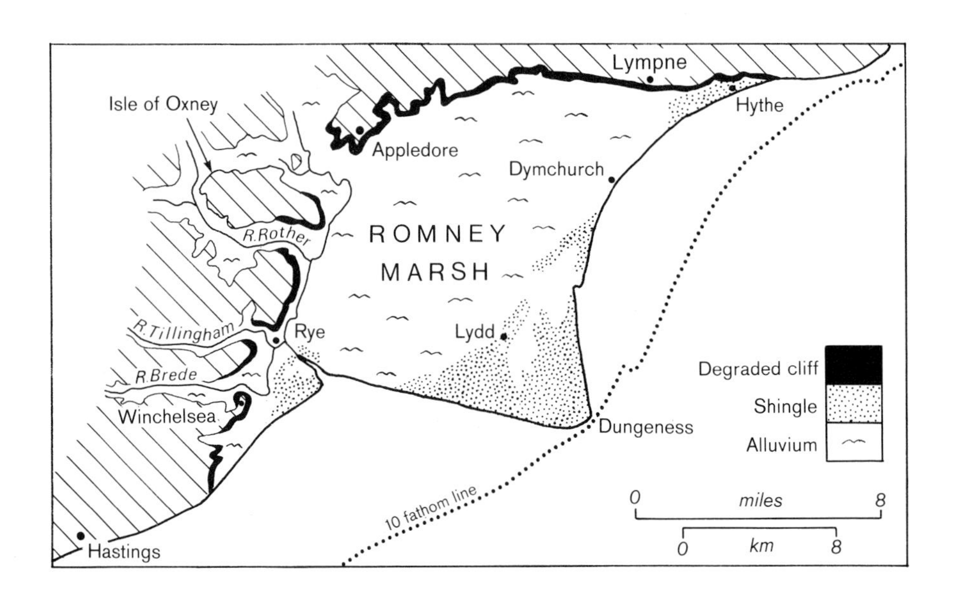

6.12 The former sea cliff which was abandoned as a result of the formation of Romney Marsh behind a major barrier of sand and shingle (after Jones, D.K.C., 1981).

unloading following the melting of the Devensian ice-sheet has resulted in crustal rebound that has exceeded the rate of postglacial sea level rise, thereby producing raised beaches and abandoned clifflines. Some of these features on the west coast of the Scottish Highlands are sufficiently young and well developed to be the sites of continuing rock instability.

By contrast, sea level change in coastal locations has been of paramount importance. The present coastline is the product of long and complex evolution and the landslides located along it have, in many cases, complicated histories of development. Each lowering of sea level resulted in the abandonment of the coastline, **as it then was,** so that the clifflines freely degraded through landsliding. Each phase of inter-glacial higher sea level saw the sea eventually reoccupy part or all of the former coastline, with renewed wave attack steepening cliffs and regenerating failures. On the 'soft' rock coastlines of extra-glacial Southern Britain, the former shoreline cliffs had probably become much degraded during the several thousand years of harsh arctic climate. Each inter-glacial stand of high sea level would have removed the debris and steepened up the clifflines, as has been achieved along the south coast of England over the 8,600 years since rising water levels caused the sea to penetrate northwards through the Dover Straits. Using contemporary rates of coastline retreat (0.1–1.0 m per year) suggests that each inter-glacial high sea level could have achieved up to a few kilometres of coastline retreat before once again withdrawing. By contrast, on the 'hard' rock coastlines of the west and north, very much more limited change was achieved. Finally, along those coastlines where soft rocks were affected by glacial erosion and deposition, it is likely that each phase of high sea level had to recommence the task of creating a coastline in the recently deposited, relatively weak materials. This was normally achieved with ease, thereby accounting for the rapid rates of coastline retreat recorded along the entire length of the Holderness coastline *(Figure 6.13).*

The most recent phase of sea level rise, the Flandrian Transgression, was produced by the decay of the Late Devensian ice sheets. This transgression was characterised by very rapid sea level rise from –100m at 14,000 BP to around –20m by 8,000 BP, after which the rate of change slowed, with near present levels being achieved at about 5,000 BP (a more complex pattern is evident in areas of glacial unloading such as Scotland, where isostatic rebound has caused sea levels to appear to generally fall over the same period). It is likely that the present pattern of landsliding around the coast was initiated in the period 8,000–3,000 BP, particularly the large landslide complexes found on the south coast of England such as Folkestone Warren, the Isle of Wight Undercliff (within which the town of Ventnor lies), Stonebarrow and Black Ven *(see Chapter 2).* In contrast to most inland areas, coastal landslides and rockfall clifflines initiated up to 8,000 years ago have remained sites of instability throughout the majority of the intervening period, except where protected by the accumulation of sediment (e.g. Pendine cliffs, South Wales; *Figure 6.14*) or engineered defences.

Detailed examples of the development of coastal landslide complexes through the Flandrian are very rare because marine erosion usually destroys the evidence. An exception is provided by the St Catherine's Point landslide on the south coast of the

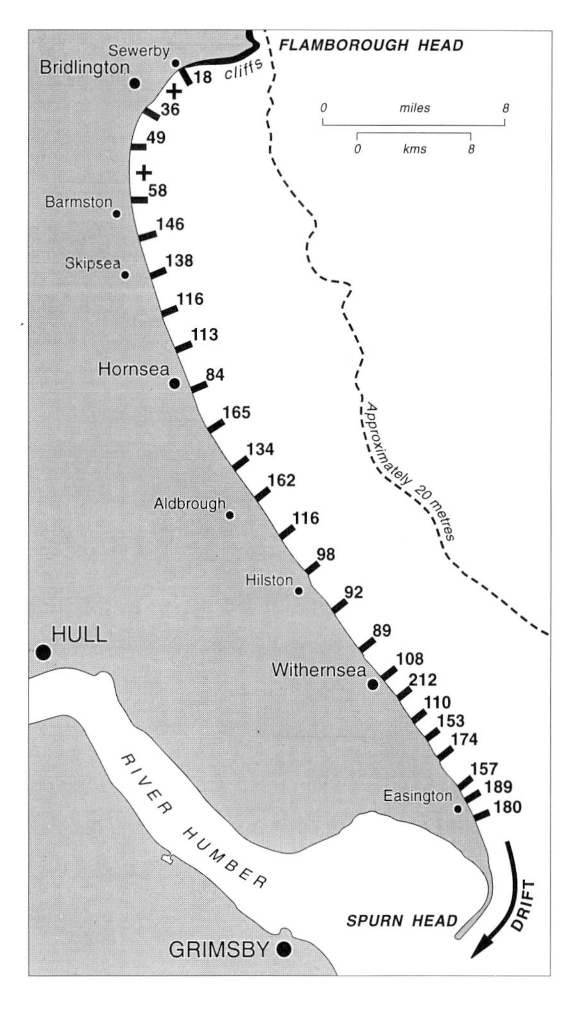

6.13 The recession of the Holderness coastline, in metres, between 1852 and 1952 (after Valentin, 1971).

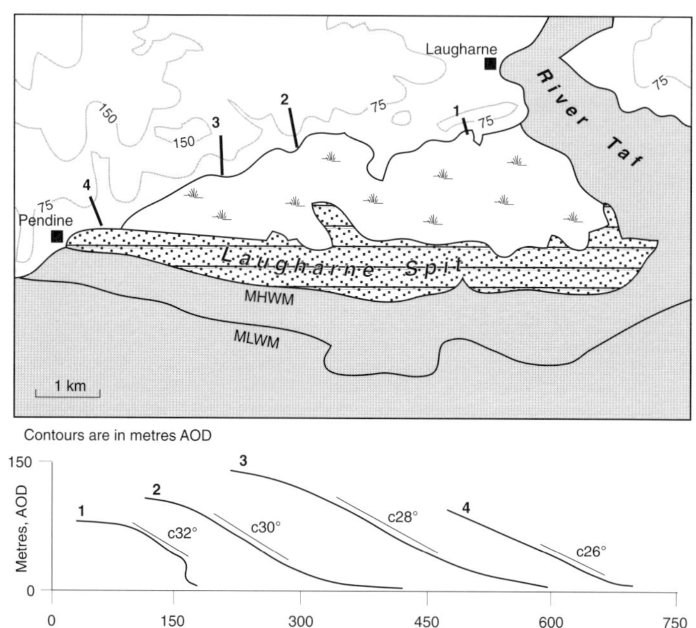

6.14 Degradation of an abandoned sea cliff between Laugharne and Pendine, South Wales (after Savigear, 1952). The cliffs were successively abandoned from the west.

Isle of Wight[33] *(Photo 3.10 and Figure 3.36)*. Here it has been established that strong periglacial activity during the Devensian produced considerable masses of predominantly fine debris through solifluction and shallow landsliding. This material buried the remnants of any earlier landslides and resulted in spreads of debris extending southwards over the exposed former sea bed. As sea levels rose during the early Flandrian, these accumulations (debris aprons) were attacked by marine erosion, and ultimately removed together with any remnants of earlier landslides.

Continuing marine erosion cut into the in situ Lower Greensand cliffs to produce the now buried shore platform and former sea cliff preserved at St Catherine's Point *(Figure 3.36)*. The levels of the landward margins of this shore platform, around 9m below present mean sea level, suggest that it may have been formed around 7,000–8,000 BP, possibly during the still-stand in sea-level at approximately this elevation identified elsewhere[34]. The main debris aprons at St Catherine's Point (over 13m of chalky material), accumulated immediately after this phase of activity, burying the landward portion of the shore platform[35]. This phase of landslide activity was followed by a period of relative stability during which soil development took place on the debris aprons. Yew logs incorporated in these soils were radiocarbon dated at around 4,000 BP, indicating that this relative stability occurred during the sub-Boreal period. These soils were subsequently buried by the accumulation of a second debris apron which probably developed in response to the climatic deterioration of the early sub-Atlantic period, around 2,500–2,000 BP.

A third phase of landsliding is believed to have occurred, involving the formation of the deep-seated compound slides. This view is based on a comparison of the chalky debris found between the linear ridges *(Figure 3.36)* and the debris aprons, with the finer infill between the ridges suggesting a later formation than the aprons. The morphology of the debris aprons is also indicative of a separate phase, as it is buckled and distorted, probably by the thrust of the compound slide which resulted in the formation of Ridge 'S'. This phase of movement remains undated, but could have occurred around 1,800 BP, as suggested by a study of chalky hillwash on the cliffs behind the landslide[36].

Continued marine erosion over the last 2,000 years has maintained marginal stability at St Catherine's Point, and other sections of the Isle of Wight Undercliff. Indeed, numerous small movements and occasional large failures (e.g. The Landslip, in 1810 and 1818; *Photo 3.7*) have been reported, often coinciding with periods of heavy winter rainfall or intense frosts[37].

Human Activity

Human activity has had a fundamental effect on the stability of both inland and coastal slopes, either by initiating first time failures, or increasing the rate of coastal cliff retreat or, more commonly, reactivating previously undetected, degraded ancient landslides. Human impact on the environment has been of major importance for at least 2,000 years and there are signs of significant human disturbance as early as 6,000–8,000 years ago, especially forest clearance. However, the scale, extent and intensity of disturbance have increased dramatically in recent centuries and there have undoubtedly been much greater numbers of human-induced slope failures over the last 200 years or so. This may be seen to be a direct consequence of rapid urbanisation, industrialisation, infrastructure development and dramatic changes in rural land use.

Land use change and **vegetation clearance** are widely regarded as the major factors in promoting slope failure in Great Britain, although few specific examples have been reported. An exception is the demonstration that there has been a marked increase in debris flow activity in the Scottish Highlands since around 1700[38] *(Figures 6.15 and 6.16)*. This pattern was recognised to be the result of over-grazing by sheep and, more significantly, burning of the moors in the late 19th and early 20th centuries. Burning was used to improve sheep grazings and latterly deer grazings, and destroys the surface moss layer that acts as a major barrier to the infiltration of water into the soil. It is considered likely that prior to widespread burning, rainfall events that now produce debris flows had little effect because the moisture was retained in the moss layer, preventing the saturation of the soil.

Other evidence for human-induced landsliding following forest clearance or land use change remains largely circumstantial. Truncated soils in upland regions, dramatic increases in sedimentation rates in lakes (one Brecon Beacon example revealed a thirteen-fold increase post 5,000 BP) and evidence from floodplains in

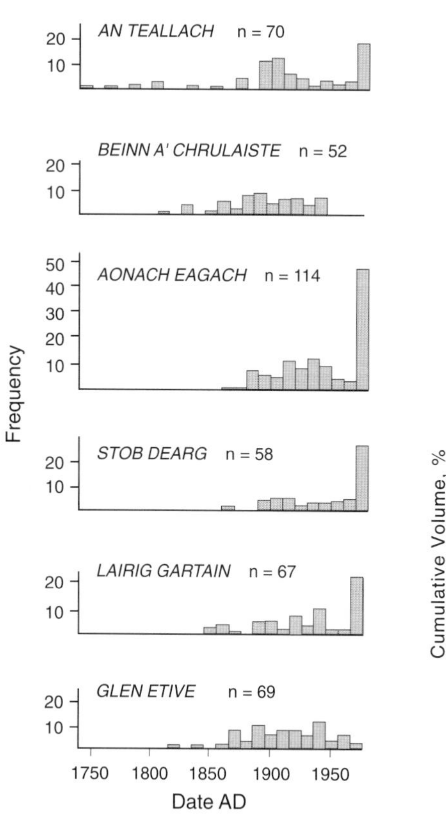

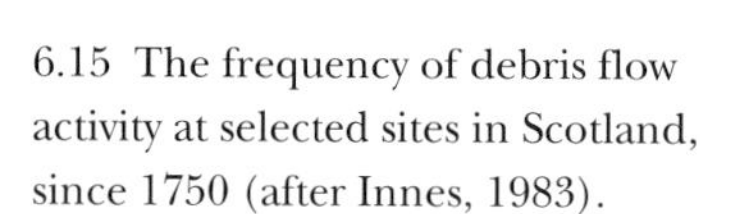

6.15 The frequency of debris flow activity at selected sites in Scotland, since 1750 (after Innes, 1983).

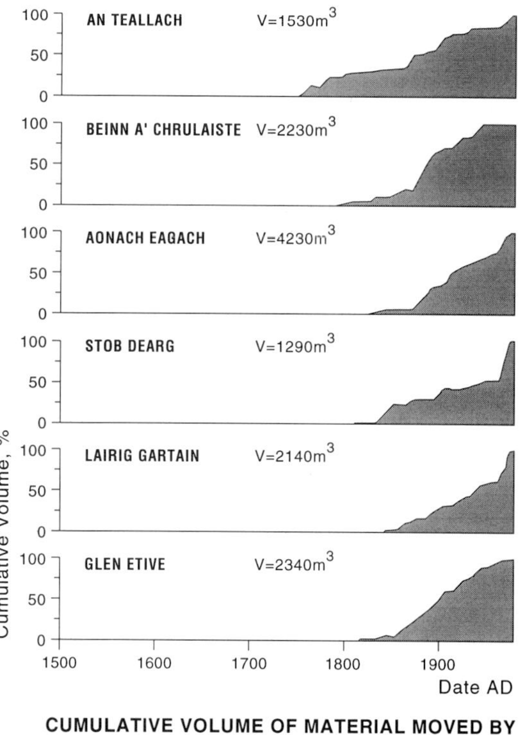

6.16 The cumulative volume of material moved as a result of debris flow activity at selected sites in Scotland (after Innes, 1983).

Lowland Britain of influxes of sediment to valley floors post 8,000 BP, are all indicative of accelerated erosion, but the relationships with human intervention and landsliding remain unclear. Similarly, the presence of failures on cleared land (i.e. fields, pastures, etc.) does not necessarily point to vegetation clearance as the main cause, for there may have been landsliding prior to or during woodland cover. Nevertheless, it does seem likely that woodland clearance did exacerbate existing slope failure and lead to new movements. For example, throughout the Cotswolds there is evidence of shallow landslide activity since the Middle Ages, when 'ridge-and-furrow' cultivation expanded during what is known as the "little optimum" of AD 750–1,300[39]. Many of these systems which were developed on Lias clay slopes have since been disturbed by shallow mudslides (e.g. at Woolstone Hill and Crane Hill near Bishops Cleeve). At Ebrington Hill and Leckhampton Hill, for example, several areas of medieval fields have been disrupted *(Photo 6.2)*. Although many of the slides at Ebrington probably were active before the field enclosures of around 1810, three large mudslides have disturbed the field boundaries indicating more recent activity *(Figure 6.17)*.

Mineral extraction, especially underground coal mining, has been recognised as a contributory cause to recent failures in a number of areas. Subsidence resulting from the collapse of shallow pillar-and-stall workings, and possibly deep longwall

6.17 The pattern of post-enclosure (c1810) shallow landslide activity on the northern slopes of Ebrington Hill, Gloucestershire (after Butler, 1983).

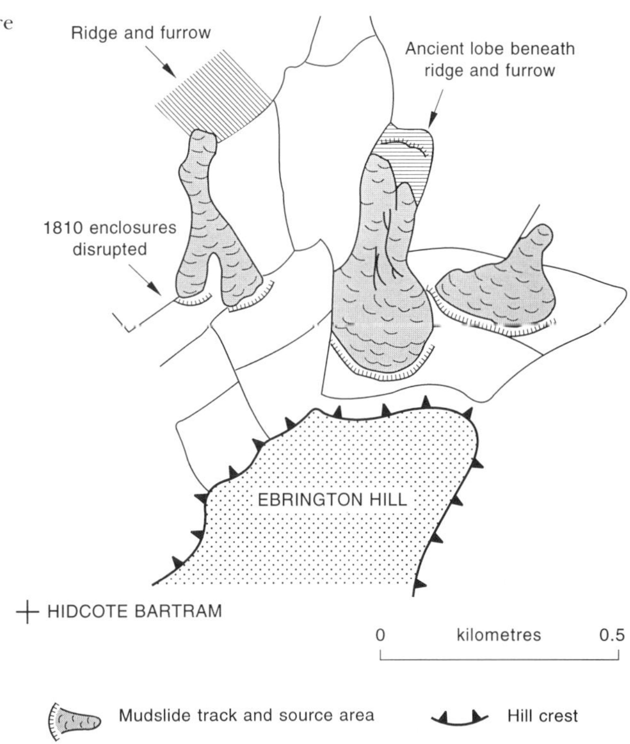

6.2 Medieval ridge and furrow system disrupted by later shallow mudslide activity, Leckhampton Hill, Gloucestershire (Cambridge University Collection: copyright reserved).

methods, is considered to have initiated or reactivated slope failures in South Wales[40]. However, recent study has shown that whilst a number of slides in South Wales could be attributed to subsidence-related problems (e.g. the East Pentwyn landslide in Gwent), there was either no evidence or inconclusive evidence for many other contemporary failures[41]. The disposal of colliery waste on hillslopes has, however, caused many landslide problems, notably the Aberfan disaster[42] and the reactivated slide at Blaina[43].

Coastal instability has been exacerbated by the disruption of sediment transport patterns, as occurred at Folkestone Warren following the construction of the harbour[4] *(see Figure 4.16)*. Similar problems were experienced at Ventnor on the south coast of the Isle of Wight, associated with attempts to build a harbour in the 1860s, when construction of breakwaters in the bay involved the removal of a rocky headland which had previously protected the shore from waves driven by westerly winds. Removal of this headland resulted in rapid depletion of the beach and erosion. As was noted at the time "the blue slipper [the Gault Clay] soon became exposed and eroded, and the land along the sea front began to slip"[44].

There are many examples of landslides associated with the construction of the railway network in the 19th and early 20th centuries, especially failures of cut slopes. However, probably the most important sector of development affected by landsliding has been the major road building programme of the 1960s and 1970s, not only in terms of the cost and inconvenience caused to specific projects, but also for the stimulation of scientific research. Examples of landslide problems associated with road construction include the Waltons Wood landslide in 1961–1962[45], the Sevenoaks Bypass ground movements in 1966[11], the problems along the M4 near Swindon in 1969[46], the Nags Head landslides on the M5 in Devon in 1968–1969, the failures along the A45 Daventry Bypass in the late 1970s[47] and the slip along the A55 at Pwll Melyn in Clwyd in 1976[48].

Housing development has also initiated landslide problems in a number of areas, through cut-and-fill operations (e.g. the Bury Hill landslide of 1960 in the West Midlands[49]) and as a result of drainage problems. The Hedgemead landslide in Bath, which occurred in the early 1870s, was the result of increased water input into superficial materials from leaking sewers, water supply pipes and storm water soakaways[2]. The first signs of failure occurred only ten years after the area had been built on, and movements continued periodically until the end of the century.

NOTES

1. Soutar, 1989
2. Hawkins, A.B., 1977
3. Douglas, 1985
4. Hutchinson, 1969
   Hutchinson et al., 1980
5. Halcrow, 1989
   Jones, D B et al., 1991
6. Sherrel, 1971
7. Grainger, 1983
8. Conway et al., 1980
9. Jones, D K C, 1981
10. Skempton, 1953
11. Skempton & Weeks, 1976
12. Bowen et al., 1986
13. Goudie, 1983
    Lowe & Walker, 1985
14. Price, 1983
15. Holmes, 1984
16. Smith, D I, 1984
17. Siddle et al., 1987
18. Henkel & Skempton, 1954
    Culshaw, 1972
    Denness, 1977
19. Chandler, R J, 1976
    Chandler, R J, 1977
20. Blackham et al., 1981
21. Hutchinson & Gostelow, 1976
22. Starkel, 1966
23. Tallis & Johnson, 1980
24. Muller, 1979
25. Chandler, R J, 1971
26. Johnson, R H, 1987
27. Coombs, 1971
28. Hutchinson et al., 1985b
29. Hutchinson, 1965a
30. White, 1789
31. Kellaway & Taylor, 1968
32. Jones, D K C, 1981
33. Hutchinson et al., 1985a
    Hutchinson, 1987a
    Hutchinson, 1991
    Bromhead et al., 1991
34. Tooley, 1978
35. Hutchinson, 1987a
36. Preece, 1980
37. Hutchinson, 1965b
    Chandler, M P, 1984
    Lee et al., 1991a
38. Innes, 1982
    Innes, 1983
39. Goudie, 1983
40. Franks, C A M, 1986
    Franks, C A M, et al., 1986
    Halcrow, 1989
    Jones, D B, et al., 1991
41. Halcrow, 1989
    Jones, D B, et al., 1991
42. Bishop et al., 1969
43. Gostelow, 1977
    Pullan, 1986
44. Whitehead, 1911
45. Early & Skempton, 1972
46. Hutchinson, 1984a
47. Biczysko, 1981
48. Subramaniam & Carr, 1983
49. Hutchinson et al., 1973

CHAPTER 7

Landsliding as a Hazard

Coastal Landsliding

On Sunday February 21, 1977, a school party from Warlington in Surrey were studying the geology of Lulworth Cove on the Dorset coast, when they were buried beneath a sudden rock slide. Despite rescue attempts by local ambulance men, the schoolteacher and a pupil had been killed and two more pupils seriously injured, one of whom died later in hospital. At the subsequent inquest the deputy coroner for south-east Dorset warned of the dangers posed by the Dorset cliffs and commented on the possibility of a recurrence of the tragedy.

Indeed, this had not been an isolated incident. In 1925, eight workmen, busy extending a road from Bascombe to Southbourne, were buried beneath 100 tonnes of rubble from a rockfall, three of whom died. On August 28, 1971, a nine year old girl was hit on the head by falling rock whilst walking on the beach at Kimmeridge, and later died of her injuries. At Swanage, a schoolboy on a field course was seriously injured by a rock fall in February 1975, and a year later (April 1976), a young boy was killed after being hit on the head by falling rock. In July 1979 a woman, sunbathing on the beach near Durdle Door, was killed when a 3m overhang collapsed. These incidents, and others, led the Chief Inspector of Wareham police to coin the phase "killer cliffs", highlighting the serious danger that rock falls and landslides posed to tourists and educational parties.

Danger posed by unstable cliffs is not confined to the Dorset coast, for fatal accidents and serious injuries caused by rock falls and slides have been reported from many tourist resorts. For example, four small boys were buried by sand in a cliff fall at Alum Bay, Isle of Wight, during September 1959, and although quickly uncovered by nearby holidaymakers, one child died. A similar accident occurred near Newquay, on the Cornish coast, in June 1986, when an Australian lifeguard was killed when a beach hut was hit by falling mud and rocks. Indeed, several isolated pillars of rock (stacks) on the coast of Thanet, north Kent, were removed in the 1970s because of the danger they posed to tourists.

Tragic though these events may be, the losses of life are small compared with other causes of sea-side deaths, such as drowning or falling off cliffs. However, while drowning and falling are perceived as acceptable hazards associated with human choice, being hit by falling debris is generally considered unacceptable. This is a common attribute of all losses resulting from the impacts of events that are considered 'natural hazards'. Thus relatively insignificant minor impacts caused by hazard events often cause considerable local concern, especially if they are associated with threatening phenomena that are poorly understood or occur with unexpected suddenness. Coastal falls have these attributes.

Isolated deaths may not be especially significant, but they do point to the potential for larger-scale and far more costly impacts on local communities. There are four grounds for making this statement. First, known examples of coastal landslides include the largest and most dramatic failures in Great Britain *(see Chapter 2)*. Second, coastal zone land use has been under considerable development pressure in recent decades, so that there is an ever growing potential for loss. Third, recorded erosion rates on many stretches of coast in Lowland Britain are so high that losses of roads, houses, buildings and even entire villages will be inevitable, unless such eroding coastlines are defended or systematically abandoned. Finally, comparison with the magnitude and frequency of other natural hazards, indicates the likelihood of a small number of very costly landslide events, given sufficient time. These serious impacts, often referred to as 'disasters', may be the consequence of large-scale (catastrophic) displacements, as occurred at Bindon in 1839 *(see Photo 2.6)*, or vulnerability compounded by chance.

Unfortunately, very few studies have attempted to quantify the actual cost of coastal landsliding in financial terms, although it is likely that many local authorities are regularly forced to commit large sums towards preventing and stabilising landslides. A recent review of the impact of landsliding in Dorset[1], estimated that an average of £2.5 million per year was spent as a result of coastal slope failure. An indication of the scale of expenditure commonly associated with coastal landsliding can be gained from the protection works recently undertaken at West Cliff, in West Bay. Development of West Cliff has coincided with a period of very rapid cliff retreat, at an average rate of 2.8m per year since 1965, which has resulted in a series of landslide problems. During February 1978, for example, a rotational slide removed a portion of the Esplanade, and required nearly £1 million of repair works. This failure was attributed, in part, to a burst water main discharging substantial amounts of water into the cliff, and also the low beach levels in front of the sea wall. Initially, the local authority refused to stabilise the retreating cliff, as this action lay beyond the limits of their responsibility for cliff protection. However, after a series of slides had removed substantial portions of the cliff, West Dorset Council commissioned a comprehensive coastal protection scheme involving regrading the slope, the installation of drainage and the construction of a sea wall and promenade, at a total cost of £1,048,000[1].

Particularly serious problems occur where rapid coastal erosion threatens the continued existence of villages or isolated homes. Runswick, for example, on the North Yorkshire coast, has had a long history of landslide problems. One of the earliest recorded landslides occurred in 1682 and destroyed almost the entire village:

> "The whole village, except a single house, sunk down in one night, the ground on which it then stood ... having suddenly gave way. It is stated as a most providential circumstance that most of the inhabitants were that night waking a corpse, ... they not only escaped themselves but alarmed the rest so that scarcely any lives were lost. The houses with their contents were for the most part buried under masses of earth or stone, and sunk down towards the shore"[2].

In 1829 the entire village of Kettleness, 2km south of Runswick, slid into the sea and fortunately the inhabitants were rescued by ships lying offshore waiting to take on cargoes of alum. Slope movements have continued regularly throughout the last 100 years, causing substantial damage to many properties in Runswick, such as the Rose Cottage slide of February 1977[3] which blocked a road, cut off the water supply to the village, and damaged electricity and telephone cables[4]. In 1961 the only access road to the village was abandoned because of repeated ground movements, and a new road had to be constructed. Figure 7.1 shows a zonation of Runswick, produced by students and staff from the Engineering Geology Unit of Newcastle University, which highlights the vulnerability of parts of the village to continued instability.

The Isle of Portland Naval Base lies within an area of ancient coastal landsliding and has had a long history of ground movement problems[11]. In 1941 the dockyard railway and adjacent road was blocked by landslide debris. A number of small slides have occurred since 1959 threatening the safety and stability of existing buildings such as the HMS Osprey mess and barracks accommodation. Because of the high value of the property at risk, a monitoring system was established in 1976 to determine whether any overall movements were taking place and to detect the likely onset of further landslides. This system comprised a network of survey stations throughout the landslide complex and a series of inclinometer installations to detect movement at depth.

Such dramatic examples of damaging coastal landsliding have been concentrated along the eastern coasts of England, where weak 'soft' rocks (clays) and weakly

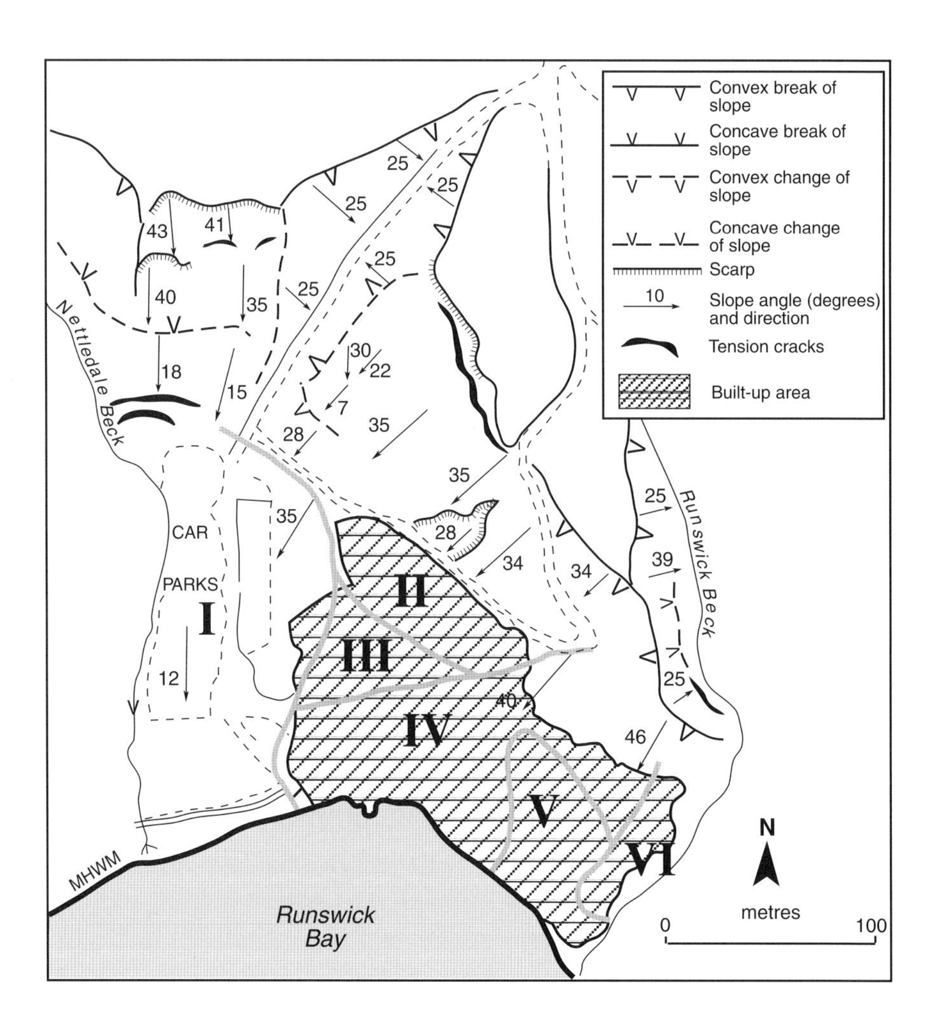

7.1 Landslide hazards in Runswick Bay, North Yorkshire (after Money, 1979).

consolidated superficial deposits are readily undermined by wave attack *(see Figure 2.3)*. The most spectacular case is the Holderness coast of Humberside, where rapid rates of coastal retreat *(see Figure 6.13)* have resulted in the loss of 200km^2 of land since Roman times (equivalent to a continuous strip 4km wide) and the destruction of 26 settlements – the so-called lost towns of Yorkshire although most were, in reality, very small. Erosion has also been severe along the coasts of Norfolk and Suffolk. A large hotel and several houses had to be demolished at Overstrand in the 1940s, Pakefield (near Lowestoft) suffered considerable destruction in the first half of this century *(see Photo 2.1)* due to very rapid rates of coastal retreat (up to 159m in the period 1905–1947), and the town of Dunwich, which was a thriving port in the Thirteenth Century, has all but disappeared.

However, the nature of the problem along the south coast is rather different to that in Humberside and East Anglia. The generally slower rates of erosion on the sedimentary rocks (as compared with glacial deposits), often means longer intervals between failures and therefore greater sense of security. In addition, mass movements will tend to be slower, creating damage and disruption rather than wholesale destruction. Under these circumstances, quite long-lived communities can come to be threatened, which tends to heighten the sense of local concern and outrage. Nevertheless, locally significant rates of erosion have also been recorded along the south coast and coastal landsliding has been widely reported as affecting many coastal communities including Sandgate[5], Barton on Sea[6] Charmouth[7], Lyme Regis[8], Swanage[1], Torbay[9] and Downderry[10] in Cornwall, in addition to the 6 examples discussed in detail below.

The Scarborough area also has a long history of major instability. In 1737 the town's first spa was demolished as an acre of coastal cliff collapsed onto the beach. The town was, of course, the scene of one of the most dramatic landslide events in recent decades when a sudden cliff failure in June 1993 made national headlines by destroying the Holbeck Hall Hotel (*Plate 4 and Photo 1.7*). Guests awoke on the morning of Friday, June 4th, to discover that the hotel gardens had largely disappeared, the cliff top having advanced by around 60m, overnight. The hotel was evacuated before breakfast with many guests forced to leave their luggage in the rush. The slide continued to develop, culminating in the collapse of the east wing on the Saturday evening (*Plate 48*), after which it assumed a temporarily stable position. The hotel was later demolished and the resulting insurance claim is thought to be as high as £3M. The landslide also overran the sea wall at the base of the cliffs, promoting fears of further movements. Scarborough Borough Council and their consultants, Rendel Geotechnics, were on site within a day and immediately began monitoring the adjacent slopes; first by routine surveying, later by electronic tiltmeters installed behind the slide. Rapid geomorphological appraisal and a ground investigation programme involving twelve boreholes was followed by the design of emergency works to protect the toe of the debris. These works included rock armour, placed to protect the landslide toe, and landscaping to achieve a stable profile at an estimated cost of £1.6M–£2M.

The small village of Fairlight, East Sussex, is to some extent threatened by the retreat of 30m high cliffs which have endangered part of the community *(Figure 7.2; Plates*

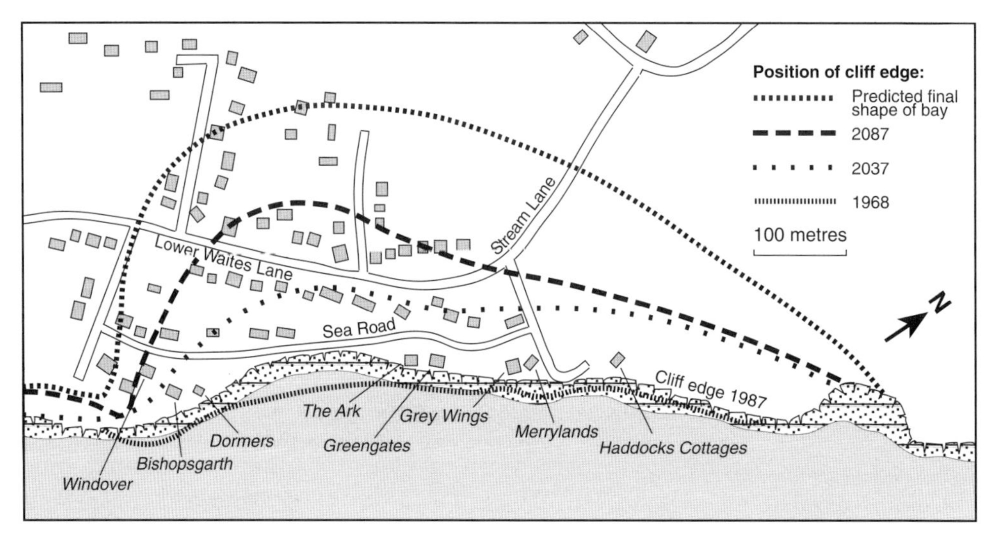

7.2 Coastal cliff retreat at Fairlight Village, East Sussex.

34 and 35). Sections of the cliff have been breaking off at up to 3.5m per annum in the 1980s, threatening 11 houses with destruction by the year 2000. It has been feared that 46 properties could go within a century. Indeed, one bungalow, The Ark, had to be abandoned in March 1987 after the owners awoke to find that their whole garden had fallen into the sea after a heavy storm. In an attempt to prolong the lifespan of properties, two houses, Grey Wings and Merrylands, have been moved inland bodily on steel runners. At Merrylands, nearly 12m of land had been lost in 4 years, and the relocation of the house 8m inland at an estimated cost of £18,000 was the only alternative to being ordered to leave by Rother District Council. In view of the seriousness of the problems facing the community, the Fairlight Coastal Preservation Association was set up to protect the interests of home owners, and lobbied for a coastal protection scheme involving sea defences to stop the erosion. This has now been undertaken at a cost of £2.5 million, involving the construction of a 550m long, 30m wide, concrete armoured berm in front of the cliffs, and the import of many hundreds of tonnes of large granite boulders from Norway.

Similar problems have occurred at Luccombe, on the Isle of Wight, where ground movements between early November 1987 and the end of January 1988, resulted in considerable damage to a number of houses in the village *(Plate 28 and Photo 4.2)*. These movements involved the reactivation of an ancient landslide system, and are the latest in a sequence of events that have affected the village since it was built in the 1930s. Fears of further movements have led to the formation of the Luccombe Residents Association, which has campaigned for a solution to the problem *(Plate 36)*. The matter was raised in the House of Commons by the local MP, Mr Barry Field, and as a result the Department of the Environment commissioned a study to review possible courses of action which could reduce the risk to the local residents. A number of the recommendations of this study have been taken up by the local authority, South Wight Borough Council, who have developed a series of low-cost remedial measures, including a first-time sewerage scheme[12].

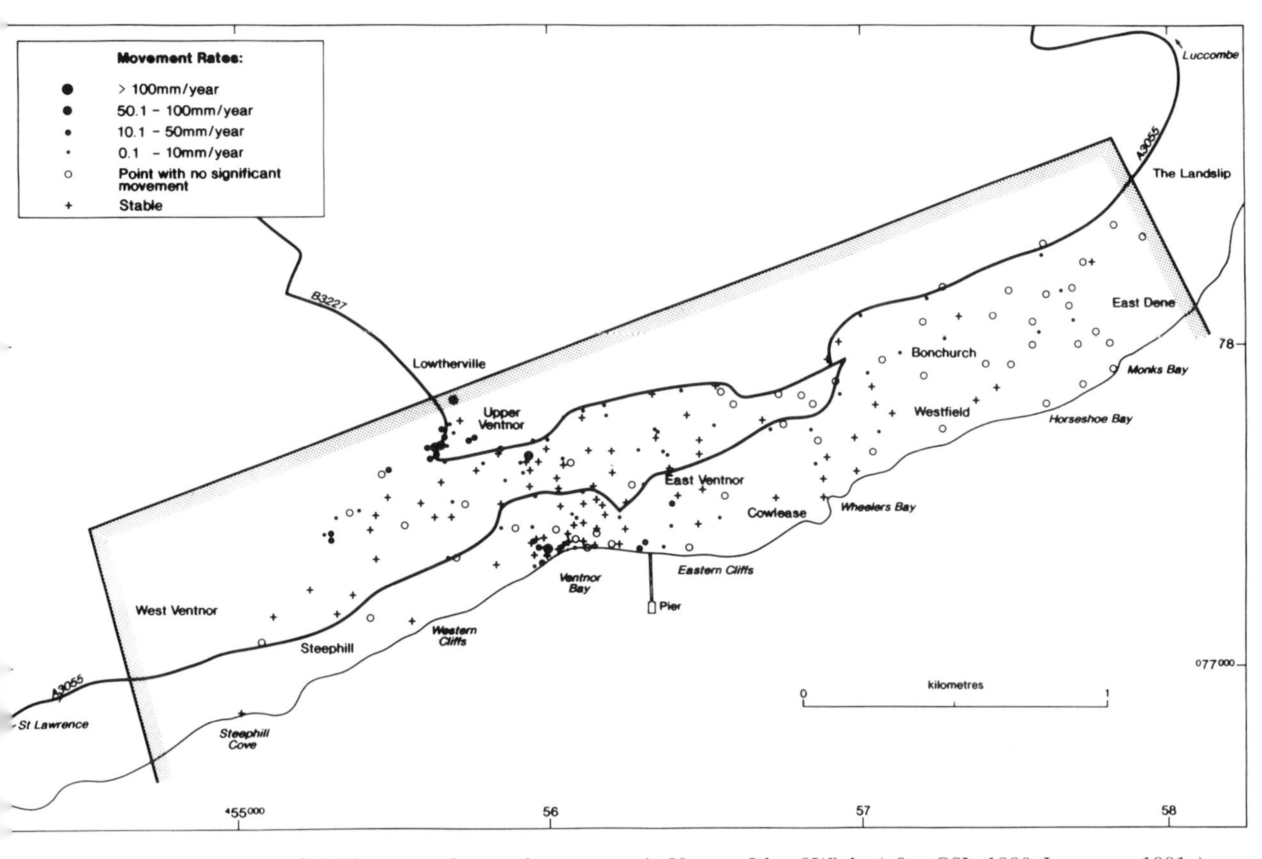

7.3 The rates of ground movement in Ventor, Isle of Wight (after GSL, 1990; Lee et at., 1991a).

Without a doubt, the most extensive coastal landslide problem in Great Britain is at Ventnor, also on the Isle of Wight, where the whole town has been built on an ancient landslide complex. Contemporary movements within the town have been slight *(Figure 7.3)*, in comparison with those at nearby Luccombe. Whilst there are a number of areas which experience, on average, over 10mm of annual movement, many parts of the town appear to have remained 'stable' or have been affected by imperceptible movement of less than 1mm a year. However, because movement occurs in an urban area with a permanent population of over 6,000, the cumulative damage to roads, buildings and services has been substantial. Over the last 100 years about 50 houses and hotels have had to be demolished because of ground movement. In addition, from a consideration of the accumulated damage, including an assessment for repeated road repairs by the Isle of Wight County Council, the total losses in the Ventnor area during the 20 years prior to 1980 have been estimated as exceeding £1.5 million[13] *(Plates 37 and 38; Photo 7.1)*. The scale of the difficulties experienced within the town, together with the innovative nature of the methods currently being employed to combat the problems[14], require that the example of Ventnor be discussed at rather greater length.

The problems of ground movement have long been recognised within Ventnor. Evidence given by Aubrey Strahan of the Geological Survey, at the Royal Commission on Coast Erosion and Afforestation in 1906, highlighted the slow creep experienced in many parts of the town: 'observations made upon houses, sewers and the flights of steps, ... shows that the mass is still subject to slow intermittent movement'. The view held at that time was that the problem of recurrent movement was directly related to coastal erosion. Indeed, attempts in the 1860s to build a harbour in Ventnor Bay by removing a rocky headland (Collin's Point), resulted in beach depletion and increased coastal erosion, causing slippage along the coast *(Chapter 6)*.

7.1 A series of deep fissures (vents) which opened up during 1954, Ventnor, Isle of Wight (Isle of Wight County Press).

In 1954, deep fissuring along Whitwell Road, inland of the rear scarp of the Undercliff, gave an indication that the problems may be more complex than previously perceived *(Photo 7.1)*. Ventnor Urban District Council asked Edmunds and Bisson of the Geological Survey to inspect these fissures (or vents) and to assess their significance. In a confidential report, they suggested that the vents may be the initial stage of the development of a major landslide, although such a failure has not occurred in the succeeding years. At the same time, Toms of British Railways was asked to view the site and he recommended that a comprehensive monitoring programme should be established within the town in order to determine the scale of the problem. Following Toms' recommendations, the Local Authority has collected records of structural damage within Ventnor since 1954.

Further dramatic movements occurred during the winter of 1960–1961 following extremely high autumn rainfall totals, when many cliff falls, collapsed walls and subsidences were reported in November and December 1960. Cracks appeared along Bath Road (to the west of Ventnor Bay; *Figure 7.4*), with the road surface dropping several centimetres: the most rapid movement observed was c. 2cm per day for seven days, contributing to an eventual 30cm drop in level. Damage was caused to a number of properties and further movements in late December resulted in these properties being temporarily evacuated by Ventnor Council, with help from the Air Ministry. Along the Esplanade, west of the pier *(Figure 7.4)*, a number of hotels were damaged by heave, two of which were declared unsafe and some of the occupants evacuated. Major subsidence also occurred in Upper Ventnor and many houses and retaining walls were damaged.

People affected by the movements were advised by the local council to apply for financial relief to the Mayor of Newport's Flood Relief Fund. Approaches were also made by Ventnor Council and the local Member of Parliament to the Ministry of Housing and Local Government, who offered financial assistance to private persons affected. This offer of assistance was described by the Ministry as neither "an acceptance of a continuing liability to make good any further damage which might occur nor a guarantee that the properties could or would be made stable"[15].

Affected parties were required to apply for a grant, but before any substantial amount could be spent, it was necessary to establish whether the repairs/improvements were worthwhile. Grants were available for the full cost of restoration of damaged homes; damage to or loss of furniture or clothing up to £300 a household; damage to or loss of stock in trade of a personal business and damage to business premises up to £2,500. No grants in any circumstances were to be paid in respect of losses or damage covered by insurance. A total of 115 claims were received

7.4 The locations of recorded landslide events in Ventnor since 1800 (after GSL, 1990; Lee et al., 1991a).

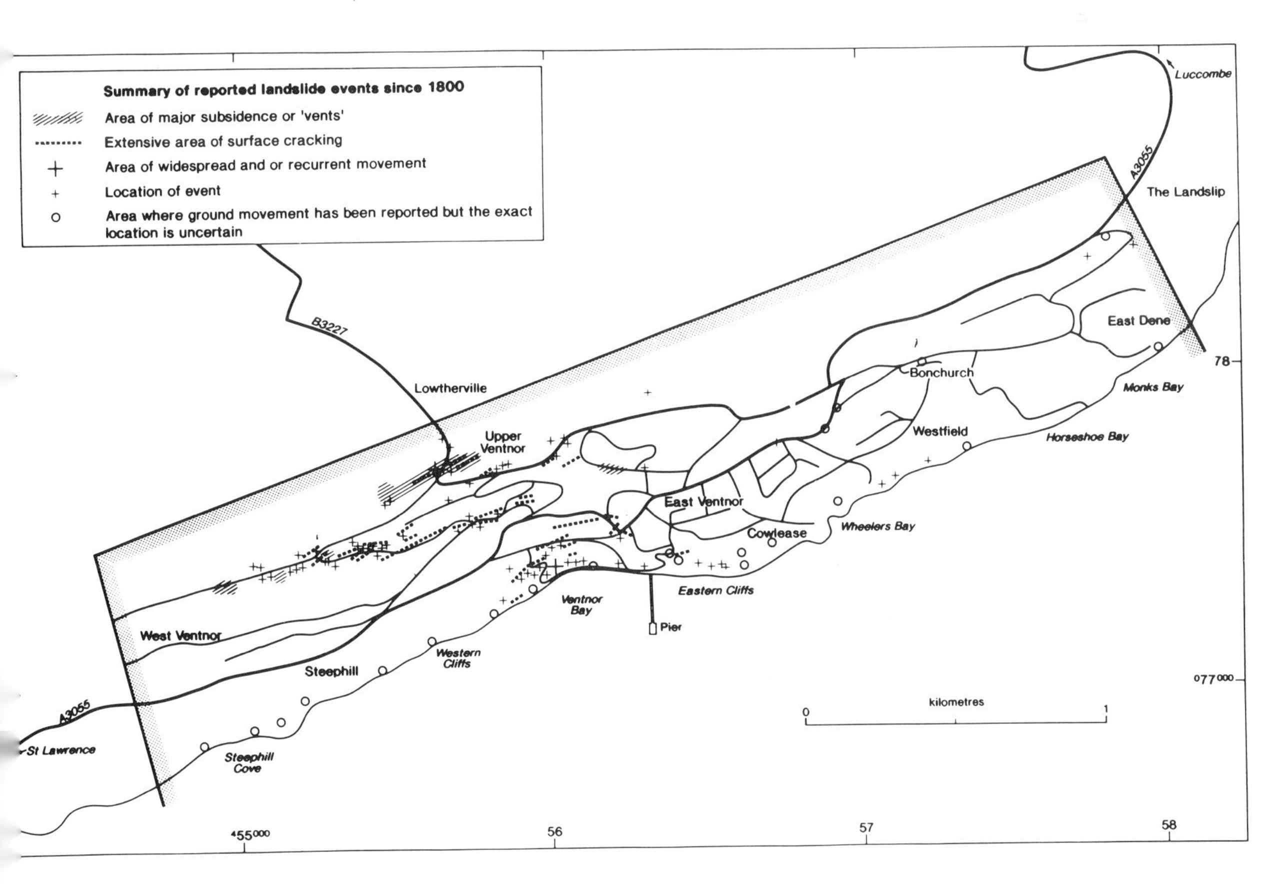

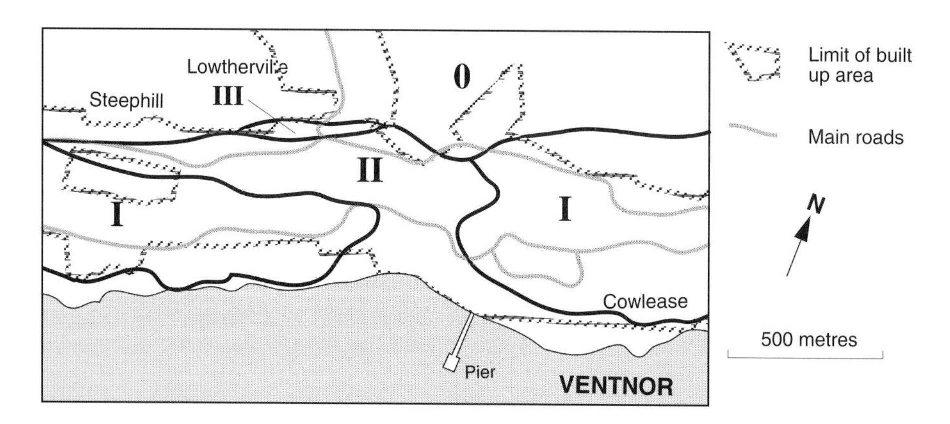

ZONE	RELATIVE PROBABILITY	DESCRIPTION
0	Negligible	Zone landward of rear scarp of pre-existing landslides, probably affected by ancient slight movements.
I	Low	Zone seaward of rear scarp of pre-existing landslides, either apparently free of slope movements, or exhibiting some of the conditions for intermittent instability. Includes renewals of movement in the old slide masses, involving slight relative subsidence or heaving and the opening of fissures and cracks.
II	Moderate	Zone affected occasionally by movements of small to moderate amplitude; or generally affected by diffuse movements; or exhibiting many of the conditions for intermittent instability. Includes minor rockfalls, gentle renewals of movement in the old deep-seated slides, and possible shallow sliding on coastal cliffs where sea defences are delapidated or absent.
III	High	Zone affected periodically by movements of moderate to large amplitude; or affected frequently by movements of small to moderate amplitude; or exhibiting many of the conditions for general instability. Comprises large subsidences and openings at fissures caused by continuing deep-seated renewals of movement in the pre-existing slide masses.

7.5 Preliminary landslide hazard map of Ventnor, Isle of Wight (after Chandler, M.P. & Hutchinson, 1984).

of an overall value at the time of £78,000, of which around £69,000 was authorised for payment.

The potential scale of the problems faced by this urban area stimulated the production of a preliminary landslide hazard assessment *(Figure 7.5)* which attempted to divide Ventnor into zones reflecting the likelihood and nature of future ground movements[13]. The highest hazard was deemed to be associated with the comparatively recent development of a graben-like feature in Upper Ventnor, which has resulted in severe subsidence to the main Newport Road and rendered a number of houses unsafe *(Plates 37 and 38)*. The recent movements in the graben area, have heightened public awareness of the potential threat and created insurance problems for many householders, as was described in an article in the Isle of Wight County Press (15.7.1988):

> 'Just ten years ago subsidence was not a major issue and most mortgages and the subsequent building insurance cover was arranged without much thought given to it. However, an increasing number of claims were made in the late 1970s and early 1980s and so, rightly, insurance companies began to give the subject more thought. Unfortunately for many of us, some companies have reacted by pulling out of the Island, and, although they might still be covering some properties, they will not extend that cover if the occupant decides to sell up, obviously leaving that householder with a major problem on his hands.

> Some have taken a more considered view and although they are aware of risks in certain areas, they will look at each case on its merits, sometimes insisting on a full structural survey themselves, and then make a final decision. Yet others fall between the two extremes and while they operate a no-go policy in certain areas, do not consider most parts of the island to be at risk. Commercial Union admit they operate a selective policy, refusing to cover property in certain areas although they would not disclose which areas and how extensive they were. The policy has been formulated in the last ten years with no-go areas decided by the number of claims they have paid out in that time'

Local concern about landslide problems faced within Ventnor not only include the risk of structural damage to property and difficulties in arranging insurance cover, but also many fear for the safety of local inhabitants. This fear was dramatically portrayed in an article in the Isle of Wight County Press in the summer of 1987, entitled "Isle of Wight landslide could kill". In this article it was argued that to evacuate a town of 6,000 people would be prohibitively expensive but might be required if the movements accelerated, and included the statement "most people know it will move but are prepared to live with the risk".

Although the pattern of displacement established for the landslide complex over the last 200 years *(Figure 7.4)* fails to provide a precedent for this headline, it does serve to highlight the need to carefully assess the significance of reported movements in the context of the known behaviour patterns of the landslide system. This can only be achieved by an improved understanding of both the nature and causes of the hazard. The article also highlights the current scientific view that hazards, such as landsliding, can no longer be regarded as unusual phenonema or 'Acts of God', but should be considered as explicable and predictable events that carry costs, both to individuals and communities[16]. With this changing appreciation of landslide hazard comes a greater responsibility for relevant authorities and interested parties to respond to such matters.

In the past there appears to have been an ad hoc response to specific landslide events in Ventnor, concentrating on repairing buildings, condemning properties and emergency action. Such crisis management responses 'after the event' are common reactions to infrequent problems throughout the world (e.g. the technical response to the Catak landslide in Turkey in June 1988[17]). However, in areas such as Ventnor, where ground movements are a recurrent problem, there is a clear need for a coherent and systematic strategy for landslide hazard amelioration. This need was recognised by the Department of the Environment which, in 1988, commissioned a detailed investigation of the problems. This study was undertaken by Geomorphological Services Limited and completed in 1991 at a cost of £250,000. The approach adopted, the techniques used and the products of this investigation, are all fully described in Chapter 8.

Inland Landsliding

The proceeding discussion on the problems of coastal landsliding has focussed attention on the fact that numbers of fatalities are a very poor measure of hazard impact. Better appreciations can only be achieved by attempting to measure the costs of impacts in financial terms. Such measures should include the direct and indirect cost of each impact (i.e. costs of repair as against cost, of delay or disruption), the costs of defensive measures (stabilisation schemes, retaining walls, sea defences, provision of insurance) and the costs of investigation and decision making *(Table 7.1)*. Many of these are difficult to obtain and some are equally difficult to value. For example, what is the value of one life, or delay, or anxiety. And of equal importance are the questions concerning who actually pays and the extent to which they can bear thc costs. Thc lack of such data clearly restricts the extent to which the significance of inland landsliding can be assessed. The review that follows must, therefore, depend on freely published "conspicuous failures" and "obvious costs" while recognising that the actual significance is greater and the consequences more widespread.

Table 7.1 **A range of costs commonly associated with landslide problems in Great Britain.**

Personal Costs;	Fatal accidents; Injuries; Psychiatric problems.
Immediate Costs;	Evacuation and provision of temporary or replacement housing; Mobilisation of relief workers and emergency services; Transport delays; Costs of investigation; Cost of repair.
Indirect Costs;	Compensation; Increased insurance premiums; Depreciated property or land values; Costs of legal actions; Costs of public inquiries into causes and responsibilities.
Cost of Prevention;	Research into the nature and extent of landslide problems at Universities; Formation of planning policies related to development on unstable land; Coastal protection schemes; Design and construction of preventative measures including drainage and regrading; Costs of monitoring potentially unstable slopes.

In general, the numerous degraded ancient slides which occur in many parts of inland Britain present only a minor threat to life, as movements, when they occur, usually involve only slow, minor displacements. Even when large displacements occur, the rate of movement tends to be gentle and not dramatic, as was reported graphically for a slide near Lympne, Kent, in 1725 where a farmhouse sank 10–15m overnight, "so gently that the farmer's family were ignorant of it in the morning when they rose, and only discovered it by the door-eaves, which were so jammed as not to admit the door to open"[18].

However, sudden dramatic failures can occur inland and may lead to extensive damage. For example, the numerous debris slides and flows triggered by severe thunderstorms in the Scottish Highlands in May 1953 caused widespread damage to forestry lands in the area[19]. Considerable acreages of recent plantings were lost and in many places older trees needed rebrashing because of the deposited debris. Great lengths of fencing to keep deer and stock out of the plantations were damaged, and had to be replaced. Damage to roads and interference to traffic was inevitable, as all the main lines of communication lie along the floors of the narrow, steep-sided valleys. In Argylshire, damage to both major and minor roads amounted to over £130,000 at current values, and made considerable claims upon labour from the Roads Department, whilst in Invernesshire many minor roads were blocked for a week. Similar problems followed the debris flows which occurred in Snowdonia during September 1983, blocking the A5 to the south of Bethesda[20]. Emergency site clearance cost Gwynedd County Council almost £10,000, and subsequent remedial works a further £46,000, with the costs of short-term improvements undertaken to reduce the debris flow hazard estimated at £90,000.

These examples merely indicate the order of costs incurred where debris falls onto or blocks a road. Sometimes the mere threat of falling material is enough to close a road, as happened in the case of Portway through the Avon Gorge at Bristol which was declared unsafe in 1974. But roads are also disturbed, deformed and sometimes cut by movements within the slope, thereby necessitating higher levels of maintenance, repair, rebuilding and even relocation. No comprehensive data is as yet available on the true levels of these costs, but it is well known that numerous long-established major roads cross areas of unstable ground, and, as a result, suffer repeated damage: examples are widespread and include the A855 Portree to Staffin Road on Skye which crosses the massive Storr landslide; the A46 in the Swainswick area of Bath; the A38 at Chudleigh, Devon, which was built in 1971 across an ancient slide; most of the roads that cross the Cotswold escarpment, including the A44 at Broadway Hill; the various roads that traverse the Lower Greensand escarpment in the Weald, including the A286 near Fernhurst and the B2042 near Ide Hill; and many roads in West Dorset including the A35 near Charmouth. It was local instability on the last example that led to a detailed study which revealed how local deformation could be explained by relating the specific **site(s)** of existing movement to the general **situation** viz a viz ancient landslide distribution over the slope as a whole[21]. The widespread occurrence of ancient landsliding *(see Chapter 6)* indicates great potential for disturbances to the road network, and problems can only increase in the future due to heavier traffic volumes, increased axle loadings and growing numbers of road widening schemes.

Probably the most acute road stability problem developed progressively at Mam Tor in the High Peak, Derbyshire, eventually resulting in closure of the A625 *(Figure 7.6, Plate 39)*. The A625 was one of only three cross-Pennine routes in Derbyshire and therefore functioned as an important link between the cities of Manchester and Sheffield. The road had been built in 1802 as an alternative route to the narrow, steep, Winnats Pass. However, the new road was inadvertently constructed across one of the largest landslides in the High Peak with the result that it has been repeatedly

7.6 The Mam Tor landslide, Derbyshire

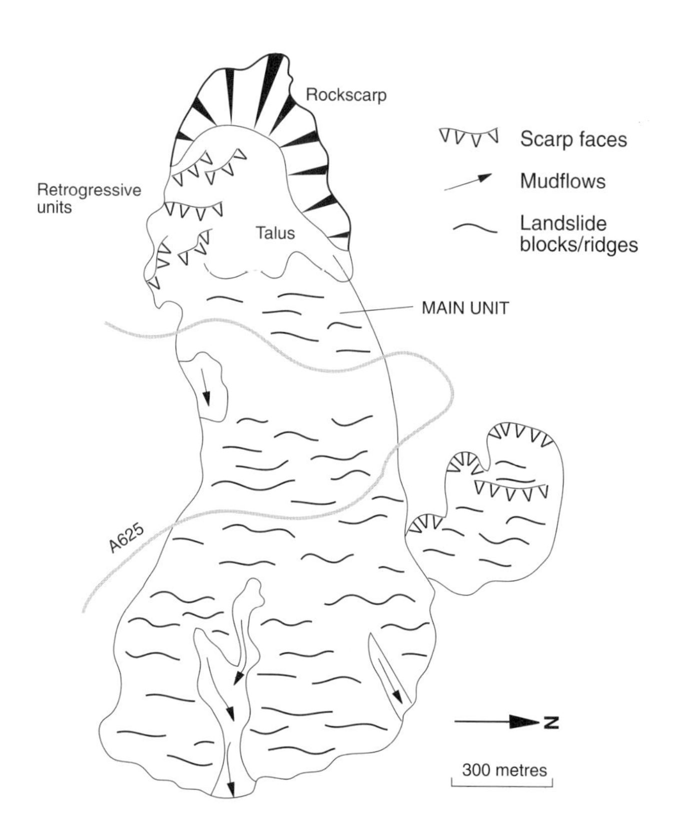

affected by cracking and settlement[22] *(Table 7.2)*. Maintenance costs were always high as work had to be carried out on the slide every year. Despite the problems, Derbyshire County Council attempted to keep the road open, at considerable expense (e.g. £13,500 on major works in 1950; *Table 7.2*), rather than adopt an alternative route. However, in February 1977 the road was closed by a large

Table 7.2 **Summary of recorded landslide damage to the A625 at Mam Tor, Derbyshire** (*after Priestley, undated; Skempton et al., 1989*).

1909	Road cracked and repaired.
1912	Road badly cracked and sinking , later repaired.
1915	Fractures and settlements of up to 30cm.
1918	Subsidence to road.
1919	Continued movement.
1920	Fractures and holes, later repaired.
1929	Road affected by movement; alternative routes suggested.
1930	Continued movement.
1931	60m long crack on road.
1937	Considerable settlement.
1939	100m long crack, 25cm subsidence.
1942	Settlement of up to 20cm.
1946	Major road works needed, including realignment of the road, costing £8000.
1948	Renewed subsidence.
1949	Slip recorded (no details).
1950	Road damaged by movement, major road works carried out costing £13,500.
1955	Large scale movement requiring £20,000 of repairs.
1965	Road closed after movement, later repaired and reopened.
1977	Road closed due to movement.
1978	Road reopened for single lane traffic.
1979	Final closure of the road, traffic diverted.

movement, and even after extensive repair works the route could only be opened to single-lane traffic. The road was closed permanently in January 1979 when a section slumped down by 2m. Owing to the high expenditure on earlier unsuccessful repairs, Derbyshire County Council decided not to undertake any further remedial works until the problem could be solved on a long-term basis. Heavy traffic was diverted via the B6049 and B623 passing through Brough, Bradwell, Peak Forest and Sparrow Pit, causing considerable congestion on the narrow roads *(Figure 7.7)*. Light vehicles were allowed to use the Winnats Pass road, despite opposition from the National Trust because it is a Conservation Area unsuitable for busy traffic. The cost of constructing a new route via Pindale and Dirtlow *(Figure 7.7)* was estimated at £2.25–2.5 million at late 1970s prices. No such route has been built to date.

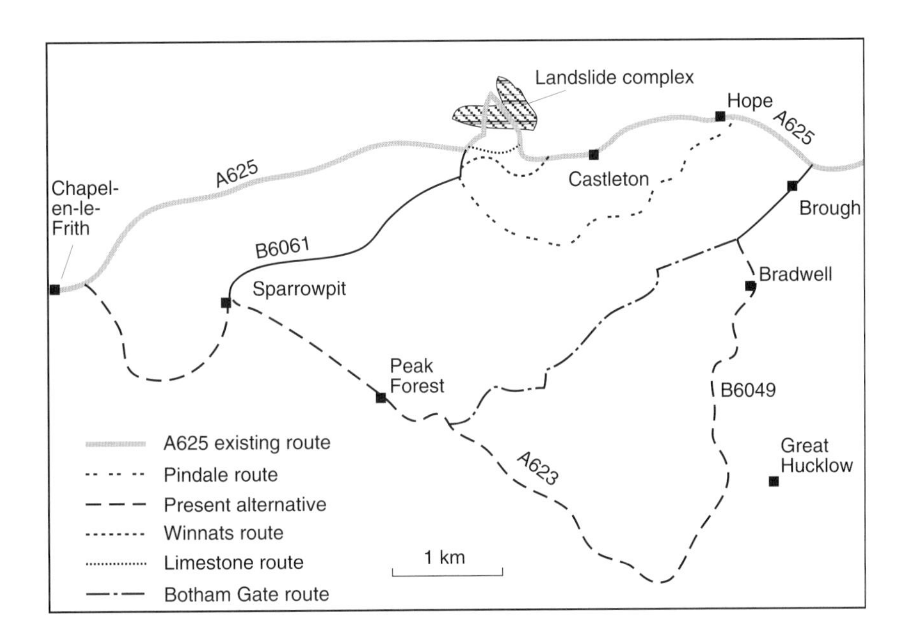

.7 Possible alternative cross-Pennine
outes avoiding the Mam Tor landslide
Derbyshire.

Even the new road building programme since the early 1960s has been affected by instability problems. The classic examples, which include the A21 Sevenoaks Bypass (1964–1966), the M6 at Walton's Wood, Staffordshire (1961–62) and the M4 near Swindon (1969), have already been described in earlier sections. The actual costs of these problems has never been publicised, although £600,000 for remedial works is quoted for Walton's Wood.

These problems encountered during construction naturally stimulated research into improved methods of mapping, sub-surface exploration and geotechnical investigation with a view to reducing the chances of future repetitions. All new road schemes now employ such methods, so the mere threat of slope instability has increased the cost of road construction by a small amount per kilometre. Herein lies a good example of the costs of avoidance. Where instability is known to be extremely problematic, or ancient landsliding is recognised to be extensive, the costs of such measures may become significant. For example, the necessity for the A470 Taff Vale Trunk Road to cross the toe of the extremely large, menacing Taren Landslide *(see Figure 2.10)* required nearly £3 million worth of investigation and remedial measures before the slope was stabilised sufficiently to allow the road to be built.

Roads are not the only form of infrastructure to be affected by slope instability. The railway network has had its fair share of failures both during construction and subsequently. Construction of the Stromeferry Bypass in the Scottish Highlands triggered a landslide in May 1969 which blocked the adjacent railway line for five months[23] *(Plate 25)*. The remote rural railway line through Glen Ogle in the Scottish Highlands was actually closed in 1965 following a major failure *(Plate 19)*. However, arguably the most dramatic example of failure affecting railway operations is Folkestone Warren, where the massive rotational movements of 1915 moved portions of the Folkestone-Dover railway line seawards by up to 50m *(Photo 2.3)* and the line has subsequently been kept open only through major investment in structural stabilisation methods and heavy maintenance *(Figure 4.16)*.

Other elements of infrastructure at risk include buried telecommunications, power lines, gas pipelines, sewers and water-supply networks. All of these are undoubtedly affected but costs are not available. Perhaps the most vulnerable is water supply infrastructure, because old water pipes are liable to fracture through movements and once fractured, provide a basic ingredient for further movements. Pipeline leakage is a major component of water demand, sometimes reaching 30–40% of the total supply. To what extent this is due to ground movements remains unknown but the potential certainly exists. Slow slope movements are known to have deformed the trunk supply pipeline on Bredon Hill, Hereford and Worcester, which had to be abandoned in the early 1980s following investigations costing about £30,000. Similarly, a proposed reservoir at Cowleigh, Malvern, had to be abandoned in 1978 when excavations reactivated a major landslide, causing disturbance to a gas-supply regulator. The 1.5 million gallon service reservoir project had to be written off at a cost of £250,000[24]. As the total cost of the scheme had been estimated at £350,000, the significance of landslide reactivation becomes clear.

While infrastructure may be especially prone to landslide problems, it is impact on buildings, and especially housing, that tends to cause the greatest levels of public concern. High cliffs are considered particularly threatening for obvious reasons. The 50m high overhang of Bwlch y Gwynt at Blaenau Ffestiniog is a focus of local concern although stabilisation is beyond the town's resources. The same was not the case in Dover where the vertical cliffs on the eastern side of the town were scaled in the mid-1960s and loose flints and lumps of Chalk removed so that they would not fall and cause death, injury or damage. But it is sliding which tends to be the major problem in many urban areas. This often arises when a development is affected by instability adjacent to the site. For example, the Cliff landslide on the banks of the River Irwell in Salford has caused recurrent damage to the road and property immediately upslope over the last 100 years[25]. The first reported movements occurred in February 1882 when a large slide was recorded by the City Engineers Department, whose response was to tip fill at the head of the slope! By May of the same year, cracks had appeared in the road immediately upslope of the slide, and subsequently a drainage scheme was installed. Continued ground movement problems resulted in the tram service along the road above the slide being discontinued in 1925, and a section of the road closed to motor traffic in 1926. Further movements have occurred intermittently and have caused damage to a row of terraced houses at the head of the slide. Of interest, these houses were subject to a special clause in the Salford Corporation Act 1933 which permitted the City Engineer to evict without notice if the houses became in imminent danger of collapse. Although the terrace remains standing today, it is now only 10m away from the slope crest.

Damage to property, or additional costs of development, also arise when the development itself initiates slope instability problems which can ultimately threaten a wider area. The Hedgemead landslide in Bath was caused by leaking sewers and water pipes installed when the area was developed between 1860–1875. Over 2.5ha of land were affected by the periodic movements and at least 135 houses were destroyed or damaged[26]. In Devon, landslide problems have recently been encountered in Tiverton and Exeter as housing development have extended onto steep valley slopes[27]. At the Exwick Farm housing development on the outskirts of Exeter, excavation of a road cutting reactivated a relatively deep landslide which required expensive stabilisation measures, including drainage and supporting the houses on reinforced concrete piles sunk into undisturbed bedrock. There are several known examples of building developments currently being affected by unforeseen stability problems, but the costs of such events are, quite understandably, not publicised.

Perhaps the fullest documented range of urban landslide problems have been identified in South Wales, and it is from here that some measure of costs can be obtained. One of the areas where landslides have caused major problems is the Ebbw Fach Valley at Blaina, where a number of slides have threatened properties *(Figure 2.11)*. The East Pentwyn slide developed in January 1954 and by March of that year the rapidly advancing toe area had reached within 9m of a row of twelve cottages. The cottages were evacuated and later demolished, and fears were expressed by consultants that the nearby council properties were also at risk. Surface drainage was installed and a drainage blanket placed in front of the advancing slide. After a period of relative inactivity there was a major reactivation of the landslide in October 1980. The local authorities commissioned Sir William Halcrow and Partners to investigate the problem, at a cost of £150,000[28], who suggested that stabilisation may be possible but at an estimated cost of over £1 million. A total of 28 houses were identified as facing immediate risk from the slide, and Blaenau Gwent Borough Council took responsibility for helping to rehouse the occupants. At the same time, warning and emergency evacuation plans were agreed with the police, and extensive monitoring was carried out[29].

The scale, extent and media profile of landslide problems in South Wales led the Department of the Environment, through the Welsh Office, to fund the South Wales Coalfield Landslip Survey[30] which was essentially an inventory of 585 landslides identified in the valleys. These landslides were classified and presented in a series of five, 1:50,000 scale maps, together with information on their types and activity *(see Chapter 8)*. Such basic data gathering exercises are essential, for sound decision-making can only be undertaken if there is adequate, accurate information to base decisions upon.

7.8 The pattern of landsliding in the Rhonda valleys, South Wales (after Halcrow, 1986).

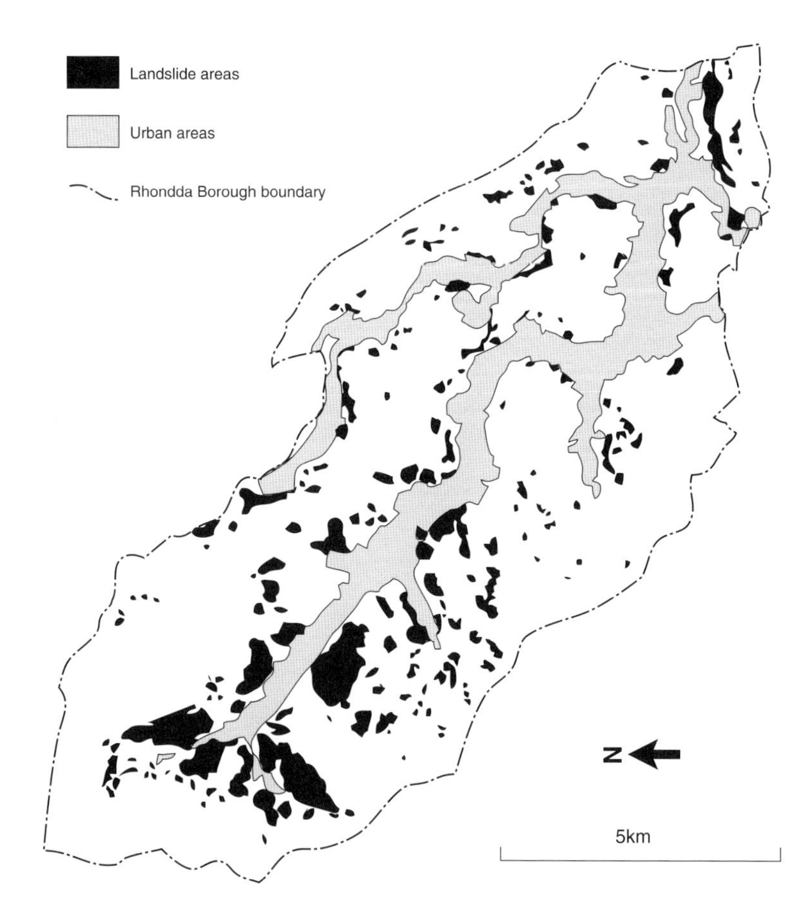

7.9 Common causes of landslides in the Rhondda Valleys (after Halcrow, 1986).

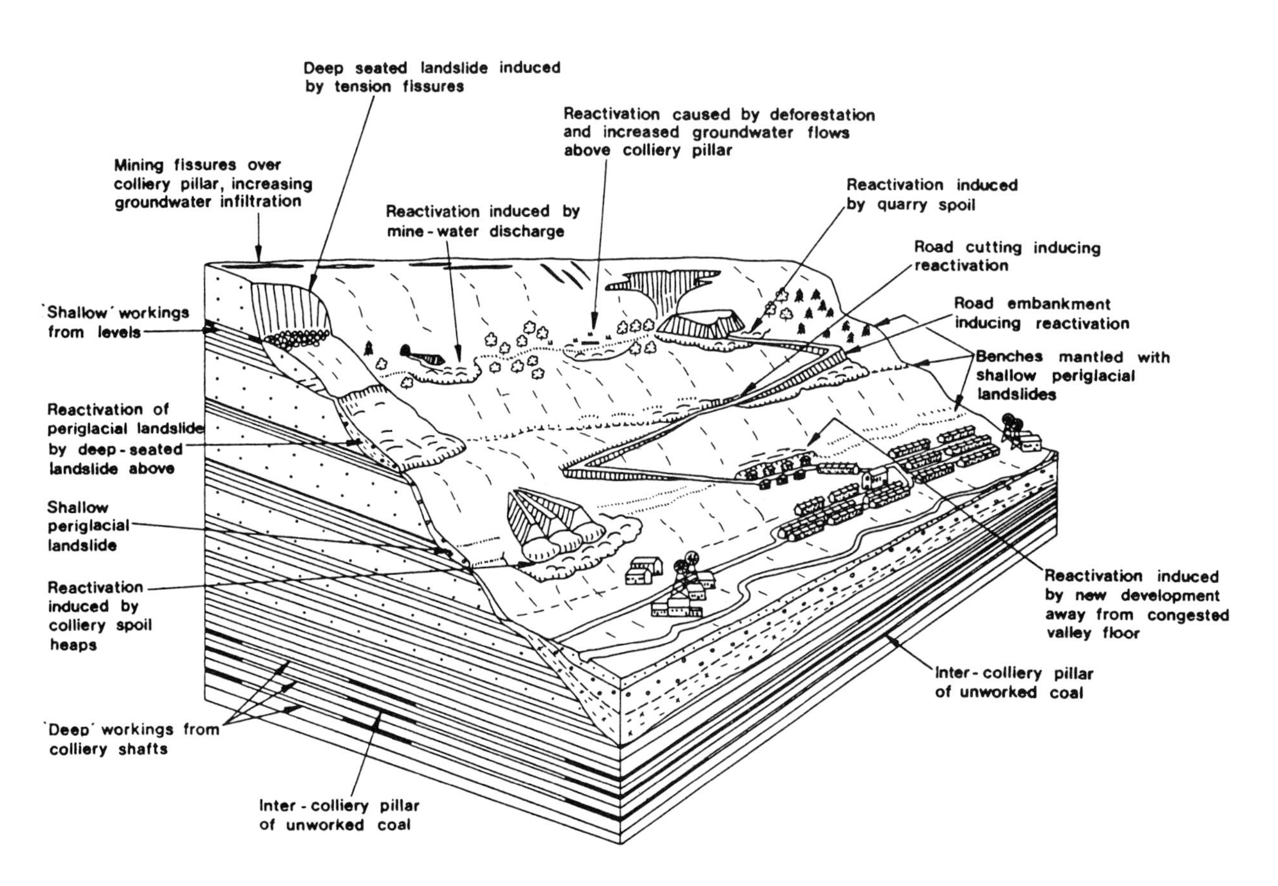

The natural progression from the South Wales Coalfield Landslip Survey was a detailed investigation of a single area of concentrated landsliding, with a view to establishing a methodology by which landslide hazard assessment could be incorporated in land use planning. Rhondda Borough was chosen and the DOE subsequently commissioned Sir William Halcrow & Partners to undertake a Landslip Potential survey in the mid-1980s. As described in Chapter 2 (*see The DOE Landslide Survey*), this detailed survey[31] added considerably to known numbers of instability features, by recording no fewer than 346 slides covering 915ha, or 9% of the Borough *(Figure 7.8)*, and ranging in size from 0.3ha to 65ha. The overwhelming majority are reported to be shallow translational failures. Many of the instability features appear to have ancient origins but have suffered reactivation during the last 120 years or so, due to human activities associated with industrialisation and urbanisation *(Figure 7.9)*. A detailed review of **recorded** damaging landslide events over the last 100 years (from newspaper sources, etc) revealed no fewer than 82 separate events *(Table 7.3)*, including:

(i) a flowslide in spoil at Pentre in 1909 which destroyed 5 houses and killed one person (the Pentre Disaster, Rhondda Leader, 13 February 1909);

(ii) the dumping of spoil at Pentre reactivated an old translational slide in 1916, destroying a billiard hall and several houses;

(iii) houses in Pont-y-gwaith were affected by a slide in 1961, an event which was described in the South Wales Echo (6.1.1961) as "the mountains slipped forward and tons of liquid yellow clay slid into the houses";

(iv) a landslide at Penrhiwgwynt, Porth in 1965, caused by disposal of waste from new water tanks onto an old landslide, threatened properties and had to be treated at a cost of c.£80,000 at prevailing prices.

Table 7.3 **Recorded damaging landslide events in the Rhondda Valleys.** *(after Halcrow, 1988)*

Damage Class	Description	No. of Events
A.	Failures where the estimated present-day cost of investigation, stabilisation and repair lies between £100,000 and £3.5M. Typically, these might include landslips which damage more than two houses.	26
B.	Failures where the estimated present-day cost of stabilisation and repair lies between £5,000 and £100,000. Typically one or two houses are affected.	24
C.	Minor failures, treated by expenditure of less than £5,000. Typically these include events where only clearance of material is necessitated or repairs to retaining walls.	32
		82

Analysis of these failures over time reveals a very unusual distribution *(Figure 7.10)* with the greatest numbers of occurrences in the 1960s and 1980s. This may, in part,

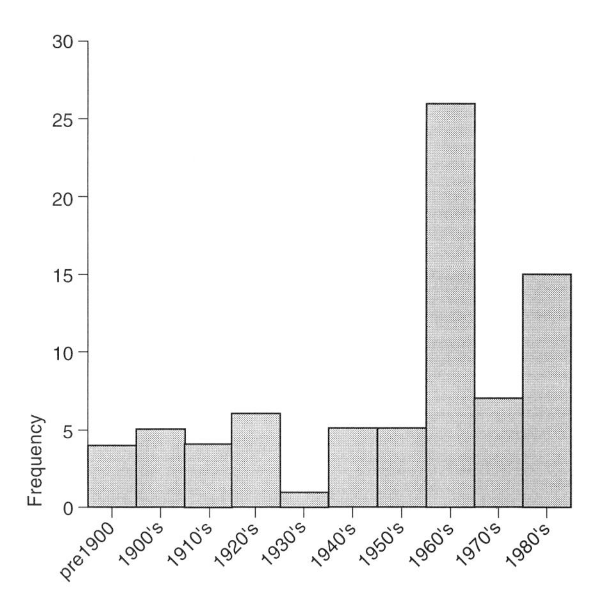

7.10 The frequency of recorded damaging landslides events in Rhondda Borough (compiled from Halcrow, 1988).

reflect the greater awareness of landsliding in recent decades but could also be the result of mining industry problems and practices in the 1930s and 1940s.

The heavy emphasis on landslide hazard assessments within the South Wales Coalfield is undoubtedly due to the fact that the most horrific landslide tragedy to befall Great Britain also occurred in this area. The Aberfan disaster of 21 October 1966 occurred when a colliery spoil tip failed at around 9.15 a.m. and the debris engulfed part of the village below[32]. The Pantglas Primary School was partially buried and 107 out of 250 children were killed together with 5 teachers. Rescue of surviving children by parents and locals began almost immediately, but no child came out alive after 10.30 a.m. that day, and bodies were still being recovered several days later. In total, 116 children and 28 adults were killed by the flowslide, and many people suffered severe emotional problems after the disaster.

In the first four years after the disaster, Gaynor Lacey, a consultant psychiatrist at Merthyr Tydfil Child Guidance Centre, saw 56 children who had developed behavioural problems since the disaster[33]. The most common problems were sleeping difficulties, nervousness, insecurity, enuresis and unwillingness to go to school. Many children lost all their friends, so there was little point in going out, and once having stayed in, it was difficult to start going out again. Educationally, the development of many children appeared to stand still for a while, until they began to overcome the trauma. However, it is likely that many of the Aberfan children will always be affected by the tragedy, and some may experience psychiatric problems in later life. The sense of guilt and bereavement which affected the whole community, was initially directed as aggression towards the Coal Board, the Local Authority and the Government. Intense feelings were caused by the Government's apparent unwillingness to promise to remove the tip, which remained a constant source of fear to many people. These feelings only began to abate once the tip was moved entirely. However, parents of the children killed in the disaster will never get over their loss.

Even before the Aberfan disaster, landsliding had long been recognised as a hazard in the South Wales valleys. In 1927 Knox presented an important review of the causes and impact of landslides such as the failures at Bournville, Blaina and Pentre which were, at least in part, related to human activity[34]. There are now known to be numerous relict slope failures on the steep valley slopes, many of which are either intermittently active or marginally unstable and present a significant hazard to the communities *(Figures 7.8 and 7.9)*. However, it took the Aberfan disaster to promote the investigation of tips and the hillsides, and raise the whole issue of instability in the minds of those responsible for the welfare of the area, even if not in all the inhabitants themselves[35]. Further, it took the horror of 21 October 1966 to convince various authorities of the dangers posed by conical spoil heaps and that there were better, safer and visually less intrusive methods of spoil disposal. Thus, the National Coal Board and the Welsh Development Agency subsequently tackled the problem of unsafe tips through the expenditure of at least £50 million on regrading. This activity was carried out throughout Britain's coalfields and must be considered to have resulted in considerable environmental improvement.

Discussion of colliery spoil heaps leads naturally on to consideration of failures in other artificial mounds and embankments. The huge numbers of such features indicates that failures have probably been numerous but mostly fairly small or failed to result in significant impact. However, some have been problematic.

The safety of china clay spoil heaps in Cornwall has recently been the subject of public concern after two large slides during February 1990. The first occurred at Fraddon Downs, near St Dennis, when 10,000m^3 of spoil flowed 70m across a road and partially buried a pensioner's house[36]. A second slide took place within a week, this time at the Penrose tip near St Austell[37]. No-one was injured by either slide, but the events were particularly worrying as English China Clay International had never previously experienced any major failure of their spoil heaps. Indeed, since the Aberfan disaster there had been strict control on the design and inspection of all waste tips, under the Mines and Quarries (Tips) Acts of 1969 and 1971. Local fears were calmed by the subsequent investigation which suggested that the St Dennis failure was caused by blocked drainage and not because of any inherent design faults which may be applicable to other tips. However, the Health and Safety Executive recommended that there should be a full and urgent assessment of all waste tips in Cornwall and Devon, especially those located within 100m of property and essential services.

In February 1992 a 21m high sludge tip failed at Yorkshire Water Services's Deighton sewage works, near Huddersfield[38]. An estimated 100,000m^3 of sewage sludge and chemical waste, from the neighbouring ICI plant, slipped 200m, destroying a pump house and putting 5 out of 8 treatment tanks out of use. The slide also blocked the river Colne over a 150–200m length, raising fears of flooding which were abated by digging a channel to divert the river around the slide. This failure has raised concerns about the stability of sludge tips, which are currently not covered by safety legislation.

Attention has also been focussed, in recent years, on the design and construction of embankment dams, following the failure of the Carsington Dam in June 1984. The dam had been designed to provide a storage reservoir for water pumped over 10km from the River Derwent in Derbyshire. The whole scheme involved a 1,250m long earthfill dam and a tunnel and pipeline aqueduct to the Derwent, at an estimated cost of £36 million (at 1980 prices)[39]. Construction of the dam took place between 1982 and 1984, and was almost complete when a slide occurred on the upstream face. Despite efforts to stop the failure by toe weighting, a major slide developed resulting in up to 10m of settlement of the dam crest and toe heave of around 2.5m. The slide was not entirely unexpected, as doubts had been roused about the stability of the dam during construction[40]. Indeed, in November 1983, the contractor had produced a report which said the dam could fail and that redesign was essential.

Detailed investigations into the cause and responsibility for the failure were carried out by Severn Trent Water Authority and the Department of the Environment, both of whom concluded that the collapse was the result of progressive failure *(see Chapter 4)*. This view was disputed by the designer who considered that the failure was due to an increase in the height of the dam, which had caused unforeseen overstressing, possibly in combination with the triggering effect of a small earth tremor or high pore-water pressures due to heavy rainfall. The Water Authority brought a legal action against the designer, alleging both breach of contract and negligence. The claim was expected to be around £50 million, but was settled out of Court with the designer paying out £3.25 million to the Water Authority[41]. However, no responsibility was accepted and the sum merely represented the limit of the designer's professional indemnity insurance cover. In February 1989, construction recommenced on the redesigned dam at a cost of £17.8 million. The scheduled completion of the whole scheme in 1992 represents a delay of eight years, with the costs rising from £36 million to £96 million, very largely a direct result of the dam failure.

The Carsington failure is merely the latest in a long history of problems associated with the construction of embankment dams. Earlier examples include the failures of the Muirhead dam in 1941, and of the Chingford dam in 1937[42]. However, the most disastrous failure occurred in Yorkshire during March 1864 when the Dale Dyke Dam collapsed[43]. The dam had been completed in 1863 and was almost full by March the following year. During the afternoon of March 11, a crack was observed along the downstream side. At 11.30pm the same evening, the dam suddenly collapsed with no further warning. Almost 700 million gallons of water was released into the Loxley valley, a tributary of the River Don. Nearly 250 people were swept away as the flood destroyed the villages of Low Bradfield, Damflask and Little Matlock. Further downstream, parts of Sheffield were completely demolished, with many buildings simply washed away with their occupants still inside. An inquest into the disaster held later that month suggested that the failure was due to seepage and erosion. However, the most eminent water engineers of the time, Hawksley, Bateman and Simpson, considered that the dam collapsed as the result of being disturbed by the reactivation of an ancient landslide which lay on the downstream side of the site. In their view, movement had been triggered by heavy rainfall and led to the distortion and collapse of the dam. Recent speculations about the cause of the failure have suggested that there may have been a slide on the dam face itself, or that seepage and piping within the dam core was the most likely explanation[44]. Whatever the cause, the impact on the community was enormous. Over and above the tragic deaths, there were 7,300 compensation claims lodged by flood victims. Many were fraudulent; a rat catcher, for example, unsuccessfully claimed damages for his lost livelihood because so many rats had drowned! In other instances there were clear cases of injustice; one couple were awarded just £25 for the loss of three young children. The total cost to the water company in terms of compensation, repairs and fees, was more than £420,000. This sum was raised by a 25% water levy over 25 years, and led to public outrage that the victims of the flood were left to foot the bill for the disaster.

Conclusions

The most obvious conclusion to draw from the above discussion is that this is not an opportune time to assess the costs of landsliding in Great Britain. The numbers of deaths and injuries to humans are clearly very small: so small in fact as to only warrant attention in those few localities where high unstable cliffs, scars or gorge faces present potential threat. Conversely, the economic implications appear to be widely distributed through many levels in the economy, including individuals, local communities, local government, national government and various sectors of industry, but the scale of impacts is generally small. This makes the estimation of total costs impossible with presently available knowledge. Certainly there are grounds for considering that landsliding is a significant hazard in Great Britain, and that the true costs of ground instability, including estimations for delay and disruption, are undoubtedly greater than most believe. However, these costs are not distributed equally over the country or within the economy.

The prime need is for more information at both the national level and at the local scale. Much more needs to be known about the true scale of landslide impacts in the transport sector, in housing development and in water supply provision. But there is also a need to focus attention on those areas where the instability problems are most obviously great, or where future development pressures are likely to increase landslide hazard loss potential in years to come. Many of these areas have been identified in the DOE commissioned national review, as was explained in Chapters 5 & 6. These are the areas which require detailed studies to not only establish the true extent and costs of past slope failures, but also the most practicable methods of planning for local development and land use change against a background of potential instability. These matters will be discussed in greater detail in the next chapter.

Finally, it has to be stressed that no phenomenon perceived as hazardous is wholly detrimental to society. There are undoubted gains from slope instability. The environmental improvements consequent upon the regrading of over 1000 colliery spoil heaps was, in part, a consequence of the Aberfan tragedy. Some landslides are of particular importance to earth science research and training, and have been recognised by the nature conservancy agencies (English Nature, Countryside Council for Wales, Scottish Natural Heritage) as **Geological Conservation Review** (GCR) sites. All GCR sites are regarded as of national importance and have been, or could be, designated as **Sites of Special Scientific Interest** (SSSIs) under the Wildlife and Countryside Act 1981. At present some 27 landslide sites in Great Britain have been identified (Table 7.4), most of which lie within existing SSSIs.

Table 7.4 **Mass movement sites of conservation value in Great Britain.** *(Source: English Nature)*

Site	Grid Reference	Descriptions
Spot Lane Quarry* Kent	TQ793542	The eastern face of this quarry exhibits a cross-section through a series of cambered blocks and intervening loess-filled gulls. The features are formed in sandy limestone of the Hythe Beds, and slipping has taken place over underlying Atherfield Clay. The site provides the best cross-section through a series of cambers and gulls currently visible in Britain.
Coire Gabhail Highland	NN167556	A massive collapse of a cliff in Precambrian metamorphics has resulted in a huge debris cone mantling the lower part of the slope. The debris cones close off the valley. Upstream is a coarse-grained alluvial flat, consisting of material of which the downstream progress has been impeded by the debris cone.
Hob's House Peak District	SK173710	This site shows a set of large blocks of Carboniferous Limestone standing on a low-angled shelf halfway down the otherwise steep slope of Fin Cop. The blocks, which are up to 8m high, and 20m long and broad, are backed by an 18m high vertical cliff face in the limestone bedrock. The blocks have moved away valleyward from this cliff, by bedding-plane slip.
Bath University Avon	ST767646	A site showing Great Oolite, which at this point is cambered towards Smallcombe Vale, a tributary of the Avon valley. The section shows a very clear example of dip-and-fault structure, where joint-bounded blocks of oolite have individually tilted downslope.
Lud's Church Staffordshire	SJ987656	Lud's Church is a vertical open fissure in Roches Grit (Upper Carboniferous). It is 4–5m wide and, including all its side passages, totals 200m in length. All but the upper third of the slope has slipped forward towards the River Dane.

Table 7.4 **Mass movement sites of conservation value in Great Britain.** *(Source: English Nature)* *(continued)*

Site	Grid Reference	Descriptions
High Halstow Kent	TQ778754	This site exhibits low-angle sliding on London Clay, at a site which does not have basal removal of material by the sea. The average slope is about 8 degrees, which is close to the angle of ultimate stability in inland London Clay slopes. A succession of rotational landslips has taken place, the results of which are visible as a series of ridges and small scarps crossing the slope.
Bucklands Windy Pit North Yorkshire	SE588828	Buckland's Windly Pit is a network of roofed-over, fissures in Jurassic Lower Calcareous Grit. The fissures are 0.5–3m wide, extend to 38m below the level of the entrance and total 366m in explored length. This is the longest and most complex fissure network in Great Britain. They are related to valleyward movement of blocks of the highly-jointed rock, over the underlying Oxford Clay but possibly also over interbedded clay bands in the Corallian succession itself.
Cum-Ddu Dyfed	SN812742	Cwm-Ddu was produced by a very large rotational slump in the Silurian Llandovery Series. It is remarkable for its size (500m wide, 1km long and 250m in vertical range), its spectacularly cliffed backface, and a classic debris tongue stretching out from the slump amphitheatre (cwm) across the low-angle footslope and the narrow floodplain of the Ystwyth.
Blacknor Dorset	SY678714	The cliffs on the western coast of the Isle of Portland exhibit probably the best British examples of slab failures. The island is traversed by numerous NE-SW trending joints in the Purbeck and Portland beds, many of them widended by slipping over the underlying Kimmeridge Clay and Portland Sand.
Llyn-Y-Fan Fach Dyfed	SN801215	The best British examples of debris flows are found on the Old Red Sandstone (Devonian) scarp of the Black Mountain, above Llyn-y-Fan Fach. A number of deep gullies are cut through the vegetated scree surface.
Creigiau Eglwyseg Clwyd	SJ221452	The most impressive set of limestone screes in Britain is found at Creigiau Eglwyseg. Carboniferous Limestone presents a west and south-west facing multiple cliffline witih an extensive scree accumulation below each cliff.

Table 7.4 **Mass movement sites of conservation value in Great Britain.** *(Source: English Nature) (continued)*

Site	Grid Reference	Descriptions
Black Ven Dorset	SY347927	This is a complex and active multiple landslip. The site is important for its demonstration of movement in arenaceous flows of cohesionless remoulded material due to copious water supply from the Cretaceous strata. The addition of more water supplies as flows progress downslope ensure that they reach the beach at the foot of the cliff.
Peak Scar North Yorkshire	SE530884	Peak Scar is a 30m high cliff in horizontally-bedded Lower Calcareous Grit (Corallian, Upper Jurassic) on the south side of Gowerdale. Downslope from the cliff is a pronounced ridge of Lower Calcereous Grit, with strong downslope dips. The sequence of slope forms is the best British example of toppling failure developed in large individual rock units.
Glen Pean* Highland	NM905897	The best British example of multiple toppling failures of steeply dipping rectangular rock units is found on the southern slope of Carn Mor. The failure stretches from Lochan Leum an t-Sagairt to Coir' a'Bheithe on the valley floor (about 75mm OD), and reaches almost to the summit of Carn Mor (892m). The rock is mica-schist, massive psammitic schist and magmatised schist.
Postlip Warren Gloucestershire	SP000260	The Inferior Oolite (Jurassic) hilltop of Postlip Warren shows the best example of "ridge-and-trough" features in Great Britain. The Inferior Oolite has broken up into separate blocks along the line of pre-existing joints, and these blocks have moved valleyward over the underlying Lias Clay.
Ben Attow Highland	NH008185	The south-west face of Ben Attow is remarkable for a set of large-scale scarp ridges, running parallel to the slope, the rocks of which appear not to have been tilted downslope. The ridges are commonly over 10m in height, and one is traceable for a distance of more than 2km. They represent the best example in Great Britain of "uphill-facing scarps".

Table 7.4 **Mass movement sites of conservation value in Great Britain.** *(Source: English Nature)* *(continued)*

Site	Grid Reference	Descriptions
Rowlee Bridge Peak District	SK147892	At this site Edale Shales and Mam Tor Sandstones on the southern bank of the River Ashop are thrown into sharp, symmetrical straight-limbed folds. These are valley bulge structures and are probably the best examples in Great Britain.
Axmouth - Lyme Regis Devon	SY234899	This is the most renowned area of landslipping in Great Britain. The very large Bindon Landslip of 1839, in the centre of the area, brought the site to public prominence. The features of the Bindon slip are now much obscured by vegetation, but the major topographic elements are clearly discernable from the air.
Folkestone Warren Kent	TR255384	This area of coastal landslides has been more intensively studied than any other of comparable size in Great Britain, because the main Folkestone-Dover railway line was displaced by slipping in 1915. The site has suffered twelve major slips since 1765, and is now protected by a complex of coastal defence works.
Alport Castles Peak District	SK138914	The whole of the valleyside of Alport Dale, from crest to the river, has been involved in a massive slip, leaving a high vertical backface, a tall pinnacled ridge and a massive flat-topped detached sandstone mass, with back-tilted rotational slumps to its rear.
Mam Tor Peak District	SK134836	This large, active rotational slip in Carboniferous strata is famous for its very conspicuous and very high arcuate back-scarp, and its repeated disruption of the main road which crosses the slumped material. This is the best example of an inland, large-scale rotational slip affecting hard rocks in England, and one of the best known landslips in Britain.
Trimingham Norfolk	TG278390	The Norfolk coast shows two areas of particularly impressive rotational slumping affecting Pleistocene deposits. The Trimingham coast is the finest site of slumping of weak, unconsolidated sediments in Britain. Huge collapses of the cliffs continue to occur, in places breaking through an elaborate set of coastal defence works.

Table 7.4 **Mass movement sites of conservation value in Great Britain.** *(Source: English Nature)* *(continued)*

Site	Grid Reference	Descriptions
Hallaig (Raasay) Highland	NG587391	This very large, highly unstable area of landslips, involves a thick series of Lower and Middle Jurassic rocks on the eastern side of the Isle of Raasay. The east-facing cliff at the rear of the landslips is about 70m high; at its foot lies a hollow some 500m in breadth after which the land rises as a broad ridge transected by a labyrinth of open fissures of unknown depth. The toe of the landslip is being actively eroded by the sea, and movements have been recorded during the present century.
Quirang (Skye) Highland	NG443665	The Quirang Landslide is the largest in Great Britain, extending 2,130m from the back-face to the coast. It is formed in Tertiary lavas and underlying Jurassic sediments. In front of the back-face is a cluster of high lava pinnacles, and further downslope are numerous slipped lava masses forming hills up to 120m high.
The Storr (Skye) Highland	NG504537	This post-glacial slide in Tertiary lavas is 1,500m long, with a back-face over 600m high and its development is seen to be related to the dolerite sills, which cut the underlying Jurassic strata. Four distinct segments of sediments and overlying lava have successively moved seawards; those furthest upslope from ridges bearing lines of spectacular lava pinnacles up to 48m high, the outstanding examples of this kind in Britain.
Canyards Hills Peak District	SK251949	This site possesses the most impressive examples in England and Wales of "ridge-and-trough" features. Beneath a 10m high cliff, the north-facing valleyside above Broomhead Reservoir is a chaotic mass of subparellel ridges, separated by intervening narrow areas of marshy ground.

Table 7.4 **Mass movement sites of conservation value in Great Britain.** *(Source: English Nature)* *(continued)*

Site	Grid Reference	Descriptions
Warden Point Kent	TQ961736	At Warden Point, a series of particularly impressive, deep-seated, rotational landslips, bench-shaped in plan, occur in London Clay. Characteristically each slip extends along the coast for distances between four and eight times the cliff height. The back-tilted blocks produced by failure are broken down by shallow slides and mudflows, the debris being removed by marine erosion. This in turn results in progressive steepening of the cliff, and thus in further landslipping. The best locality in Britain to observe the cycle of landslip and coastal erosion typical of soft coasts.

Note:
All sites were identified by the Geological Conservation Review and, with the exception of those marked*, lie withiin designated SSSIs.

SSSIs receive protection through restrictions on the type of land use or operations that can be carried out. Owners and occupiers are notified by the nature conservancy agencies of the range of operations that could damage the site, such as drainage or excavation. In general, written consent or agreement is required from the relevant agency before a potentially damaging operation (PDO) can be carried out. However, agreement is not required when works are carried out in emergencies or where the operation is authorised by planning permission. The planning system, outlined in Chapter 8, is also an important instrument in safeguarding SSSIs, by refusing planning permission to developments that could lead to degradation or damage to sites.

Landsliding can also provide scenic attractiveness. The distinctive landscapes of many valued upland areas, (e.g. National Parks, Areas of Outstanding Natural Beauty (AONBs) in England and Wales; National Scenic Areas in Scotland) are, in part, a reflection of past instability. Sites such as Mam Tor (*Photo 2.10*) and Alport Castles (*Photo 2.11*), for example, are sites where landform is of considerable tourist interest. On the cost, cliffs are shaped by, and dependent on, landslide processes. They are amongst the nation's greatest landscape assets with many safeguarded by the protection afforded by inclusion within National Parks and AONBs or through their status as heritage coasts[45]. At present around 1525km of coast in England and Wales has heritage coast status, with public enjoyment encouraged by the provision of recreation activities that are consistent with their conservation, natural scenery or heritage features[46].

Many hard rock cliffs are renowned for providing prime breeding grounds for seabirds, with cliffs from Flamborough Head north to Dunnet Head, Cape Wrath to Land's End and the Northern and Western Isles containing the bulk of Europe's breeding seabird population. Indeed, over 20% of the world's population of razorbills nest on the Great Britain coast. Many active coastal landslides give rise to important biological habitats as colonising species thrive on the environmental change produced by instability. Numerous threathened species are found in such settings; hoary stock (*Matthiola incaria*) found only on eroding chalk cliffs, the Scottish primrose (*Primula scotica*) on cliff tops in Orkney[47].

Cliffs are also of great value because of the exposures of geological features. This is reflected in the large number of geological SSSIs, especially on the eroding soft rock cliffs of England from Flamborough Head to the Exe estuary. Such sites include international reference localities for lengthy periods of geological time, such as the Bartonian Stratotype between Highcliffe and Milford Cliff in Hampshire, and provide valuable opportunities for teaching and research. The oil industry for example, uses these exposures as training grounds for their geologists who have to recognise similar underground oil field structures from sparse borehold and geophysical data[47].

Landsliding is a necessary process in certain areas and needs to be accepted, despite the loss of land. This was clearly shown in a study of the East Anglian coastline[48] *(Figure 7.11)* which revealed that sediment supplied to the shoreline by cliff failures between Great Yarmouth and Cromer was essential for beach nourishment in adjacent areas. Stabilise the coast on this stretch and the reduced nourishment of beaches will result in increased potential for erosion elsewhere which will only be negated by further expenditure on coastal defence works. The real lesson of such

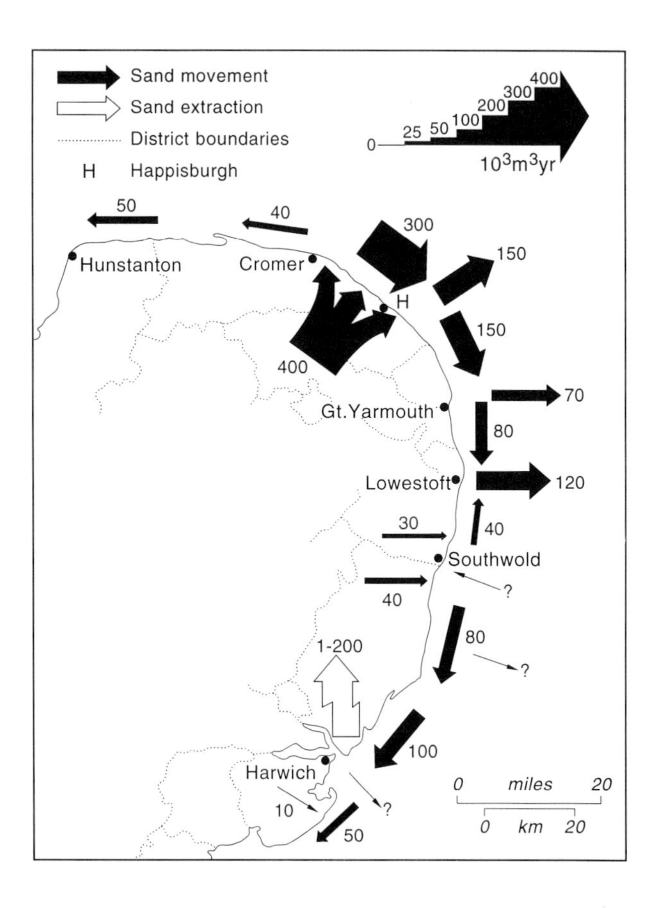

7.11 Annual net sand transport along the East Anglian coast, and administrative boundaries: the mismatch is clear (after Clayton, K.M., 1980).

studies is that there is no point in attempting to eradicate all slope instability through some ill-founded belief in the power of engineering technology, but rather that society should learn to co-exist with slope failure, avoiding the most problematic sites and stabilising only in those cases where the results are clearly beneficial.

NOTES

1. Clayton, D J, 1988
2. Young and Bird, 1822
3. Rozier and Reeves, 1979
4. Money, 1979
5. Topley, 1893
6. Clark, M J et al., 1976
7. Denness et al., 1975
8. Conway, 1979
   Pitts, 1979
9. Doornkamp, 1988
   Lee et al., 1988
10. Coard et al., 1987
11. McLaren, 1983
12. Lee and Moore, 1989
    Moore et al., 1991a
    McIntyre and McInnes, 1991
13. Chandler, M P and Hutchinson, 1984
14. Geomorphological Services Ltd, 1991
    see et al., 1991a
15. Anon, 1961
16. Jones, D K C, 1992
17. Jones, D K C., et al., 1988
18. Gostling, 1756
19. Common, 1954
20. Addison, 1987
21. Brunsden and Jones, 1972
22. Priestley, undated
    Skempton et al, 1989
23. Smith, D I, 1984
24. Malvern Gazette and Ledbury Reporter, 1981
25. Harrison, C and Petch, 1985
26. Hawkins, A B, 1977
27. Crofts and Berle, 1972
    Grainger and Harris, 1986
28. Pullan, 1986
29. Pullan, 1986
    Jones, D B and Siddle, 1988
30. Conway et al., 1980
31. Halcrow, 1986
    Halcrow, 1988
32. Miller, 1974
33. Lacey, 1972
34. Knox, 1927
35. Siddle et al., 1987
36. New Civil Engineer, 1990a
37. New Civil Engineer, 1990b
38. Greenman, 1992
    Russell, 1992
39. New Civil Engineer, 1990c
40. Bromhead, 1986
41. Middelboe, 1988
    Hayward, 1990
42. Penman, 1986
43. Youdale, 1989
44. Binnie, 1978
45. Lee, 1993
46. Countryside Commission, 1992
47. Stevens, 1992
48. Clayton, K M, 1980

CHAPTER 8

Dealing with the Problems

Introduction

It has already been shown in Chapter 7 that landsliding in Great Britain has not posed a significant threat in terms of loss of life or injury. In historic times, many were undoubtedly killed or maimed by ground movements while accomplishing the major feats of engineering that characterised the Industrial Revolution (e.g. building the canals and railways), but contemporary society suffers few casualties: far fewer in fact than are caused by other natural hazards such as radon, flooding, windstorms, etc. However, the human dimension of landslide hazard cannot be dismissed, for the potential for disaster certainly exists, most especially along crumbling coastal clifflines, within upland areas with precipitous bluffs and overhangs, and in the vicinity of large artificial slopes in cut and fill. The harrowing scenes at Aberfan, in 1966, highlight the unexpected suddenness of catastrophe and warn against complacency. But forecasting such events is notoriously difficult (see **Ground Management,** later) and dependent on the identification of preliminary signs of impending movement (e.g. tension cracks) and subsequent detailed monitoring. Fortunately, most inland failures either produce preliminary diagnostic symptoms or move relatively slowly (e.g. East Pentwyn), so that normal emergency action procedures (emergency services, evacuation, etc.) are usually adequate to cope with the vast majority of instability events. Nevertheless, there remains the need for careful monitoring of threatening situations, together with the management of slopes that are recognised to pose a danger to the public through the employment of measures such as reinforcement, retainment, drainage and regrading.

Conversely, landsliding in its various forms is known to have caused considerable damage to buildings, structures and infrastructure. Damage due to instability can lead to expensive remedial measures or, where repair is considered uneconomic, the abandonment/loss of property. The permanent closure of the A625 Manchester-Sheffield road at Mam Tor[1] in 1979 *(Plate 39)* and the decision not to reopen the railway link to Killin following the Glen Ogle rockslide[2] *(Plate 19)*, are the best known examples, but there are numerous instances of destroyed and damaged houses (e.g. Hedgemead, Bath[3]), dislocated roads and railways (e.g. Folkestone Warren[4]) and abandoned infrastructure.

Even more significant in terms of publicity, is the long and ever-growing list of schemes that have been delayed, redesigned or even abandoned due to the effects of instability during construction. Problems encountered by engineering works associated with the programme of motorway construction and trunk road improvements in the 1960s and 1970s *(see Chapter 7)*, merely brought into sharp focus difficulties that had been experienced by property developers over the last two centuries. Thus the instability generated by the post-war housing developments at Bury Hill[5] and Brierley Hill[6], in the West Midlands, at Ewood Bridge[7] in the Irwell Valley and at Gypsy Hill in south London[8], simply represent extreme examples of a phenomenon more generally characterised by frequent localised impacts associated with smaller-scale developments. Indeed, the only significant difference between these recent examples and the problems encountered at Hedgemead in the 1870s, is that the effects of slope remodelling and loading were exacerbated in the latter case by seepage from leaking sewers and water-supply pipes.

Landsliding, therefore, does result in important impacts that are widely distributed throughout society, although rarely achieving sufficient scale in Great Britain to be classified as disasters. The cumulative costs of landsliding to the British economy must, therefore, be 'significant', although no figures are available at present. There are five main reasons for this ignorance. First, landsliding affects such a wide range of interests, that information is held in the files and databases of a multitude of separate organisations and has yet to be systematically consolidated and synthesised. Second, the huge majority of failures are relatively small and of local interest only, so they rarely get publicised by the media or recorded in proper fashion. As a consequence, perception of landslide impact is generally poor. Third, the unexpected occurrence of instability during engineering works is not a matter that most firms consider should be publicised, due to problems of litigation and blame and because such publicity could adversely affect the firm's image. Thus some instability events remain shrouded in secrecy. Fourth, actually ascribing costs to individual slope failures is difficult and rarely attempted *(see Chapter 7)*. Lastly, many minor displacements or slow-moving forms of mass movement are misinterpreted as subsidence, because of the general misconception that landsliding is violent and rapid[9]. This mislabelling naturally considerably reduces appreciation of the distribution, extent, variety and significance of landsliding.

The above points have been reiterated here merely to emphasise the need for consideration of landslide potential in developments involving sloping ground. Clearly geological factors will determine the threshold conditions for slope failure *(Chapters 4 and 5)* and these must be weighed against proposed alterations to slope conditions *(Chapter 4)*. However, it has long been claimed that landsliding is one of the potentially most predictable of geological hazards[10], with very high estimated potential for loss reduction (up to 90%) and very favourable cost-benefit ratios of costs of investigation/protection measures as against savings in terms of construction costs / damage / disruption / destruction[11]. Thus it may be concluded that unnecessarily large costs are being incurred by the British economy because of too limited consideration of landslide potential: costs that are, for the most part, hidden within the overall financing of development projects and maintenance programmes.

The recent report on site investigation practice, by the Ground Board of the Institution of Civil Engineers, suggests that in any civil engineering or building project the largest technical and financial risk lies in the ground conditions[12]. Delays, claims and overspending resulting from unexpected ground conditions (of which landsliding is one of many potential problems) are very common: the National Economic Development Office found that 40–50% of all industrial and commercial developments were delayed because of ground problems[13]; the National House Building Council pays out £5–11 million each year for claims related to domestic housing, over 50% of which are related to geotechnical problems[14]; the final cost of large highway construction projects was on average 35% greater than the tendered sum, with half this increase directly attributable to inadequate site investigation[15]. These examples certainly lend support to the view that much time, effort and money could be saved by undertaking fuller site investigations prior to the commencement of construction programmes.

Obviously it is in a developers' interest to determine whether a site is on unstable land, as any future movement will affect the value of the site and its development costs. If there are reasons for suspecting instability problems, the developer should instigate appropriate investigations to determine whether:

- the ground is capable of supporting the loads to be imposed;
- the development will be threatened by unstable slopes within or adjacent to the site;
- the development will initiate slope instability which may threaten its neighbours.

The required assessments of landslide potential and the associated dangers are tasks that require careful professional judgment. Developers should seek expert advice about the likely consequence of proposed developments on sites where landsliding is known or may be anticipated. This advice should ideally involve a series of phases beginning with some form of **preliminary assessment** as to the presence or absence of landsliding and the type(s) of movement that has occurred, followed, if necessary, by **detailed investigation** into slope stability conditions as part of site investigation procedures. This will allow decisions to be made directed to limiting the potential for adverse impacts, either through **planning strategies/land management strategies** which seek to limit vulnerability by avoidance or zoning (most valuable elements located in least unstable sites thereby minimising risk), or **ground management strategies** which seek to limit adverse effects on the site or neighbouring land through measures designed to reduce the likelihood of slope failure. These aspects will be briefly examined in the following sections, first with respect to site specific problems and then with reference to areally extensive land management and planning.

Preliminary Assessment

The land planning and development control frameworks currently operated in Great Britain mean that the majority of questions relating to ground stability are site-specific and not areally extensive (e.g. the whole of the Weald or the entire Southern Uplands of Scotland). Under these circumstances, two approaches to preliminary assessment are generally applicable: **desk study** and **ground inspection.**

The desk study[16] approach has been traditionally employed in ground engineering over the past few decades and involves collecting, collating and synthesising all available information on the site of interest and similar sites nearby. The growth in soil mechanics and ground investigation during the past three decades has resulted in something of an explosion of information regarding stability characteristics. However, it has to be stressed that this information is strongly focussed on those areas with either the greatest recognised problems or where there has been recent development involving extensive foundation investigations. Thus, the data is highly variable in quantity and quality, and there remain surprisingly large areas for which extremely limited information exists.

The problems of the desk study approach have been considerably eased in recent years by data collected together during the DOE commissioned **Review of Research on Landsliding in Great Britain**[17] and the subsequent updating process. These data now form the National Landslide Databank[18], a computerised consolidated archive of all available published sources in the public domain*. Although this represents the single most complete data set on the nature, occurrence and extent of landsliding in Great Britain, it is wholly composed of reported occurrences and therefore an artifact of knowledge rather than a true representation of the actual distribution of landslipped ground *(see Chapter 2)*. An absence of reported landslides at a particular location does not, therefore, imply the absence of unstable or potentially unstable ground. This is the fundamental shortcoming of any database involving **reported** occurrences, and although the level of information will improve as more reports are included, it will never approach the accuracy of a database produced through systematic survey.

Accessing the databank can also provide an indication as to whether unstable ground could be anticipated in an area by comparing the site geology with the list of bedrocks which have high recorded densities of known landslides on their outcrops *(see Table 5.4)*. For example, landslide problems have been widely recognised on the Lias clays and so any proposed development on these materials should take into consideration the potential hazards of reactivating previously unrecognised slides or initiating first-time failures. This is a rather crude approach, however, for it must be fully appreciated that instability is frequently the product of specific site conditions or localised characteristics/changes in lithology or hydrogeology *(see Chapter 5)*.

In those areas where landsliding is known to have posed a significant threat to local communities or infrastructure, attempts may already have been made to systematically identify, delimit or classify instability features. The resulting maps and accompanying reports contain much greater levels of information than are contained on standard geological maps, and higher levels of confidence can be placed on the portrayed distribution approaching the true distribution. The Environmental Geology Map series commissioned by the DOE for certain selected locations are one important source**. However, even greater levels of detail and accuracy can be attained if such surveys are specifically focussed on landsliding. These are known as **landslide inventory maps** but, unfortunately, very few have been produced to date.

The pioneer landslide inventory map in Great Britain was the result of a series of studies of the South Wales Coalfield commissioned by the Department of the

---

* The National Landslide Databank is currently held by Rendel Geotechnics who offer either a data search service or can provide copies of the databank for use. Copies are designed to run on any desk top computer that is compatible with an IBM PC machine.

** In recent years the Department of the Environment has commissioned a series of Environmental Geology Mapping studies as part of their Planning Research Programme. Those areas investigated include Glenrothes, Fife (Nickless, 1982), the Bath region (Forster et al., 1985), Morpeth, Northumberland (Jackson & Lawrence, 1990), Torbay in Devon (Doornkamp, 1988) and St Helens, Merseyside (Doornkamp and Lee, 1992).

Environment, in collaboration with the Welsh Office, and carried out by the Institute of Geological Sciences (now the British Geological Survey) over the period 1977–1980[19,20]. The final product, the South Wales Coalfield Landslip Survey[20], is a two-volume report accompanied by 1:50,000 scale maps showing the distribution of 595 landslides with respect to topography, solid geology, superficial deposits and slope angles. This study has been hailed as a major advance for all those concerned with land management and ground stability in the area[21]. For the first time the pattern and nature of landsliding in South Wales was systematically documented, thereby providing the information base that is the essential prerequisite to an informed assessment of hazard and the planning of more detailed studies targeted at resolving site specific problems[22].

It is undoubtedly true that a great benefit would be obtained by further landslide inventory surveys undertaken along similar lines, although it must be recognised that no single map will represent a perfect portrayal of reality and that there will always be marked differences in skill between surveyors (operator variance). Thus it should come as no great surprise that a subsequent more detailed investigation of the Rhondda Valleys, commissioned by the DOE through the Welsh Office and carried out by Sir William Halcrow & Partners in the early 1980s[23], raised the total reported landslides from 118 to 351 within this single area of 100km^2 *(Figure 7.8)*. Herein lies the fundamental problem inherent in maps. Once the distribution of landsliding is portrayed on a map it is naturally assumed that (i) the boundaries of landslipped ground are accurately shown, (ii) that all areas of instability have been identified and depicted and (iii) that no landslides exist in the intervening areas which must, therefore, be considered 'safe'. All these assumptions are unsound to varying degrees. The accuracy of inventory maps must be recognised to be a function of the methodology adopted, the skills employed, the detail pursued and the funding available.

Irrespective of the availability of background information, there will inevitably be a strong case for ground inspection to confirm desk study results or to provide indications as to the extent, character and activity state of existing landslips at the site, or sufficiently close to the site to pose a potential threat. There are a number of morphological and botanical indicators that can be used by trained personnel to identify, delimit and classify landslides *(Table 8.1)*. The need for skilled advice arises from the fact that: (i) many landslides are hybrids and do not display the characteristic forms described in Chapter 3; (ii) ancient landslides often have exceedingly subdued morphological expression which makes identification difficult and delimitation extremely problematic *(Plate 40)*; (iii) it is not the presence of individual indicators but associations of indicators that point to the existence of slope failure (e.g. tilted trees can be produced by many processes; cracked walls may be the result of poor workmanship or bad foundations); and (iv) landform morphology apparently characteristic of landsliding can, under certain circumstances, be produced by other geomorphological processes in a phenomenon known as convergence (e.g. old quarries and certain forms of glacial deposits can sometimes be easily confused with landslides). Thus it is only through the application of skill and experience that irrelevant signals can be filtered out. Finally,

Table 8.1 **Features indicating active and inactive landslides** *(after Crozier, 1986).*

Active	Inactive
Scarps, terraces and crevices with sharp edges.	Scarps, terraces and crevices with rounded edges.
Crevices and depressions without secondary infill.	Crevices and depressions with secondary infill.
Secondary slides on scarps.	No secondary slides on scarps.
Fresh slickensides.	No fresh slickensides.
Fractured block surfaces.	Weathered block surfaces.
Disrupted drainage.	Integrated drainage.
Pressure ridges.	Marginal fissures, abandoned levees.
Tilted trees, mainly fast growing vegetation species.	Trees tilted but with new vertical growth, vegetation cover predominantly slow growing species.

it must be recognised that such qualitative assessments cannot be infallible. Slopes may be so remodelled by human activity or natural processes that no surface indicators exist to the presence of failed masses and buried shear surfaces. Conversely, prominent tension cracks on highly irregular ground may be the result of relatively insignificant superficial slippage on generally stable landforms produced by quite different processes. Thus, irrespective of the interpretive skills brought to bear, it is often necessary to confirm visual diagnoses by subsurface investigation.

Detailed Investigation

If desk study or preliminary assessment indicates that a site is likely to have been affected by instability, then detailed investigations should be undertaken to ascertain the true extent of the hazard. Once again, two approaches are available: subsurface investigation and surface mapping. These can be employed independently but are generally considered more effective if used in combination.

The scale of necessary ground investigations will clearly depend on the character of the proposed development and the nature and extent of the perceived ground problems facing the developer. Typically, they involve a combination of topographic survey, sub-surface investigation, laboratory testing[24], monitoring of ground movement and groundwater studies *(Table 8.2)*, which are carried out with the view to establishing the nature of ground materials, the number, position and form of shear surfaces, and the areas subject to marked fluctuations in pore-water pressures, so that stability conditions can be assessed[25]. Some of this information may be obtained by remote sensing (gravity surveys, seismic surveys, earth resistivity surveys) but more usually the data is produced by trial pits, trenches and boreholes. These may be sited randomly or regularly over the site, but it is coming to be more usual for expensive investigations (boreholes) to be located with reference to either

Table 8.2 **Commonly used ground investigation techniques**

Approach	Technique	Usage
Sub-Surface and In-Situ Testing	Trial Pits/Trenches	Sampling and logging of exposures. Most useful in investigating shallow instability in soils and soft rock or locating boundaries of disturbed ground.
	Boreholes	Sampling and logging of disturbed or undisturbed core samples. Useful in investigating deeper instability problems. Variety of techniques eg. shell and auger, rotary drilling etc. allows use in all rock types, although core recovery can be a problem.
	Adits	Large excavations to establish sub-surface conditions in major, deep-seated landslides. Very expensive and are generally used as a drainage measure.
Surface Monitoring	Topographic Survey	Measurement of displacement rates between surveyed points. Problems of vandalism.
	Extensometers	Measurement of enlargement of tension cracks, building cracks etc. Problems of vandalism.
	Analytical Photography	Retrospective analysis of displacement of points on photography of different dates, usually aerial but can be hand-held. Expensive, taking considerable computing effort. Not sensitive to very small displacements.
Sub-Surface Monitoring	Inclinometers and Slip Rods	Identification of zones of movement, monitoring of displacement rates.
	Piezometers	Monitoring groundwater levels and pore-water pressures.
Hydrological Monitoring	Tracer Experiments, Pump Tests etc.	Enables groundwater flow monitoring for design and investigation of remedial measures.

predetermined development considerations (e.g. the proposed locations of buildings, structures, infrastructure, etc.) or preliminary identification of potentially problematic areas (see below).

This is not the place for a detailed discussion of site investigation procedures, as these are readily available elsewhere, and one case study will suffice. Probably the best example to demonstrate all the related methods is the recent investigation of the East Pentwyn landslide at Blaina, South Wales *(Figures 3.37 and 3.38)*. After a major reactivation of the slide in 1980 the responsible local authorities commissioned a series of investigations to identify possible remedial measures[26]. A total of 14 boreholes were sunk to establish the geological conditions and depths of the shear surfaces *(Figure 3.38)*. The position of the active shear surfaces were determined by observing where sliprods, inserted periodically into the boreholes, were distorted by movement.

Disturbed and undisturbed samples were obtained from the displaced material and debris apron, and cores were recovered from the underlying intact strata.

Laboratory tests were carried out on these samples to establish the nature and strength of the materials involved in the landslide. These tests included an assessment of moisture content, bulk density, plasticity and residual shear strength[24]. It is worth noting that a wide range of shear strength values were recorded, reflecting the natural variations in material at the site and the common problem of obtaining undisturbed and representative samples.

To establish the hydrogeological conditions within the landslide, 16 standpipe piezometers *(Figure 5.7)* were installed to determine representative groundwater levels, and rainfall was measured using a continuously recording raingauge linked to a data logger. The piezometers were positioned immediately above the several active shear surfaces identified in the boreholes, thereby allowing the accurate measurement of pore-water pressures acting on these surfaces. In addition, a groundwater experiment using a tritium tracer[27] was carried out to establish water-flow paths into the landslide and thereby ascertain the effects of abandoned shallow coal mine workings on hydrogeological conditions.

Ground movement observations were made regularly by accurately resurveying the positions of several 2m long steel pins driven into the landslide and a number of marked boulders. An example of the survey results is presented in Figure 8.1 which clearly demonstrated the pronounced seasonal pattern of movement noticeable in many inland landslides. By plotting the combined horizontal and vertical movement vectors, together with the shear surface positions observed in the boreholes, an accurate model of the failure mechanism was established for the whole landslide *(Figure 3.37)*.

The results of this investigation at East Pentwyn were used to mathematically model the stability of the slope. Because of the problems, mentioned earlier, of the wide range in measured shear strength values, an analysis was carried out to calculate the

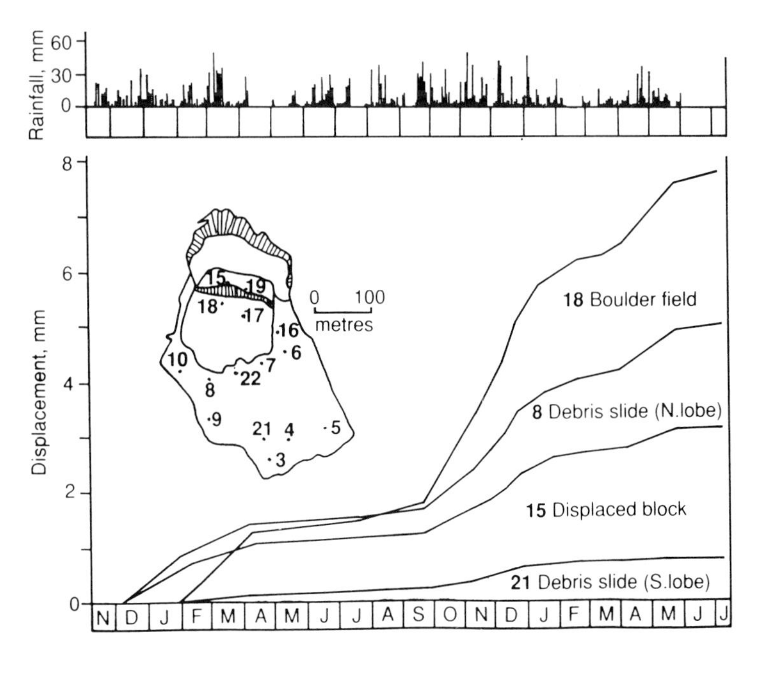

.1 Ground movements recorded at the East Pentwyn landslide using surface measurement between survey stations (after Jones, D.B. & Siddle, 1988).

shear strength that would be mobilised when the Factor of Safety of the landslide was 1.0, i.e. at the point of failure, with the water table at the typical winter level*. This strength value was then used to analyse the suitability of alternative proposed remedial measures, together with an assessment of the sensitivity of the landslide model to changes in water level. These analyses demonstrated that remedial works to improve the stability of each of the main elements of the landslide (debris apron, boulder field, displaced block and scree), and hence the whole system, were feasible but likely to be very costly.

This example, while clearly displaying the potentially varied nature of relevant investigation and monitoring techniques, is not typical of site investigations for development purposes, because it was designed specifically to establish the nature of movement in a known instability situation of some complexity and great local concern. For the most part, standard geotechnical investigations are normally restricted to establishing the nature of materials over a particular site and with depth, the presence and form of shear surfaces, and groundwater characteristics. Nevertheless, they can still prove expensive and there is often a temptation to limit the scope of site investigation and thereby avoid what is seen to be unnecessary expenditure and delay by obtaining information about an uncertain (unlikely) problem. For example, the East Pentwyn investigation cost an estimated £150,000 and took over five years to organise and complete, although most surveys are considerably simpler, cheaper and of much shorter duration. It follows, therefore, that any techniques that can be used to reduce the costs of detailed site investigation are to be welcomed. Specialised ground surveys have been the most successful to date.

Following the various problems experienced with unforeseen instability during the highway development programme of the 1960s, it was concluded that such problems could be reduced if areas of unstable ground or ancient landslides could be identified and delimited prior to site investigation. Such an approach would ensure: (i) that all areas of unstable ground were thoroughly investigated, thereby avoiding the possibility that they might be missed by site investigation programmes designed on the bases of random or regular sampling (e.g. boreholes and trial pits at regular intervals often miss particularly problematic areas); (ii) that site investigation effort is directed at problem sites thereby ensuring cost-effectiveness and limiting unnecessary expenditure; and (iii) that early knowledge of problematic areas would facilitate fine-tuning of development design so as to minimise vulnerability to instability.

As a consequence, the traditional initial input of topographic survey came to be augmented by mapping techniques designed to portray information on ground conditions of value to engineers. These **engineering geology maps** were initially fairly rudimentary[28] but have become increasingly sophisticated with time and now have come to largely replace topographic survey as the essential preliminary geotechnical

---

* This approach in determining shear strength at moment of failure, under assumed ground conditions, is commonly known as back-analysis and is a frequently used technique in analysing failures (see Bromhead 1986).

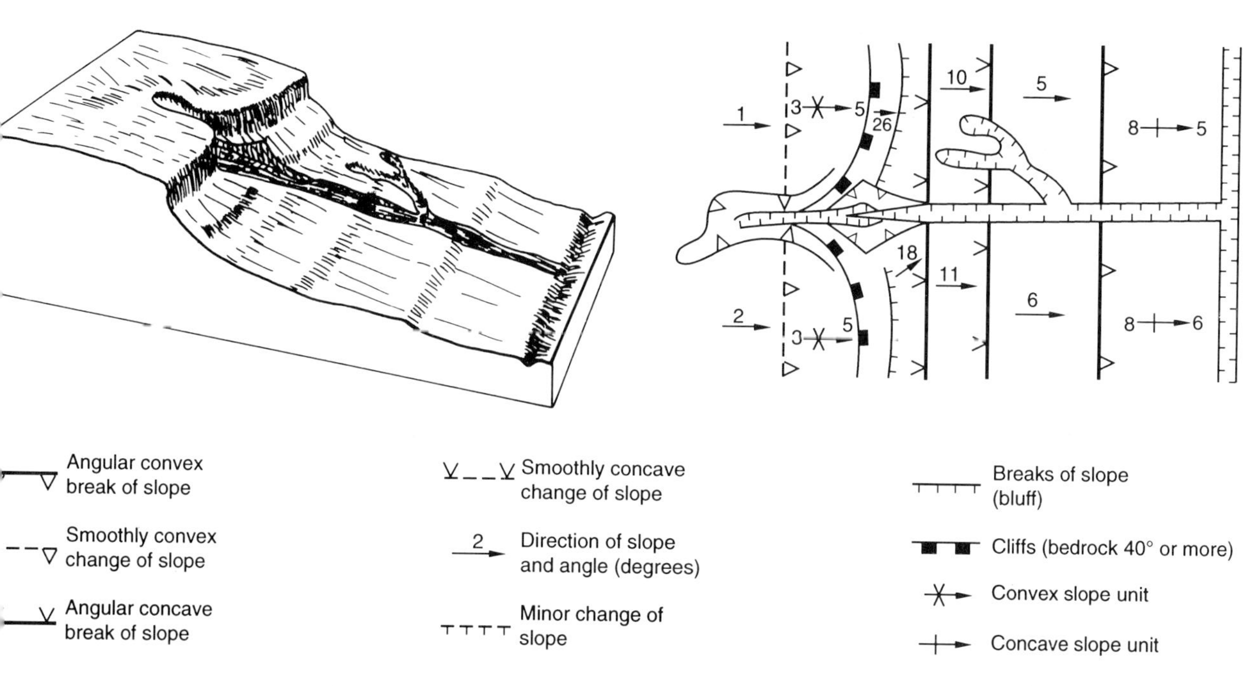

8.2 A simple morphological mapping system (after Savigear, 1965, Cooke and Doornkamp, 1990).

information input[29]. An alternative, but complementary, approach has been to use the portrayal of ground surface form as an indicator of surface materials and surface processes, including instability. This is heavily dependent on the technique of **morphological mapping**[30] which portrays surface form by means of symbolic notation. Using this technique, skilled operators can quickly and cheaply produce maps which show the ground divided into units depending on ground slope characteristics (planar, concave, convex), together with information on the direction and steepness of slopes *(Figure 8.2)*. Morphological mapping was initially developed as a means of describing surface form for academic purposes and was first used to portray a landslide in 1965[31]. The real value of morphological mapping at scales of 1:5,000 and larger is that map production requires careful scrutiny of ground form over an entire site as well as adjoining land, thereby facilitating the recording of tell-tale indicators of instability. Of even greater importance, large-scale morphological maps can be enhanced by shading of slope categories to reveal ground forms diagnostic of ancient instability. This was first shown for the Char Valley slopes in West Dorset *(Figure 8.3)*, where localised displacement of the A35 was found to be due to partial reactivation within an extensive sheet of ancient landslide debris[32]. The technique was subsequently utilised to analyse instability problems along the A470 Taff Vale Trunk road in South Wales[33], including preliminary assessment of the crossing of the Taren landslide *(see Figure 2.10)*.

Morphological mapping in itself produces very complex maps that cannot be easily interpreted by the non-specialist. While slope shading helps clarification *(Figure 8.4)*, the most useful development is to use the morphological notation as a basis for the geomorphological assessment of the ground in terms of landforms and surface materials. These **geomorphological maps** can be developments of the morphological map at the detail-scale (1:200–1:5,000), but become increasingly abstract at the

8.3 Mapping the landslides on the northern flanks of Stonebarrow Hill, Dorset

(a) a sample of the morphological map

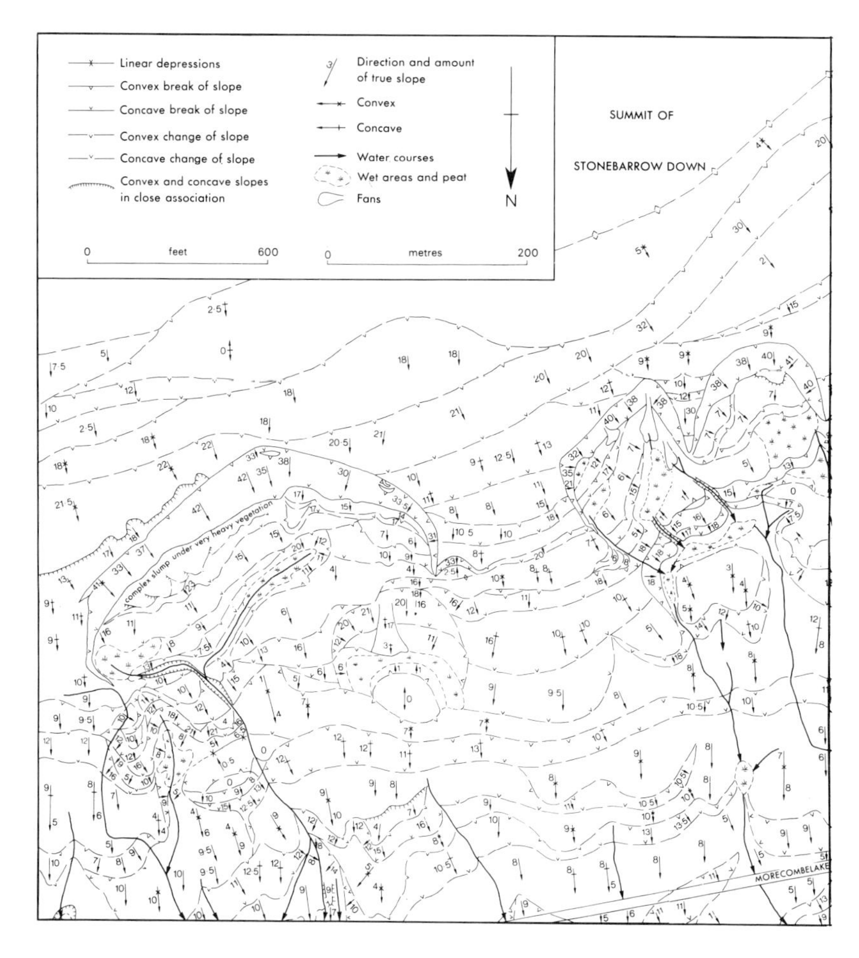

(b) a shaded slope map (after Brunsden & Jones, 1976).

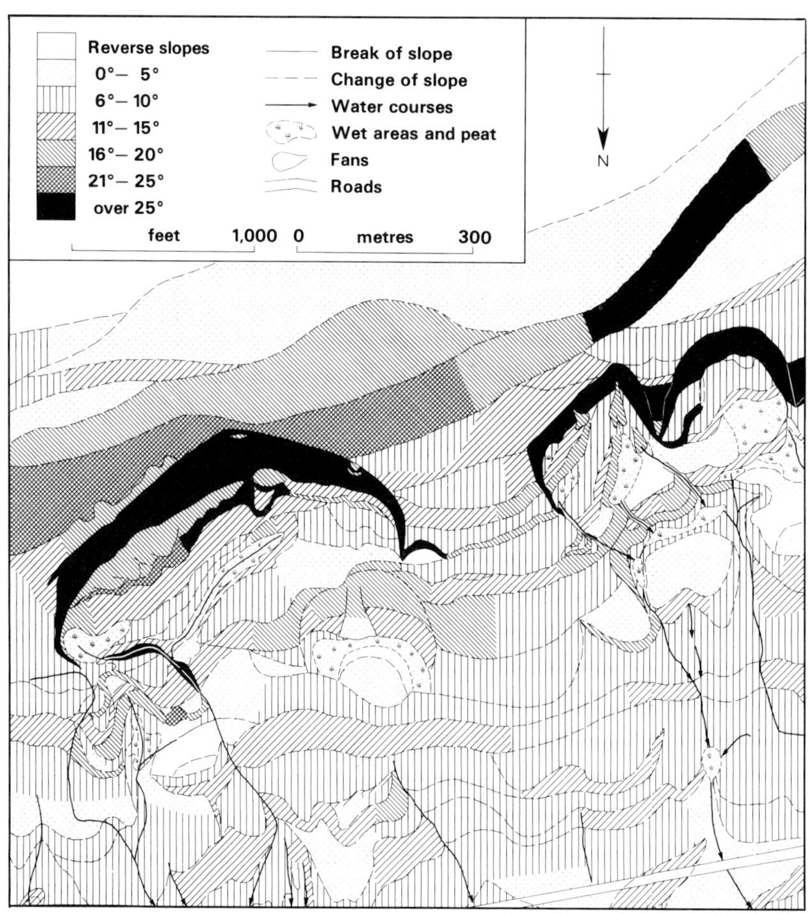

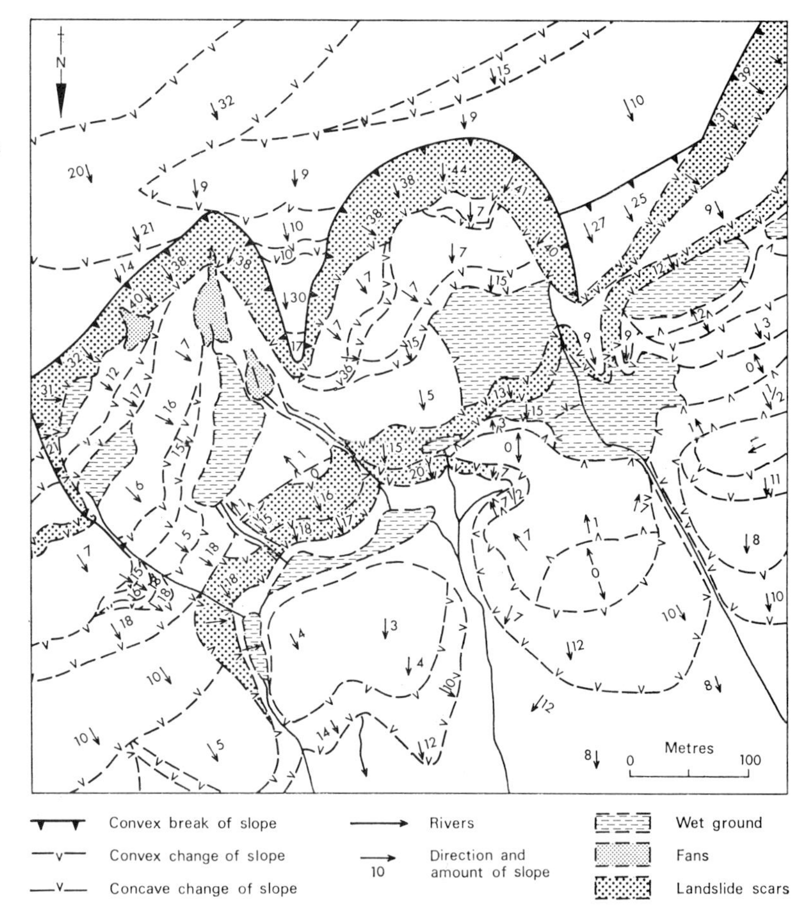

8.4 A shaded morphological map of a section of Stonebarrow Hill, Dorset, emphasising landslide backscars and areas of wet ground (after Brunsden & Jones, 1976).

large-scale (1:5,000–1:10,000) and medium-scale (1:10,000–1:50,000)[34]. Adaptations of geomorphological mapping techniques have been incorporated into engineering geological mapping to produce **engineering geomorphological mapping**[35]. This is not the place to discuss the distinctions between the various forms of mapping but merely to note that a variety of detailed, sophisticated mapping techniques involving consideration of ground form are now available and are widely used in development projects. Geomorphological mapping, in one form or another, is now often employed to evaluate alternative road alignments[36], to establish the nature and extent of landslide features in an area, and to determine the degree of threat that instability may pose to existing property and future developments. For example, geomorphological mapping was carried out as part of an investigation to establish the cause of landsliding at Luccombe, Isle of Wight, in the winter of 1987–1988, which had resulted in a number of houses being damaged beyond repair[37] *(Plate 28 and Photo 4.2)*. By carefully mapping the limits of the recent movements, together with the surface form of the surrounding area, it was possible to demonstrate that the village had been built on an ancient landslide and that the recent movements were a reactivation of parts of this feature *(Figure 8.5)*.

In addition, by analysing ground form it is often possible to improve the design of costly site investigation, both in terms of the scale of investigation and the location or type of investigation employed, as well as provide a framework within which the results obtained from samples yielded by boreholes and trial pits can be assessed. There are numerous instances where this procedure has been employed. For example, the very costly stability analysis programmes undertaken for the Taren

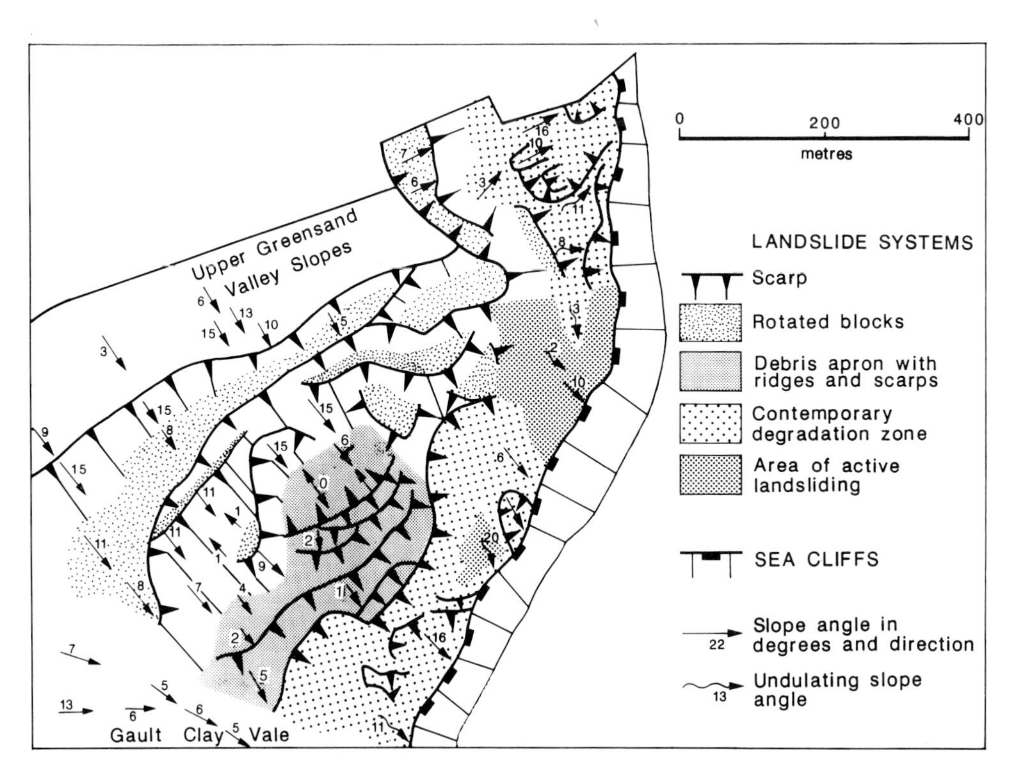

8.5 A simplified geomorphological map of the landsliding at Luccombe Village, Isle of Wight (after Moore et al., 1991a).

landslide in the Taff Valley *(see Figure 2.10)* and East Pentwyn (see earlier) were in both cases preceded by detailed geomorphological mapping relying heavily on morphological mapping techniques[38]. Indeed, the investigation of stability problems affecting water supply infrastructure on Bredon Hill in the early 1980s *(see Chapter 2)* produced such a detailed and informative geomorphological map assessment that no expensive sub-surface investigation was required to reach the decision to abandon the site.

The use of geomorphological mapping has heralded a new approach to site evaluation which may be conveniently called **geomorphological assessment,** that emphasises the systematic collection of all relevant surface information prior to geotechnical site investigation. This approach is well illustrated by the recent investigation into the landslide problems at Ventnor, Isle of Wight[39]. The work was carried out over a two year period and involved a thorough review of available records, reports and documents relating to instability, followed by a detailed field investigation comprising geomorphological and geological mapping, photogrammetric analyses, a survey of damage caused by ground movement, a land use survey and a review of local building practice *(Figure 8.6)*. The results of these investigations provided an understanding of the nature and extent of the landslide system, together with the type, size and frequency of contemporary movements and their impact on the local community. This detailed understanding of ground behaviour was used in conjunction with knowledge of the vulnerability to movement of different types of construction and the spatial distribution of property at risk, to formulate a range of management strategies designed to reduce the impact of future

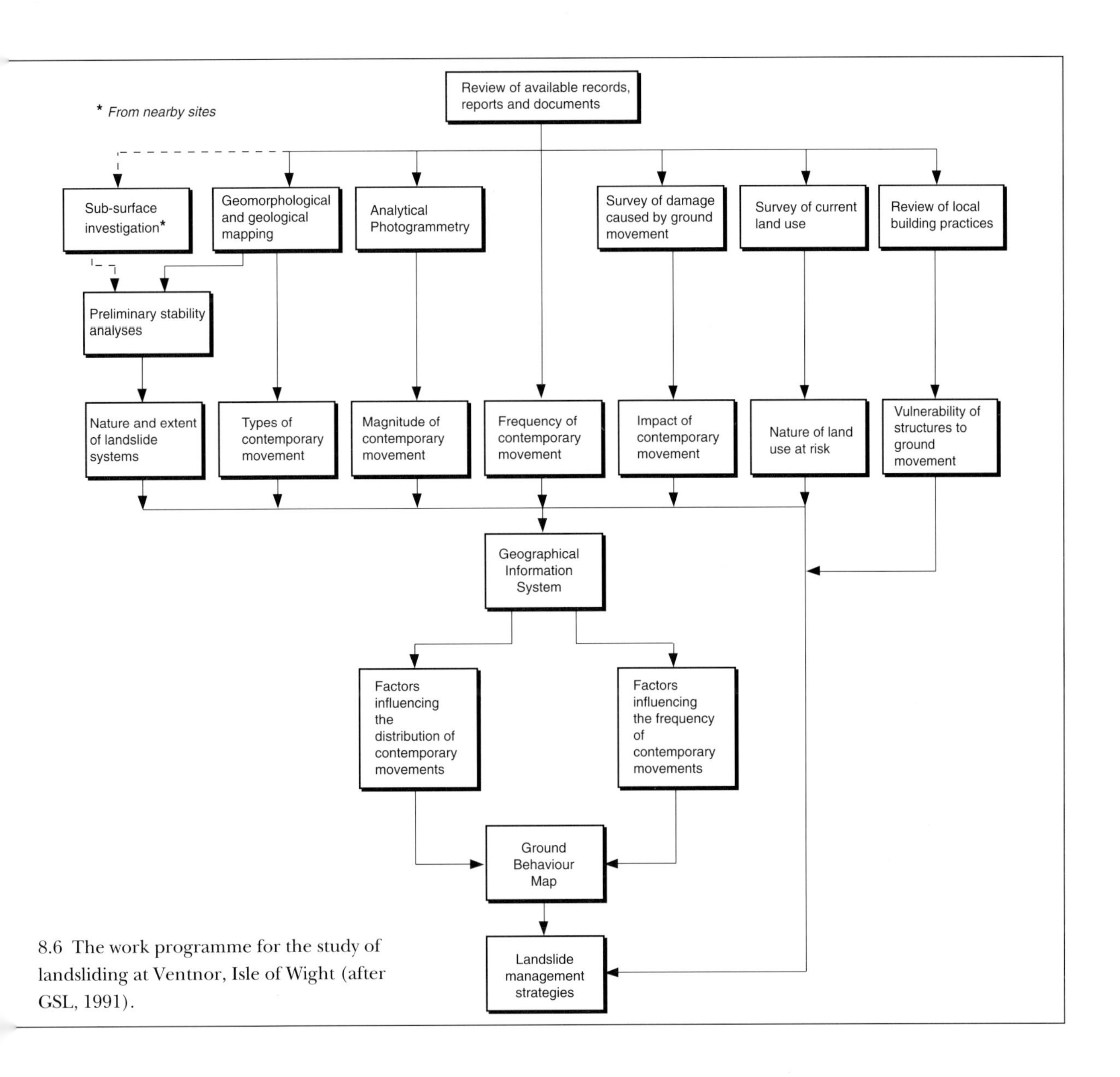

8.6 The work programme for the study of landsliding at Ventnor, Isle of Wight (after GSL, 1991).

movements. This method of investigation, although very broad in scope, was essentially a geomorphological-based approach, relying on an understanding of surface processes and features (landslide movement, property damage etc.) to arrive at a model of landslide behaviour.

While it is recognised that the Ventnor study represents a major response to an atypical problem affecting a very large site, the underlying principles apply equally well to new developments on smaller sites. It is recommended that the traditional approach of rushing to site investigation and then broadening consideration when and if problems are encountered, should be replaced by the more logical approach of obtaining surface information prior to embarking on costly sub-surface ground investigation. This 'geomorphological approach' is not meant to be an alternative to geotechnical investigation, but merely a way of reducing the costs of such investigations while increasing the flow of relevant information to those responsible

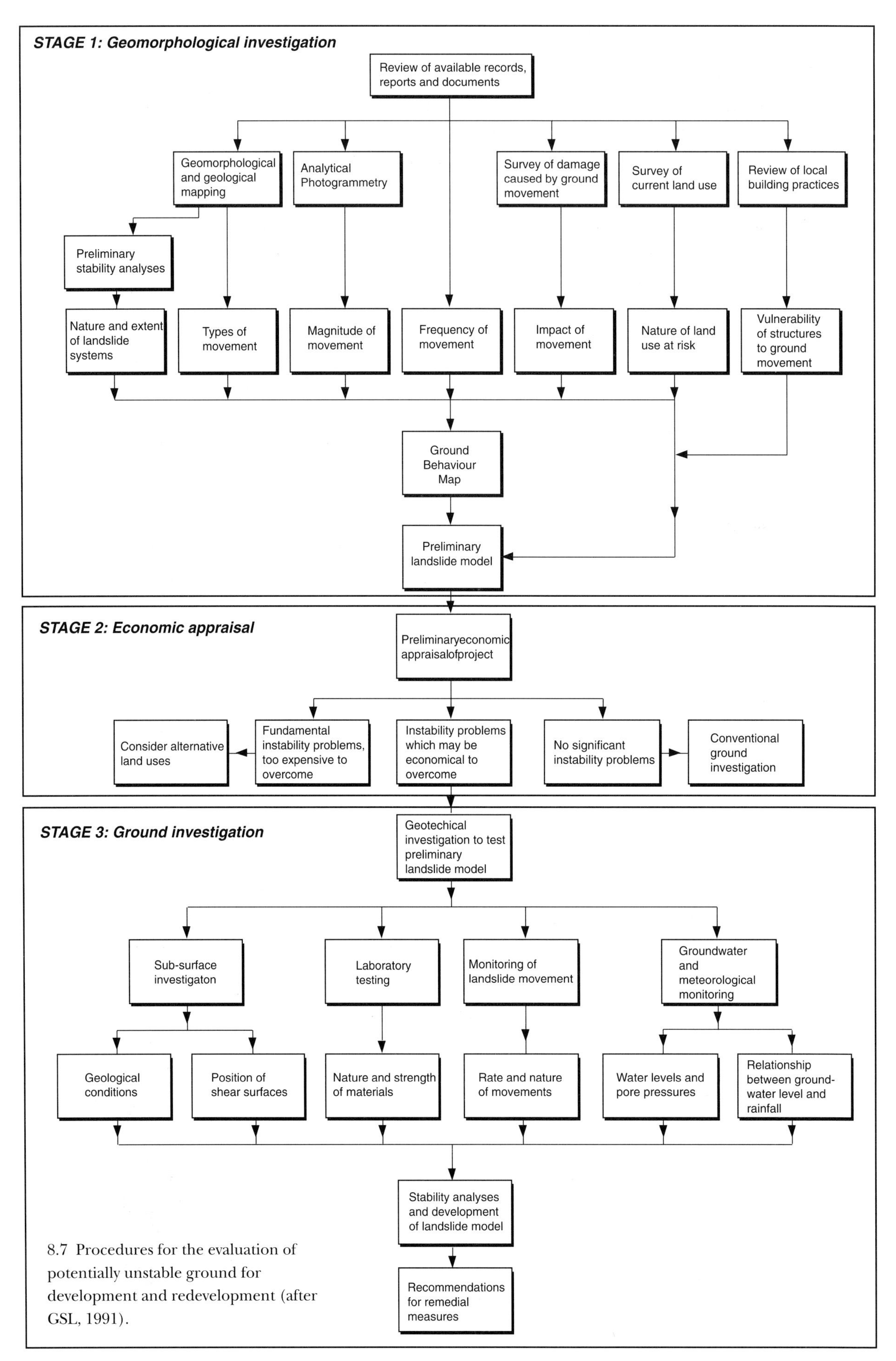

8.7 Procedures for the evaluation of potentially unstable ground for development and redevelopment (after GSL, 1991).

for construction. In addition, it should be used as a cost-effective way of helping to identify at an early stage whether instability problems at a site are likely to prove too expensive to overcome (e.g. Bredon Hill), or may be minimised by redesign prior to site investigations, which can then be tailor-made to the refined requirements.

The decision to proceed with an expensive and often time consuming ground investigation programme should only be made after an economic assessment of the project has been carried out with the benefit of the results of an initial geomorphological appraisal. This logical sequence highlights a 3-stage approach which is considered to be the most appropriate way to investigate areas of landsliding or potentially unstable land *(Figure 8.7)*; initial geomorphological investigation, economic appraisal and then a suitably designed, detailed ground investigation. The resources invested in each of these stages, and the techniques used, will obviously vary according to the complexity of the landslide problem and the nature of the hazard. For example, it would not be relevant to carry out photogrammetric analysis to determine the magnitude of contemporary movement in areas where landslides are essentially dormant, as is the case in many inland situations. The ground investigations, too, will reflect conditions at the particular site and the types of problem anticipated; for example, investigation of cliff fall hazard or debris flow hazard would obviously require only limited sub-surface investigation, but great emphasis on understanding the nature of the materials involved in movement.

Land Management

Various aspects of land management have been discussed in the previous section and need not be reiterated. To summarise, information on site conditions produced by desk-study, preliminary assessment, mapping and geotechnical investigation should be used, wherever possible, to redesign development plans so that vulnerability is reduced and instability minimised. This may involve (i) redesigning building height to reduce loading, (ii) changing the pattern and scale of cut and fill earthworks, (iii) changing the location of buildings and structures to maintain stability, (iv) incorporating special measures into the design so that the threat of future instability is reduced (e.g. flexible service pipes, employment of comprehensive drainage networks, prohibition of soakaways, etc), or (v) abandoning proposals in favour of a less problematic site. Some of these issues will be considered further under **Planning Strategies.**

Ground Management

Should investigations indicate that a development can proceed on the basis that stability problems can be overcome or the risks reduced to acceptable levels with minimal threat to neighbouring sites, then it will often prove necessary to take steps to minimise hazard potential from landsliding. In addition, instability may occur during construction or subsequent to completion, in which case stabilising remedial measures will be required. Finally, natural processes of erosion or changes in environmental conditions can cause new, repeated or reactivated instability to threaten property, thereby necessitating stabilisation. These considerations can all be included under the heading **ground management.**

A wide range of stabilisation methods are available *(Table 8.3)*. The most frequently adopted approaches include a combination of modification of slope profile by

Table 8.3

Principal methods of slope stabilisation

Approach	Methods
Excavation and Filling	– Remove and replace slipped material. – Excavate to unload the slope. – Fill to load the slope.
Drainage	– Lead away surface water. – Prevent build up of water in tension cracks. – Blanket the slope with free draining material. – Installation of narrow trench drains aligned directly downslope, often supplemented by shallow drains laid in a herring-bone pattern. – Installation of interceptor drains above the crest of the slide or slope to intercept groundwater. – Drilling of horizontal drains into a slope, on a slightly inclined gradient. – Construction of drainage galleries or adits, from which supplementary borings can be made. – Installation of vertical drains which drain by gravity through horizontal drains and adits, by siphoning or pumping.
Restraining Structures	– Retaining walls founded beneath unstable ground. – Installation of continuous or closely spaced piles, anchored sheet or bored pile walls. – Soil and rock anchors, generally prestressed.
Erosion Control	– Control of toe erosion by crib walls, rip-rap, rock armour, revetments, groynes. – Control of surface erosion. – Control of seepage erosion by placing inverted filters over the area of discharge or intercepting the seepage.
Miscellaneous Methods	– Grouting to reduce ingress of groundwater into a slide. – Chemical stabilisation by liming at the shear surface, by means of lime wells. – Blasting to disrupt the shear surface improve drainage. – Bridging to carry a road over an active slide. – Rock traps to protect against falling debris.

excavation or fill, drainage works and the construction of retaining structures. Which combination of methods are adopted and their relative importance depends on the type of landslide, failure size, the nature of ground materials, the danger posed by failure and the value of the property being defended. In general terms the scale of investment in stabilisation measures has increasingly come to be determined by Cost-Benefit Analysis which seeks to relate the cost of proposed measures to the losses likely to result from a threatening impact. In reality, this is an extremely complex question and involves consideration of the likelihood of damaging events occurring, the opportunity costs of money utilised in maintaining stability, and numerous intangibles such as pride, fear, beauty, reputation, corporate image, reliability, safety, etc.

Environmental considerations have increasingly become an important factor in the choice of suitable landslide stabilisation measures, particularly issues such as visual intrusion in scenic areas or the impact on nature or geological conservation interests. This is especially so on the coast where it is now widely recognised that protection works can have significant effects on other environmental interests, such as the disruption of sediment supply to amenity beaches and the degradation of conservation sites dependent on continued erosion. Over the last decade there has been a notable shift towards **soft engineering** solutions to coastal defence problems, which aim to work with natural systems by manipulating coastal processes to the benefit of environmental interests as well as protecting coastal communities. From a geological conservation perspective, the design of coast protection works should aim to control erosion sufficiently to protect the coast whilst allowing a small degree of erosion to maintain the geological exposure[40]. In this context, the Government has recently issued environment guidance in selecting appropriate coastal defence solutions[41].

One of the most obvious methods of stabilisation is to change the distribution of weight over a slope which will have the effect of altering the balance between displacement forces and resisting forces. However, the success of corrective slope regrading (cut or fill) or the application of a restraining load (fill or structure) is determined not merely by size or shape of the alteration, but also by position on the slope. Reference to Figure 4.15 will reveal that removal of material from the lower portion of a landslide reduces resistance to shearing (toe unloading). Similarly, while application of a load to the lower slopes increases stability by moving the centre of gravity downslope (toe weighting), application of the same load on the higher parts of the same slope will have the reverse effect if it increases the overturning moment.

In this context, the concept of a "neutral point" is useful[42]. The neutral point within a landslide cross-section is the position where a load will have no effect on the Factor of Safety. The sum of neutral points on landslide cross-sections results in a neutral line across the landslide *(Figure 8.8)*. There are, in reality, two neutral lines within a landslide: the "drained neutral line" and the "undrained neutral line" (referring to drained or undrained soil conditions; *see Chapter 4*), with the latter usually displaced downslope. Any load, such as embankment fill, should be placed downslope of the undrained neutral line in the toe area of a slide. If this is not practicable, then it must be positioned below the drained neutral line. Adding weight upslope of this line will reduce the stability of the slope and may initiate further movements of the landslide.

Probably the best known example of landslide stabilisation by means of 'toe loading' or 'toe weighting' is Folkestone Warren *(Photo 4.1, Figure 4.16)* where a massive concrete structure was built on the foreshore in 1948–1954[43] to provide passive support and stop toe unloading through wave erosion. Inland, toe embankments have been used to stabilise many large landslides such as Walton's Wood on the M6[44] and the huge Taren slide on the Taff Vale Trunk Road in South Wales[45] *(Plate 41 and Photo 8.1)*.

Where clay slopes are affected by relatively shallow landsliding, it is common practice to excavate and remove the slipped material and back fill with free-draining

8.1 Construction of the toe weighting to stabilise the Taren landslide, South Wales (Rendel Palmer and Tritton).

frictional material such as brick rubble, broken rock or gravel. This method is frequently used in the case of shallow failures in road cuttings, e.g. the A12 in Essex. A further example is at Longham Wood on the M20 immediately to the east of Maidstone, Kent, where a large cutting in Gault Clay had to be excavated and backfilled with shingle from Dungeness[46]. Other, lower cuttings along this stretch of motorway have also been flattened by excavation because regrading is often considered the easiest and cheapest method of imparting stability if there is

.8 An explanation of the neutral point oncept (after Hutchinson, 1984b).

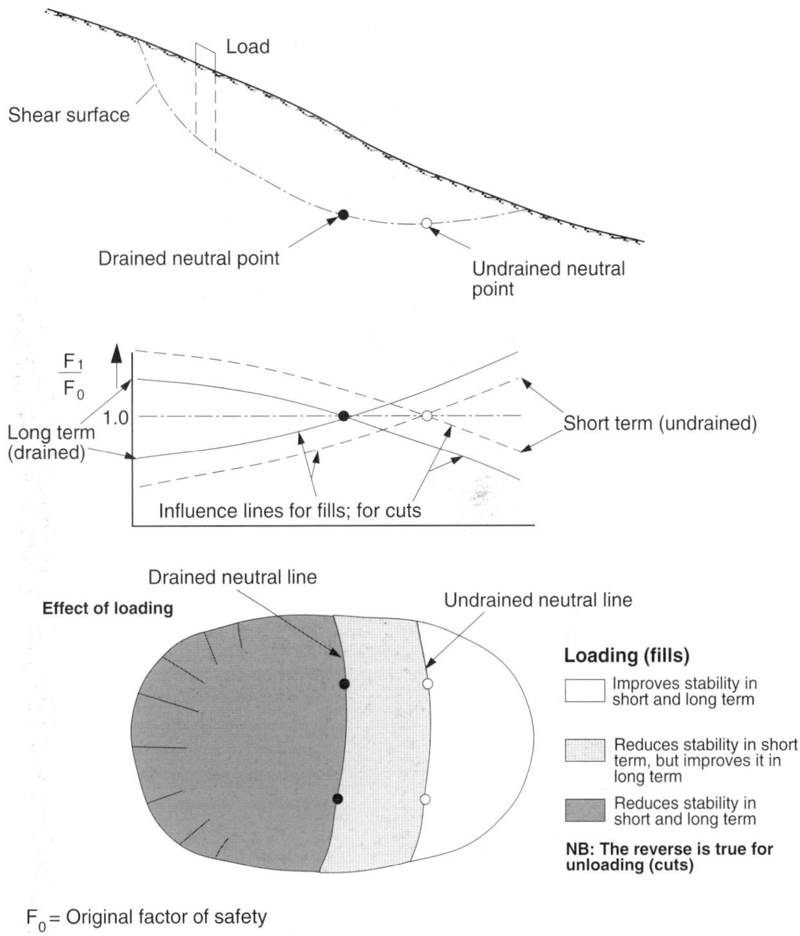

sufficient space. However, it must be noted that regrading can destabilise deep-seated failures, as happened at Marine Parade, Lyme Regis in 1962 when reprofiling of Lower Lias clay slopes resulted in the reactivation of an unrecognised, pre-existing, large slide *(Plate 26).*

The second group of measures shown in Table 8.3 – slope drainage – are often crucial to slope stabilisation because of the important role played by pore-water pressures in reducing shear resistance. Drainage systems are widely installed to collect surface water and lead it away from the sites of potential instability or to lower the groundwater table and hence reduce pore-water pressures acting along shear surfaces. Shallow drainage is usually achieved by networks of trench drains; trenches 3.5m–5.5m deep filled with gravel and usually spaced 6–29m apart[42] *(Plate 42 and Photo 8.2)*, feeding to a collector drain at the slope base which leads the water away from the site. Amongst the many documented examples of the installation of trench drainage is along the Sevenoaks Bypass[47]. Drainage of shear surfaces, on the other hand, is usually achieved by counterfort drains which are trenches sunk into the ground to intersect the shear surface and extending below it.

In the case of deep landslides, often the most effective way of lowering groundwater levels is to drive gently inclined drainage galleries into the slide mass. At Folkestone Warren, for example, a series of drainage galleries were constructed, discharging

8.2 Trench drains along the Taff Vale Trunk Road, South Wales (J S Griffiths).

just above sea level[48]. However, it is often unsafe to drive galleries into an unstable slide and, therefore, it is common for galleries to be driven into the intact rock beneath the landslide. From this position, a series of upward-directed drainage holes can be drilled through the roof of the gallery to drain the sole of the landslide.

8.3 Emplacement of reinforced anchor pads to stabilise the landslide at Nantgarw, South Wales (Rendel Palmer and Tritton).

Alternatively, the galleries can connect up a series of vertical drains sunk down from the ground surface, as was the case at Taren[49]. In instances where the groundwater is too deep to be reached by ordinary trench drains and where the slide is too small to justify an expensive drainage gallery, bored sub-horizontal drains can be used, as was the case on the Otley Bypass in Yorkshire[50]. Another approach is to use a combination of vertical drainage shafts linked to a system of sub-horizontal drains which act as gravity discharge for the vertical drains[51].

During the early part of the post-war period landslides were generally seen to be 'engineering problems' requiring 'engineered solutions' involving control by the use of conspicuous structures. This structural approach initially focussed on the construction of retaining walls but has subsequently been diversified to include a wide range of sophisticated techniques including soil and rock anchors and cantilever sheet pile walls. While there is nothing wrong with the techniques themselves, fixation with structural solutions has in some cases resulted in the adoption of over-expensive measures that proved to be less appropriate than alternative approaches involving profile modification or drainage, and there are numerous reported cases of the failure of such structures[42]. When properly designed and constructed, they can be extremely valuable, especially in areas with high loss potential, or when steep unstable rock slopes are involved or in restricted sites. For example, an active rockslope failure which threatened the new A470 Taff Vale Trunk Road at Nantgarw, South Wales, was stabilised by means of a unique array of 32 surface reinforced anchor pads which were tied into sound bedrock by 128 deep anchors[51] *(Photo 8.3)*.

The alternative, non-structural approach is exemplified by recent experimental work which has demonstrated the potential for increasing the stability of clay slopes by the addition of lime[52]. The stability of clays can be improved, both in the short-term and in the long-term, by the use of lime columns, lime piles or lime slurry injections. The addition of lime results in an almost immediate **modification** of the clay, causing a significant reduction in water content since water is used in the reaction between the lime and clay minerals. Modification is followed by **stabilisation** which results from a long-term gain in strength by the development of a cemented structure. The gain in strength can be considerable and depends upon a variety of factors including lime content, clay mineralogy, temperature and time[53]. Small landslides, natural slopes and artificial slopes can all be stabilised in this way, and the improvement in strength may allow the creation of steeper slopes on cuttings and embankments, a development that could have considerable economic implications in future road construction or widening programmes.

The remaining measures listed in Table 8.3 are relatively cheap and concerned with minimising surface effects and limiting minor impacts, e.g. wire nets to minimise the effects of stone falls etc. *(Plate 43)*. Although often of limited value when used in isolation, they can be employed in combination with slope regrading and drainage to produce low-cost, relatively low technology solutions to problems that frequently prove as effective as more impressive large-scale engineered solutions. This point will be discussed further later in this chapter.

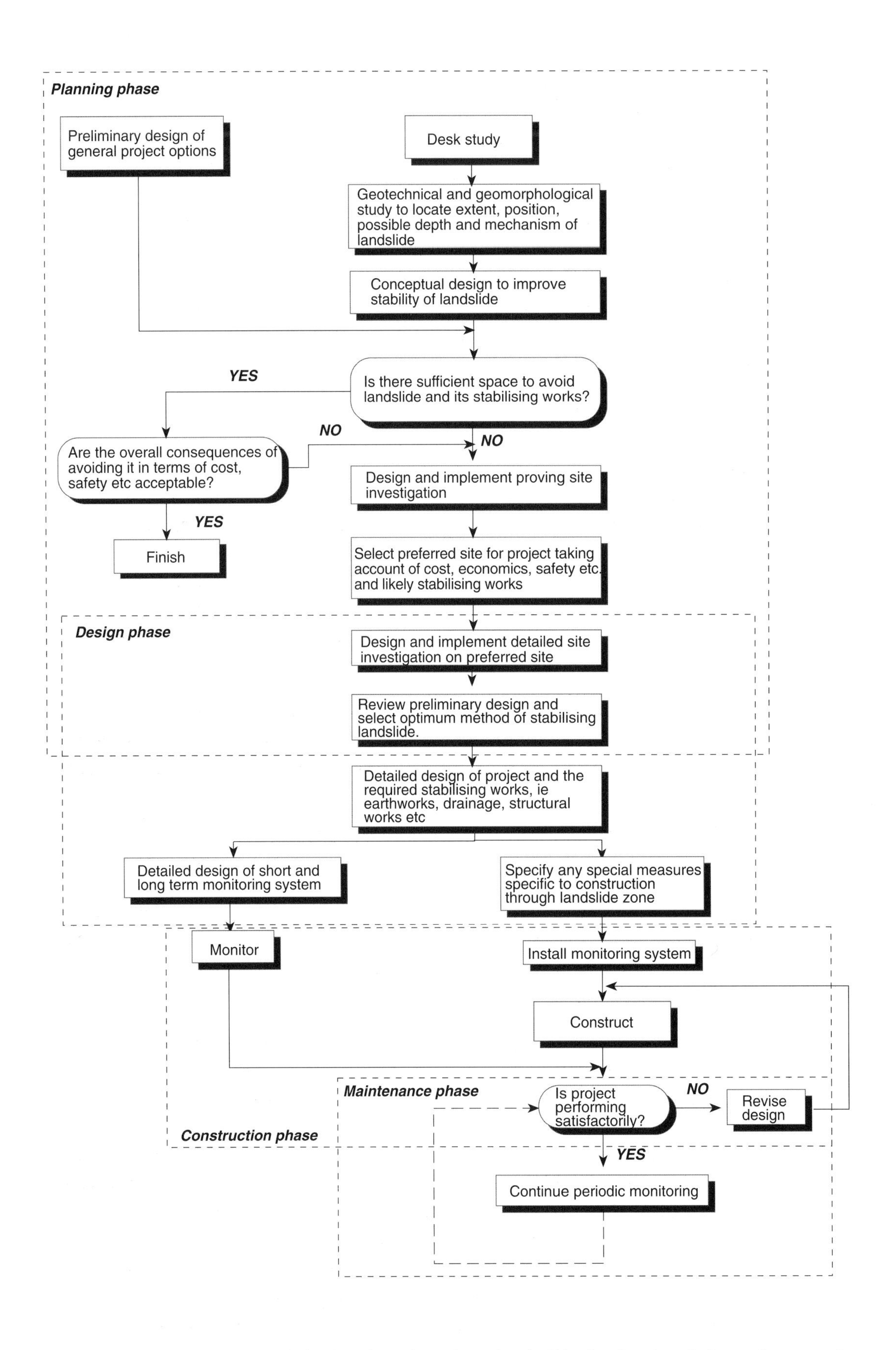

8.9 A flow diagram showing the various phases involved in the planning, design and construction of landslide stabilisation measures (after Kelly & Martin, 1986).

It is important to stress that the procedures involved in the planning, design, construction and monitoring of stabilisation works are often complex and time consuming. The sequence of activities that should be contemplated in any development/construction programme involving slope stabilisation are detailed in Figure 8.9. The wide variety of possible slope failure types and causes indicates that effective stabilisation measures can only be planned properly following detailed geomorphological and geotechnical assessment of the site directed to establishing the mix of failure mechanisms present and the interplay of causal factors. This is very important, for if the stability analyses undertaken to determine the effectiveness of proposed measures are not representative, then the solution chosen may ultimately prove to be inappropriate or unsuccessful.

In addition to mapping, sub-surface investigations and laboratory studies, assessments also have to be made of the position, pattern and rates of ground movement. There are several categories of monitoring designed for slightly differing purposes but generally involving similar techniques. **Preliminary monitoring** involves the establishment of data on pre-existing movements so that the dangers can be assessed and stabilisation methods properly designed or the site abandoned. **Precautionary monitoring** is carried out during construction in order to ensure safety and to facilitate redesign if necessary. **Post-construction monitoring** is undertaken to check on the performance of stabilisation measures and to focus attention on problems that require remedial measures.

The types of monitoring that can be undertaken vary greatly depending on the nature of the terrain, the types of failure, the scale of failure, the time available and the size of allocated funds. Distinctions must also be drawn between the monitoring of rock slopes[54] and engineering soils[55] and the techniques most suitable for inland situations as against coastal failures. A wide variety of techniques are available for monitoring surface movements including repeated accurate topographic survey, the measurement of marker positions using electro-optic distance measuring instruments, repeated accurate profiling, repeated analysis of aerial photography, extensometers, terrestrial photogrammetry[56], analytical photogrammetry[57], etc. Sub-surface techniques include borehole extensometers, borehole inclinometers, piezometers, seismometers, etc.

Unfortunately, the proper application of monitoring tends to be inhibited for three main reasons. First, satisfactory monitoring can only be undertaken after the spatial pattern of potential ground movements has been clearly established by detailed mapping and subsurface investigation. Second, meaningful results can often only be obtained from fairly lengthy periods of monitoring, especially in those instances where movements are cyclical or episodic with large return periods. Third, monitoring can prove to be a significant element of cost, not so much because of the instrumentation involved, the need for observers and the analysis required, all of which are relatively minor compared with the total expended on development projects, but mainly because of the potential for causing delay. As a consequence, there has been general reluctance to embark on monitoring with emphasis placed instead on claiming 'unforeseen ground conditions' which require 'engineering out of the difficulties' in those instances where problems have been encountered. Only in those cases where remedial measures are obviously required is monitoring widely employed. Even in these cases, records may be unsatisfactorily short because of the widespread failure to embark on monitoring exercises until their need has been proven, by which time the opportunity for a useful period of records may have been lost.

Ground investigation and monitoring together provide an assessment of the pattern of contemporary movements and the likelihood of future movements. They, therefore, provide the bases for **landslide forecasting** which is concerned with establishing the location, timing and magnitude of future movements. Landslide forecasting is an essential basis for the effective implementation of emergency action procedures and would assist greatly in establishing the need for stabilisation. Unfortunately, there are so many variables involved in landslide reactivation that it is impossible to be precise about the timing of triggering factors and the scale of subsequent failure response. Thus landslide forecasting remains highly qualitative and imprecise. As a consequence, evaluations of the future likelihood of landslide impact are dependent on **landslide prediction** which is concerned with establishing the probability of occurrence of landslide events of different magnitude. It is location and magnitude specific but not time specific and suffers from the same failing as landslide forecasting, in that uncertainty with respect to the action of the triggering factors makes calculations of probabilities imprecise *(see later)*.

However, landslide prediction is important to the calculation of **landslide risk** which is, in turn, essential to the proper assessment of cost-benefit ratios with regard to the various management options available for developments on potentially unstable sites.

Before proceeding further, it is necessary to clarify the difference between **'hazard'** and **'risk'**. The term 'hazard' refers to the likelihood of slope instability and is concerned with the location, extent, frequency, scale and rate of ground movement. The notion of 'risk', by contrast, is concerned with establishing the potential for impact on society. Risk assessments attempt to recalibrate hazard in terms of the scale of losses likely to be experienced by vulnerable human activities and artifacts. Thus, for example, two identical cliffs may have the same likelihood of rock fall (hazard rating) but pose very different levels of threat (risk) if one is located in an isolated, remote location in an inhospitable rarely visited area and the other backs a favoured holiday beach or looms above a high-cost housing development.

There are numerous and diverse definitions of risk, but the most useful with respect to landsliding is:

$$R_S = E \times H \times V$$

where R_S = Specific Risk associated with a landslide of particular magnitude;

E = Elements at Risk or the total value of population, properties and economic activities within the area under consideration;

H = the probability of occurrence of the specified magnitude of hazard event;

V = Vulnerability, or the proportion of E affected by the specified magnitude of hazard event.

It is computation of risk that is basic to the cost-benefit ratio calculations that determine what scale of stabilisation measures should be adopted as part of development design or remedial measures. However, computation of risk is not easy, for there are many aspects of Elements at Risk (E) which cannot be valued in economic terms and many components of vulnerability that cannot be quantified.

There are numerous examples of stabilisation schemes that could be used to illustrate features focussed upon in the above discussion, but probably the best are the recently completed coastal scheme at Whitby, North Yorkshire and the Castle Hill scheme near Folkestone, Kent, associated with the portals of the Channel Tunnel.

The Whitby scheme, completed in May 1990, clearly demonstrates the engineering approach to cliff stabilisation. The affected section of coast involves 750m of cliffs on the northern flanks of the town which are composed of easily eroded glacial tills and fluvio-glacial gravels overlying Middle Jurassic sandstones, siltstones and mudstones. Marine erosion has caused repeated destruction, with about 50m of cliff top recession over the last 100 years so that increasing numbers of houses have come to be threatened *(Plate 44)*. In addition, the crumbling cliffs were considered unsightly and dangerous and not in keeping with the image of a seaside resort.

Initial geomorphological investigations undertaken in 1983 identified the problem as due to the combination of basal erosion by wave attack and high groundwater tables within the slope. Following an extensive ground investigation in 1985, a series of alternative engineering solutions were designed. This led to the selection of a preferred scheme and a cost-benefit analysis to enable the scheme to be justified by the funding bodies* (MAFF and the European Regional Development Fund, Scarborough Borough Council and North Yorkshire County Council). The estimated costs of the scheme included expenditure on construction, maintenance and consultants fees; the benefits identified were protection of cliff top properties *(Plate 44)* and infrastructure, and safeguarding the tourist industry. The ratio of benefits to costs was calculated as 1.70 and resulted in scheme approval.

The 750m length of stabilisation eventually cost a total of £3 million, of which one-third was spent on sea defences and the remainder on earthworks. The scheme consisted of three main elements[58]:

- **slope reprofiling;** involving a balance of cutting and filling to create stable slopes at between 28° (cuts) and 24° (fills) *(Plate 45)*. In general, material was stripped from the upper sections of the cliff and placed on a drainage blanket on the lower slopes where it was compacted to improve the strength.
- **drainage;** interceptor drains were laid to collect water from the permeable layers of fluvio-glacial gravel. A network of trench drains was laid to control surface runoff and intercept significant springs *(Plate 42)*.

---

* Although the primary responsibility for protecting property or land lies with the owner, the Coast Protection Act 1949 empowers coast protection authorities, usually maritime district councils, to carry out works to defend the coastline. Under this Act MAFF, the Welsh Office or the Secretary of State for Scotland is responsible for making available grants towards capital expenditure.

- **rock armour wall;** design of this sea defence was based on the 50 year return period storm wave data and consists of a concrete promenade overlying fill, fronted by rock armour stones (between 4.2–6.9 tonnes each). These were brought by barge from Stavanger, Norway, in loads of c.9,000 tonnes and dumped on the beach near the low-water mark. Each rock was then carefully inspected and weighed prior to individual placement to form the defensive wall.

The heavily engineered approach adopted to combat the problems at Whitby is more than matched by the sophisticated investigations that have been undertaken in the vicinity of the Channel Tunnel terminal at Cheriton, Kent. The North Downs escarpment in this area is affected by large relict landslides in the clayey Lower Chalk and underlying Gault Clay. Detailed geomorphological mapping revealed that these features were more extensively developed than had previously been thought and underlay parts of the terminal area as well as the main portal site, thereby necessitating extremely careful investigation and stabilisation[59]. One slide in particular caused significant engineering problems.

The Castle Hill landslide *(Plate 46)* is located at exactly the point where the cut-and-cover section of the route leading from the terminal area joins the first section of bored tunnel through Castle Hill *(Figure 8.10)*. This landslide is one of a series of slides which were mapped in great detail using geomorphological techniques and then subjected to subsurface investigation. In the case of the Castle Hill landslide, geomorphological predictions of landslide form based on surface (morphological) indicators were tested by means of an adit driven into the slope along the line of the proposed tunnel and found to be remarkably accurate.

Monitoring results and historical records indicated that the slide had been moving at 1–2mm per year for the last 200 years and was therefore only marginally stable. Yet the engineering design required that two separate railway tunnels be driven through the slide mass. As a result of the careful mapping and sub-surface

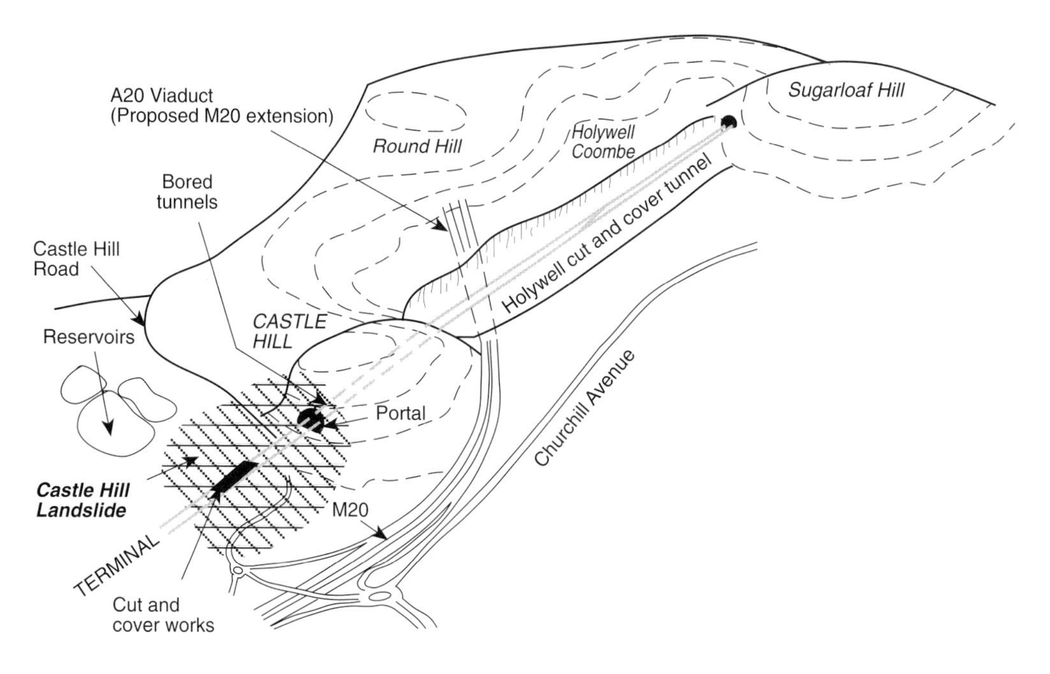

8.10 The location of the Castle Hill landslide in relation to the main Channel Tunnel Portal structures, Folkestone (after Duggleby et al, 1991).

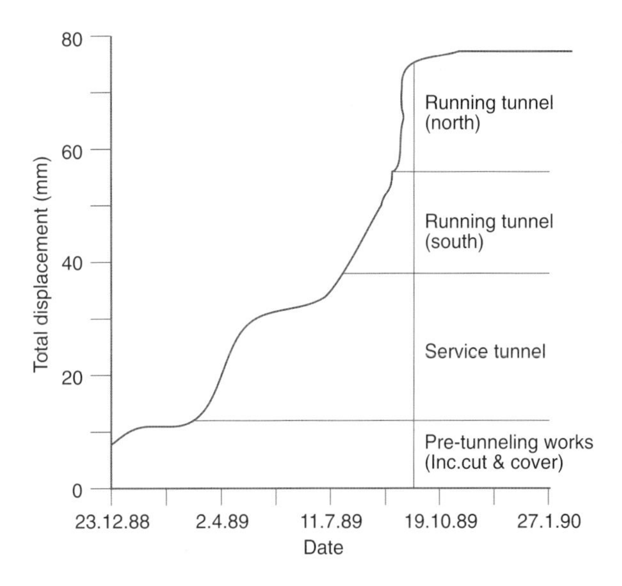

8.11 Ground movements recorded at the Castle Hill landslide during different phases of construction works (after Duggleby et al, 1991).

investigation programmes it was possible to weigh-up the alternatives of realignment (both horizontal and vertical) versus stabilisation. In this instance, stabilisation proved to be the preferred option with the necessity that there be a substantial, permanent improvement in the Factor of Safety.

In order to achieve these requirements, it proved necessary to undertake a combination of stabilisation measures and a strict construction programme. A 100,000 tonne toe weighting was placed on the foot of the slope, to maintain stability during construction and to contribute to the long-term stabilisation of the slope. In addition, three drainage adits were driven into the slide to lower the groundwater levels and thereby improve the long-term stability. Ground movements during construction were minimised by a carefully phased sequence of top-down construction (which minimised changes in overburden pressure) and controlled tunnel boring. An extensive array of inclinometers and survey pins were used to monitor the effects of construction. A typical plot of ground movement in the central area of the landslide is shown in Figure 8.11, and clearly shows how various activities contributed to displacement. Fortunately, stability returned rapidly after the end of the disturbance, and the whole project has proved to be a great success.

Planning Strategies

The previous discussion has focussed upon ground stability problems in the context of specific sites and has concentrated on identification, investigation, avoidance and stabilisation. This is the scale of consideration that has dominated past analysis of landsliding problems in Great Britain because the responsibility for safe development and secure occupancy has rested with the developer and/ or landowner. However, there are signs that controls on built development by means of planning permissions and building regulations are being progressively tightened with regard to ground instability, thereby requiring that increased knowledge of potential ground problems be obtained for extensive tracts of the land surface in addition to the detailed information produced for specific sites. This change of scale requires the employment of rather different methodologies and techniques to those already described. However, before these are discussed, it is useful to briefly examine the changing character of forward planning and development control processes within Great Britain.

The Planning System

The main control over development in Great Britain is exercised through the operation of the planning system. This system, established in 1947, seeks to control development in the public interest. A primary aim of the planning system is to facilitate acceptable and appropriate development and to strike the right balance between that development and the environment. The current legislation is contained in the **Town and Country Planning Act 1990** (as amended by the **Planning and Compensation Act 1991**) which specifies that development* may not be undertaken without planning permission. In most cases specific (express) planning permission is required upon application to the relevant local planning authority (usually the district or borough council). In Scotland, the equivalent legislation is the **Town and Country Planning (Scotland) Act 1972.**

The decision whether or not to grant planning permission must "be made in accordance with the development plan unless material considerations indicate otherwise" (S.54A). The precise nature of "material considerations" have yet to be defined and they can cover a very wide field although they must be genuine planning considerations (i.e. relate directly to the physical development) that the local planning authority or appropriate Secretary of State considers fit, reasonable or necessary.

Slope stability is not mentioned specifically in the amended 1990 Act. However, local authorities are legally required to draw up development plans which set out the land use policies and proposals that will apply in an area. There are three main types of development plan:

(i) **Structure plans,** in which shire counties set out key strategic policies as a framework for local planning. They have to have regard, inter alia, to national and regional policies. These plans have to include policies on conservation of the natural beauty and amenity of the land, as well as on the improvement of the physical environment and management of traffic.

(ii) **Local plans;** district councils and National Park authorities have to prepare plans covering the whole of their areas. These local plans set out more detailed policies for development control and proposals for specific sites. Local plans are required to be in general conformity with the structure plan.

(iii) **Unitary development plans** (UDP), prepared by London and metropolitan boroughs and combining the functions of both structure and local plans.

In advance of preparing such development plans, local planning authorities are required to undertake surveys which include reviews of "the principal physical characteristics of the area" and to produce statements on "measures for the improvement of the physical environment"[60]. While the term "physical environment" is not defined in the Act, it is generally interpreted as relating to those issues affecting pollution, nuisance, amenity, quality of life and danger. In principle, therefore, slope stability would appear to qualify as an element of the "physical

---

* 'Development' is defined in the 1990 Act as "The carrying out of building, engineering, mining or other operations in, on, over or under land, or the making of any material change in the use of any buildings or other land" (S.55).

environment" that is a "material consideration" to the granting of planning permission. Indeed, the Minister of Housing and Local Government stated in 1961 that "decisions should not be taken to permit surface development without giving due weight to what is known or can be conjectured about the stability of the site"[61]. While this advice was given in respect of surface development in coal mining areas, it is clearly applicable to other areas affected by other forms of slope instability. Ways in which instability might be treated in development plans and in considering applications for planning permission have recently been outlined by the DOE and Welsh Office in **Planning Policy Guidance Note (PPG) 14: Development on Unstable Land** (1990) and clearly point the way ahead for much greater consideration of landsliding in the context of forward planning and development control. It is important to note, however, that PPG 14 does not apply in Scotland. Indeed, it is by no means certain that this approach to addressing the planning and legal issues raised by instability problems will be taken by the Scottish Office, in the future.

The need to take account of landsliding in the planning process was reiterated by the DOE and Welsh Office in Planning Policy Guidance Note (PPG)20: Coastal Planning which states:

> "Due to the nature of coastal geology and landforms, there are risks, particularly from flooding; erosion by the sea; and land slips and falls of rock. The policy in these areas should be to avoid putting further development at risk. In particular, new development should not generally be permitted in areas which would need expensive engineering works, either to protect developments on land subject to erosion by the sea or to defend land which might be inundated by the sea. There is also the need to consider the possibility of such works causing a transfer of risks to other areas." (DOE, 1992).

Not all development requires specific planning permission. The **Town and Country Planning General Development Order 1988** (GDO) gives general planning permission in advance for certain defined classes of development set out in Schedule 2 to the Order. In National Parks, AONBs and the Broads some permitted development rights are reduced and others withdrawn completely. Important examples of permitted development which may be relevant to slope stability issues include:

- the construction of a swimming pool or hard standing within the boundaries of a property (Part 1 Class E & F);
- development under **Local or Private Acts**, or **Harbour Orders** (Part 11 Class A);
- certain development by local authorities, highway authorities, drainage bodies, the NRA, sewerage undertakers and statutory undertakers (Parts 12–17). Statutory undertakers include dock, pier, harbour, or water transport undertakings. Part 17 Class D authorises the use of any land for the spreading of any dredged material.

Such permitted development rights can be withdrawn by a direction under Article 4 of the GDO, with the Secretary of State's approval. However, it is intended that these rights should only be withdrawn in exceptional circumstances and local planning authorities are often unwilling to seek such directions because of their liability to pay compensation if planning permission for the development was subsequently refused.

Under the 1990 Act a number of activities are specifically excluded from the definition of development of land and, therefore, do not require planning permission (S.55). These include:

- maintenance, improvement or alteration of a building which does not materially affect its external appearance;
- the carrying out of road improvement or maintenance works on land within the boundary of the road;
- the carrying out of inspection, repair or renewal works (eg for sewers or water mains) by local authorities or statutory undertakers;
- the use of a building or land within its curtilage for any purpose incidental to the enjoyment of the dwelling;
- the use of land and existing buildings for agriculture and forestry.

In the context of slope instability and land management, it is important to recognise that a number of activities that do not require express planning permission described above (ie permitted development and development outside planning control), may in fact increase the risk of ground movement even to the extent of affecting adjacent property. For example, construction of swimming pools, terracing of gardens, open trench excavations, removal of vegetation from slopes, and building improvements have all been demonstrated to contribute to slope instability problems in both Luccombe and Ventnor, Isle of Wight[62]. This has led to South Wight Borough Council and the Joint Island Planning Unit to seek additional control over many such activities as part of their Undercliff Management Strategy[63] *(see Living with Landslides, below).*

Environmental Assessment

Although environmental issues have always been a factor in preparing development plans and determining planning applications, the UK is now bound by an European Community directive which requires an **environmental assessment** to be carried out before consent is granted to certain projects which are likely to have significant effects on the environment[64]. The directive contains two separate Annexes listing the type of projects which may require environmental assessment (EA):

(i) Annex I Projects for which EA is required in every case (eg a thermal power station, a crude-oil refinery, a radioactive waste storage installation, etc,);

(ii) Annex II Projects for which EA is required only if the particular project is judged likely to give rise to **significant** environmental effects (eg reclamation of land from the sea, mineral extraction, an industrial estate development, road construction, oil or gas pipelines, dams, etc,).

The planning system is the main instrument for taking account of environmental assessment in the decision making process, through the Town and Country Planning (Assessment of Environmental Effects) Regulations 1988. In preparing an EA, developers are advised to consider the possible short and long-term effects of the proposed project on slope stability[65]. However, two factors suggest that the environmental assessment procedures may not contribute greatly to the reduction of landslide hazards. First, many of the landslide problems that have arisen in recent years are the result of projects not covered by existing regulations such as small-scale housing developments *(see Chapter 7)*. Second, many activities which could have an adverse effect slope stability do not require express planning permission and, hence, there is no formal mechanism for reviewing their potential effects.

Building Regulations

Building regulations are made by the DOE and Welsh Office to secure "the health, safety, welfare and convenience of persons in and about the building" (The Building Act 1984, S.1(1)) and provide a complementary mechanism to the planning system for addressing instability issues during development. The 1991 Building Regulations drew, for the first time, the building industry's attention to the problems that landslides and subsidence may cause to a building and the surrounding area:

> "The building shall be constructed so that ground movement caused by:
>
> (a) swelling, shrinkage or freezing of the subsoil; or
>
> (b) land-slip or subsidence, in so far as the risk can be reasonably foreseen, will not impair the stability of any part of the building" (Requirement A2).

In Scotland, landsliding is not explicitly addressed within the Scottish Building Standards Regulations 1990, although Regulation 11.1 (General Loading) may be taken to consider the effects of unstable foundations.

Landsliding is a factor that needs to be taken into account before proceeding with the design of a building or its foundations. Indeed, the Approved Documents which accompany the 1991 regulations draw attention to the availability of relevant information held in the National Landslide Databank. However, the instability considerations currently required by the regulations may need to be further extended in the future, because: (i) they exclude consideration of the effects of development on adjacent areas; (ii) they exclude consideration of the cumulative effects of a number of separate decisions on the site; (iii) they exclude consideration of the impact of developments at adjacent sites ((i) in reverse); (iv) they exclude prescription of maintenance requirements; and (v) they exclude a range of development activities that can have adverse effects on slope stability such as the removal of trees, terracing, creation of hard standing with drainage, construction of tennis courts, ponds and swimming pools and the storage of materials. Thus, currently applied building regulations cannot be used to ensure full consideration of slope stability issues that may have been missed in the planning process and it was for this reason that recently issued PPG 14 has stated, unequivocally, that "the stability of the ground, in so far as it affects land use, is a material consideration which should be taken into account when deciding a planning application".

Discussion

The statements in PPG 14 and PPG 20 together with the 1991 Building Regulations clearly herald a move towards greater consideration of ground instability in development processes. The reasons for this were clearly outlined in the discussion of the variety and significance of impacts at the beginning of this chapter, to which must be added the more general requirements specified by the increasing use of environmental assessments before granting consent to major development projects. However, such a change in emphasis cannot be achieved without new and improved levels of knowledge regarding the variable nature, distribution and potential significance of ground instability and the employment of additional planning department staff with suitable expertise. It is with the former that the remainder of this chapter is concerned, most especially with the methods available for obtaining useful, spatial information on landsliding and how this knowledge is best applied in the context of land use management and development control.

Before proceeding further, however, it is necessary to recognise that attempts to incorporate knowledge of ground instability into the development control process essentially depends upon obtaining accurate spatial information on the likelihood of landsliding in the future (**landslide hazard assessment**) and the effect that this landsliding could have on society (**landslide risk assessment**). Both require accurate information on the spatial occurrence, type, magnitude and frequency of landsliding so that the landsurface can be divided into areas (**zones**) considered to pose similar levels of danger (**landslide hazard zonation**). In those cases where high levels of information are available, it may prove possible to provide a very detailed sub-division of the ground surface by a process known as **hazard micro-zonation.** The establishment of realistic hazard zones is an essential prerequisite for effective land management because it provides the spatial framework for planning controls, building regulations, site investigation requirements, insurance premiums, etc. Indeed, hazard zonation is fundamental to effective land use planning, through the avoidance of unsuitable areas, the control of development to minimise vulnerability and the removal of development from areas exposed to high levels of threat.

Landslide Hazard Assessment

Landsliding represents something of a paradox. While it is frequently possible to adequately explain failures by ex-post evaluation or hindcasting, it is a very different problem to forecast future events with respect to location, magnitude and timing. However, it has often been claimed that landsliding is one of the most predictable of geophysical hazards[10] capable of extremely high estimated potential for loss reduction (90%) with excellent cost-benefit ratios[11] or, to put it another way, landslide hazard can be effectively mitigated at relatively little cost compared with the potential scale of losses/costs. Such optimistic statements are made with particular reference to small- and medium-scale events, similar to those that occur in Great Britain, and depend upon three basic assumptions[66]:

(i) that uniformitarian principles can be applied to landslide hazard assessment, in that the conditions that led to slope instability in the past and the present will apply equally well in the future. Thus, the estimation of future instability can be based on the assessment of conditions that led to slope failure in the past;

(ii) that the main conditions that cause landsliding can be identified; and

(iii) where the causes of landsliding can be identified, it is usually possible to estimate the relative significance of individual factors. This facilitates assessment of degree of hazard by examining the number of failure-induced mechanisms present in any area.

Such statements of confident optimism should not be allowed to conceal the existence of some very real problems with landslide hazard assessment. First, while it is indeed true that earth scientists and engineers believe that the main causes of landsliding have been identified, the significance of certain elements, such as chemical changes, have not yet been fully evaluated. Second, uniformitarian principles can only be applied in general terms as both climatic change and human intervention variably influence the stability conditions of slopes over time. Third, it is not always possible to assign relative levels of significance to landslide causes with any degree of confidence, especially in those instances where complex sequences of rocks are involved in failure.

Despite these problems, numerous different approaches to landslide hazard assessment have been developed over the last three decades[67]. Three main groups of techniques can be identified:

(i) **Direct mapping,** involving the analysis of landforms and identification of existing landslides so that areas of past instability can be identified, thereby facilitating extrapolation from areas of recognised past instability to similar situations which may suffer slope failure in the future;

(ii) **Indirect mapping,** which requires the collection of data on the causes and mechanisms of landsliding so that assessments of slope stability can be made by the application of known landslide-inducing parameters;

(iii) **Land systems mapping.**

Direct Mapping

The simplest form of mapping is the **Landslide Inventory Map** which displays the distribution of recognised landslides. Methods of portrayal will vary with scale, with dots used for synoptic maps (1:100,000 scale and smaller), a mixture of dots and outline shapes for medium-scale (1:25,000–1:50,000 scale) maps and increasing detail on large-scale (1:10,000–1:5,000) maps. The information may be obtained by aerial photographic interpretation, ground survey or literature review, and additional information may be included on the type, size, age or state of activity of the recognised failures. Examples of this type of mapping include the South Wales Coalfield Landslip Survey[20] completed in 1980, the Rhondda Valley Survey (Figure 7.8) produced in the mid-1980s[23] and the nationwide map coverage produced at scales of 1:250,000 and 1:625,000 as part of the national **Review of Research into Landsliding in Great Britain**[17] *(Figure 2.14).*

The obvious next step towards the production of a landslide hazard assessment is to generalise the data in the landslide inventory map over administrative areas, geological outcrops, geomorphological regions, or through the application of isopleth or choropleth mapping *(Figure 2.17)* to provide **regional distribution maps**.

A rather different approach which places landslides in their setting and can indicate the causes of slope failure is geomorphological mapping. The nature and value of geomorphological mapping was discussed earlier in this chapter and need not be repeated here. Needless to say, detailed geomorphological mapping by ground survey or aerial photographic interpretation is not particularly cost-effective if employed over extremely large areas. However, in areas of known landsliding activity measuring up to 25km^2, geomorphological mapping can provide an excellent framework for slope instability assessment and land management. The recently completed DOE commissioned detailed geomorphological assessment of Ventnor, Isle of Wight[39], represents a benchmark for the potential value of geomorphological mapping to the planning process and is worthy of extended consideration. Indeed, rather than repeat some of this information, it is considered most effective to deal with all aspects of the Ventnor study at the end of the chapter under the heading *"Living with Landslides"*.

Indirect Mapping

The causes and mechanisms of landsliding are now thought to be reasonably well understood[66], although it has to be recognised that landsliding in any specific area is almost certainly the result of the complex interaction of a number of causes which will vary in significance over space and time. If causal factors can be identified, recorded, measured and mapped, then comparison of the distributions of different factors *(Table 8.4)* allows the identification of areas with varying potential for landsliding. The resultant element maps or factor maps may either consist of a series of overlays, each of which displays the varying strength of a particular factor, so that the cumulative results can be assessed, or they take the form of a grid or polygon mesh for the purposes of factor scoring *(Plate 47)*. Such approaches essentially entail the collection and assessment of data based on three main groups of analytical techniques:

(i) graphical methods, mainly involving the use of maps;

(ii) the analysis of empirical relationships between landsliding and individual causal factors such as slope steepness, material type, vegetation cover and rainfall intensity; and

(iii) multivariate analysis of the varied causes of landsliding.

Table 8.4 **Element (factor) maps typically produced in landslide hazard studies.**

Geological structure
Chronostratigraphy of rocks
Lithostratigraphy of rocks
Lithostratigraphy of soils
Rockhead contours
Geotechnical properties of soils
Geotechnical properties of rocks
Hydrogeological conditions
Geomorphological processes
Geomorphological history
Seismic activity
Hydrological conditions
Climate
Ground morphology
Land use
Vegetation
Pedological soils
Landslide deposits
Landslide morphometry

This approach was adopted in the Landslide Potential Mapping Programme undertaken in Rhondda, South Wales[68], which is fully described below in the section headed **"Landslide Hazard Assessment and Planning"**.

Land Systems Mapping

The land systems mapping approach occupies an intermediate position between direct and indirect mapping. Land systems mapping involves the definition of areas within which occur certain predictable combinations of surface forms, soils, vegetation and surface processes. The result is a hierarchical classification of terrain into **land systems** (large-scale divisions) composed of **land facets/land units** which are further divisible into **land elements** *(Figure 8.12)*. The technique was first developed in Australia where there was an urgent need for maps to serve as a basis for rural planning, but has subsequently been widely applied elsewhere for it "produces an extremely simple, cost-effective and versatile method for rapidly classifying large areas of relatively unknown territory, and providing a regional framework for environmental data collection and storage"[69]. The technique is well described in Ollier[70] and recent practice is discussed in King[71].

12 The relationship between land stems, land facets and land elements fter Lawrence, 1972).

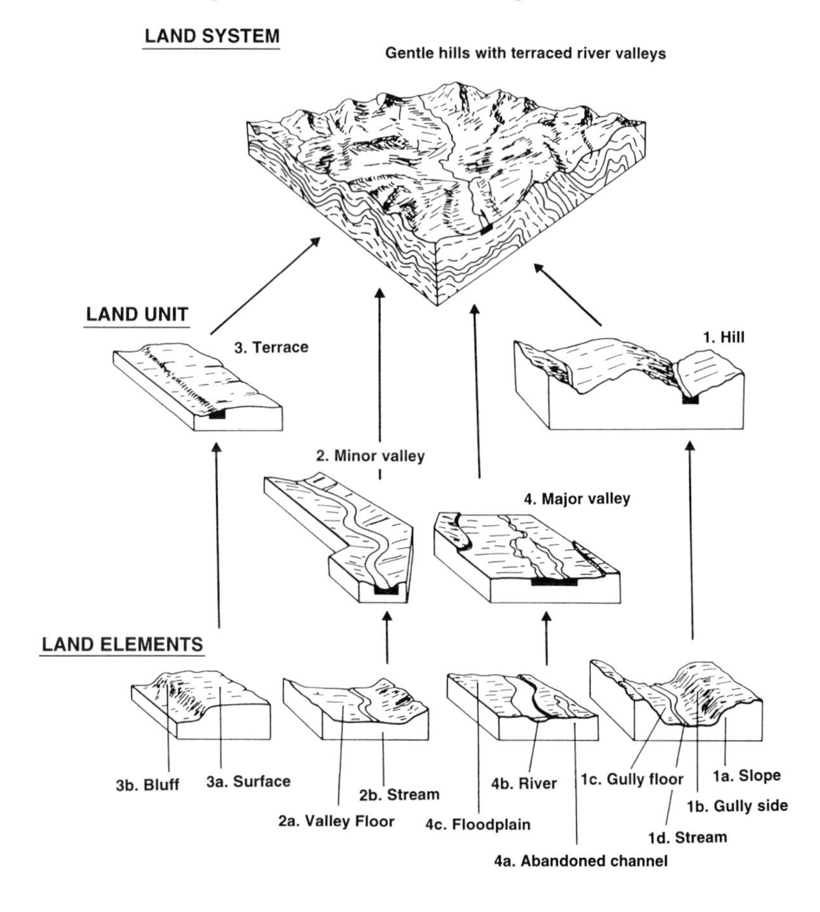

Landslide Hazard Assessment and Planning

The three approaches to landslide hazard assessment outlined above have varying applicability in Great Britain. Land systems mapping has potential for rapid preliminary assessment, especially for areas that are remote or characterised by more difficult terrain. Direct mapping has obvious applicability in relatively restricted areas with identified complex landslide problems, as will be clearly shown in the case of Ventnor. For more extensive areas, probably the greatest potential lies in the combination of direct and indirect mapping approaches, with the latter providing generalisation and extrapolation between the detail provided by the former. This has certainly proved to be the case in the Rhondda Valley study and may well indicate a possible approach that could usefully be adopted elsewhere. However, before examining the Rhondda study in a little more detail it is useful to clarify the terminology that has been developed to describe such studies.

Some confusion exists over the terms "hazard", "susceptibility" and "risk", which are often used interchangeably in the literature. There are some who argue that maps showing the distribution of landslides (landslide inventory maps) or landslide deposits, should be referred to as hazard maps. However, the term hazard only has meaning when viewed from human perspective, because hazard is culturally defined. Thus past events are only of significance if they are likely to be repeated or reactivated in the future. True hazard maps display the extent of hazardous events **(hazard zone)** together with an internal division **(hazard zonation)** reflecting the magnitude frequency distribution of the hazardous events **(hazard rating)**. Such maps must be based on evidence of past failures together with judgments as to the likelihood of future events. It is these maps that have often been referred to as **'landslide susceptibility maps'** or **'landslide potential maps'** and which are of greatest value to development planning as they attempt to assess spatial dimensions of the potential threat from slope instability, especially if modified by human intervention. They are not risk maps, however, as risk also reflects the value placed on land use and property and can vary dramatically irrespective of stability considerations.

Landslide hazard mapping techniques have been extensively reviewed elsewhere[72] and need not be described here in detail. Instead, the emphasis will be placed on the recently completed DOE/Welsh Office commissioned research programme in Rhondda, South Wales.

The research into landslide potential mapping undertaken by Halcrow[23] has utilised a fairly well established methodology *(Figure 8.13)*, albeit one that was suitably modified for the particular conditions of South Wales. Not only does it represent the first attempt to apply this approach in the context of landslide problems in Great Britain, but it pioneered efforts to translate and communicate the results in such a way that they can be utilised in planning decision-making and incorporated into local planning policy.

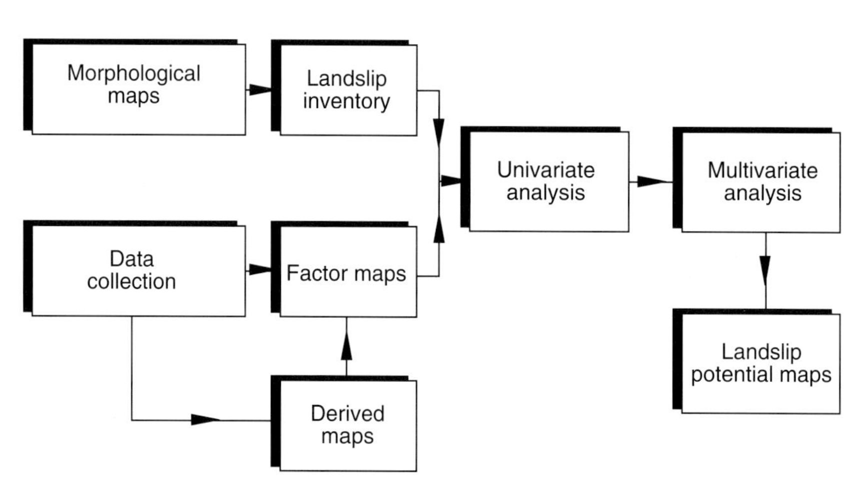

8.13 Procedure used for assessing landslip potential in Rhondda Borough (after Halcrow, 1986).

The Landslip Potential *(Plate 47)* maps were created in the following way:

(1) A landslip inventory was created by amalgamating the sites shown on existing geology maps and the maps produced by the South Wales Coalfield Survey[20] with information obtained from archival sources, reports held by consultants, contractors, statutory undertakings, local authorities, etc., and augmented by an intensive programme of direct mapping. As a consequence of this concerted effort the number of known landslides was raised from 118 to 351 **in an area that was already considered to have been adequately investigated.** This landslip inventory represents essential base-line data as to the extant problem and great care was taken to standardise the information on each landslide, including use of aerial photography to establish extent and type, and the results were portrayed on maps and entered into a computerised database[68].

(2) Factor maps were prepared for the various attributes that contribute towards slope instability *(see Chapter 4)*. A total of 20 factors were originally identified as of potential importance, but lack of data resulted in the number being reduced to 12, although subsequent analysis raised the total to 16[73] *(Table 8.5)*. Maps were prepared for each factor, sub-divided into an appropriate number of 'factor zones', each of which was characterised by a consistent set of conditions.

Table 8.5 **Factors contributing to landsliding in the Rhondda Valleys** ***(after Siddle et al., 1987)***

Intrinsic factors	**Extrinsic factors**
Slope aspect	Erosion potential
Slope angle	Ground strains due to mining
Lithostratigraphic units	Ground tilt due to mining
Superficial deposit thickness	Excavations and filling
Dip in relation to ground slope	Vegetation changes
Faulting	Seismicity
Groundwater potential	Rainfall
Joint direction/density	
Ground height	

(3) Univariate analyses were then performed to derive a rating for the degree of association between the known landslides and each factor zone. This was achieved by encoding all factor maps and landslide maps using a grid of 50m x 50m elements oriented with the National Grid. The ratings were produced by comparing the incidence of landslides in each factor zone to that of the area as a whole, i.e.:

$$\text{Rating} = \frac{\text{Proportion of a factor zone occupied by landslides} \times 100}{\text{Proportion of the entire area occupied by landslides}}$$

A rating of 100 indicated that a factor zone contained no more landslides than average but a value above 100 indicated a possible association between the factor and landsliding, while values below 100 indicated the reverse. A summary of the ratings derived from each of the original 12 factors analysed over a trial area of 22km^2 is shown in Table 8.6.

Table 8.6 **Summary of ratings for factor zones in 22 km2 trial area in the Rhondda Valleys.** *(after Siddle et al., 1987).*

Factor	Number of factor zones	Range of ratings (%)	Standard deviation
Slope aspect	9	11–136	40
Slope angle	11	6–185	68
Lithostratigraphic unit	10	8–179	50
Superficial deposit type	8	7–247	94
Superficial deposit thickness	3	41–200	70
Dip v ground slope	6	26–253	70
Faulting	3	88–130	18
Groundwater potential	6	36–251	79
Erosion potential	3	20–204	75
Mining strains	6	92–107	5
Mining tilt	6	91–194	42
Excavation and filling	5	13–145	55

(4) Multivariate analysis was then undertaken to combine ratings in a simple algorithm to define the 'landslip potential' of each 50m x 50m element. Trial and error revealed that the best results were obtained using the formula:

$$LP = \frac{RPn2 \times (RPn4 + RPn5 + RPn8)}{300}$$

where LP = Landslip Potential
RP_{n2} = Slope Angle
RP_{n4} = Superficial Deposit type
RP_{n5} = Superficial Deposit thickness
RP_{n8} = Groundwater Potential

The resultant Landslip Potential values ranged from zero to 518. As the planning authority has a relatively limited range of available responses to planning applications, it was decided to restrict the number of zones portrayed on the final maps to six *(Plate 47)*, but grouped into four main categories:

(i) areas of active landsliding;

(ii) areas of dormant landsliding;

(iii) areas of 'some landslip potential' covering the two zones with LP values of 240–518 and 121–240; and

(iv) areas with 'little landslip potential' covering the two zones with LP values in ranges 33–121 and 0–33.

The resultant Landslip Potential Maps provide a good basis for defining the relative significance of the mass movement hazard although they could be further refined with respect to nature and magnitude of potential failure. They also provide a zonation scheme that can be readily communicated to the general public, for the four zones may be clearly and simply described as: (i) "areas where landsliding is occurring"; (ii) "areas where landsliding has occurred"; (iii) "areas where landsliding may occur"; and (iv) "areas where landsliding is unlikely to occur". Such qualitative descriptions are much more suitable for publicising hazard potential within the public domain (developers, etc) than the numerical scheme which creates a false aura of accuracy and preciseness, and lends itself to misinterpretation (i.e. the larger the Landslip Potential value the bigger the likely landslide!).

However, to be of value to the planning process, always assuming that the local planning authorities begin to actively consider landsliding in development control *(see Chapter 9)*, further refinements will be required. First, the various Landslip Potential zones would have to be interpreted in terms of the hazard presented to development. For example, what is the difference, in practicable terms, between LP values of 121, 240 and 518? Second, the hazard potential presented by the different zones has to be interpreted in terms of possible planning responses regarding development control decisions, building codes and site investigation requirements. Preliminary proposals for such developments have been put forward as a result of the Rhondda study[68] *(Table 8.7; Figure 8.14)* but there remains considerable scope for further refinement.

Table 8.7 **Landslip potential categories: Rhondda Valleys, South Wales** ***(after Halcrow, 1986).***

Landslip potential categories
Minimal Landslip Potential: Areas for which the combination of factors is unlikely to adversely influence stability. Preliminary investigations for stability generally not required and costs of development unlikely to be greater than normal. Preferred for constructional development.
Slight Landslip Potential: Areas for which the combination is generally unlikely to adversely influence stability and which are therefore suitable for constructional development. Costs of development are unlikely to be greater than normal. Providing existing conditions are not radically altered by, for example, steep cuttings, external loadings or drainage modifications, investigations into stability or remedial works are unlikely to be required.
Moderate Landslip Potential: Areas for which the combination of factors may adversely influence stability, but which may be suitable for development after appropriate investigation. Constructional development requires an evaluation of slope stability by a competent person which must include appropriate recommendations to ensure stability during and after construction. Ground investigations controlled by the competent person are likely to be required and the cost of development may be increased by remedial and preventive measures. Other forms of non-constructional land use are generally suitable, providing drainage is not adversely modified.
High Landslip Potential: Areas occupied by existing landslips or those for which the combination of factors is likely to adversely influence stability and which are therefore generally unsuitable for constructional development. The cost of investigation, remedial and preventative work is likely to be high and thus preclude all but the most essential of construction. Comprehensive ground investigation under the control of a competent person will be required, including an evaluation of existing ground stability and recommendations for remedial and/or preventive measures to improve stability. Proposals for other, non-constructional land use, involving changes in topography or drainage, need to be examined by a competent person to ensure that stability is not decreased.

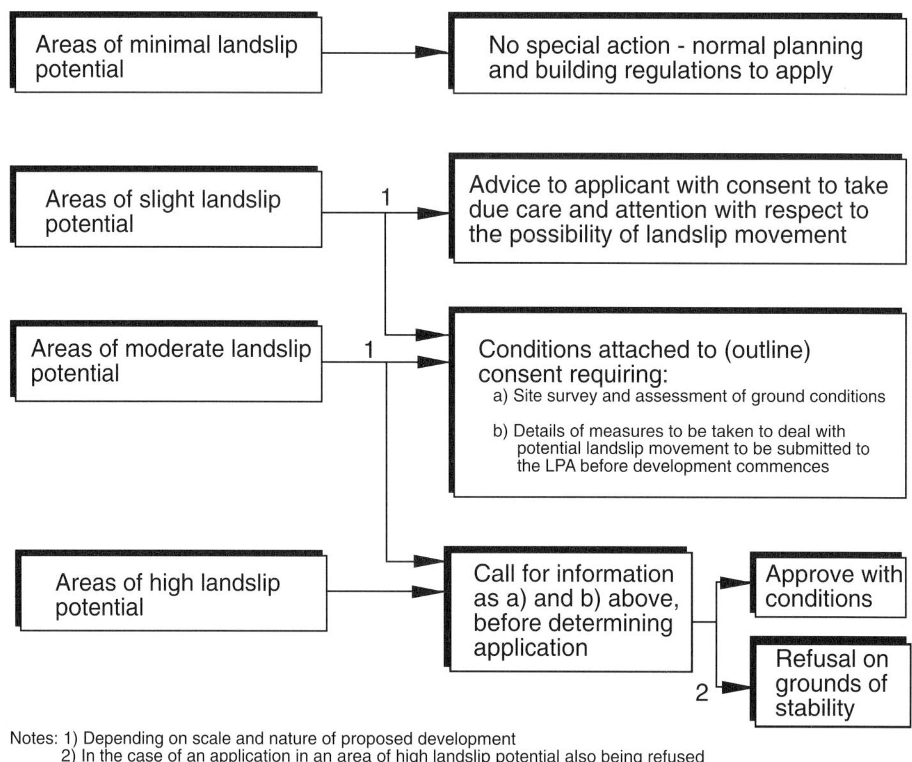

8.14 Landslip potential maps and development control for Rhondda Borough (after Siddle et al, 1987).

Living with Landslides

The Rhondda Landslip Potential Maps and other forms of hazard zonation mapping can thus be used to minimise the threat posed by instability to future development, either by assisting the adoption of avoidance strategies, where appropriate, or by ensuring adequate consideration of landslip potential on unstable or potentially unstable slopes which cannot be avoided (often for perfectly good reasons). However, there are numerous situations where historic development has resulted in intimate superimposition of urban development and infrastructure on unstable ground, often on a scale that indicates that total avoidance or abandonment are out of the question, as is recourse to large-scale and inordinately expensive engineering solutions. Under these circumstances, detailed knowledge of instability is required so that pragmatic policies can be developed to assist communities to reduce risk. This approach has been pioneered by the recently completed detailed study of Ventnor, Isle of Wight.

Ventnor is an unusual situation in that the whole town lies within an ancient landslide complex. The problems are, therefore, related to the control of the nature of development in those parts of the town which have been shown to be particularly susceptible to ground movement. This is in contrast to other landslide prone areas, such as South Wales where landslides, once identified, can be avoided by future development. The hazards faced by the local community in Ventnor have had to be defined in terms of an understanding of contemporary ground behaviour within an extensive belt of landslipped ground[39].

The methodology used in this study is shown in Figure 8.6 and highlights the importance of geomorphological mapping to the whole project. Simplified versions of the 1:2,500 scale geomorphology maps are presented in Figure 8.15 which reveal the scale and complexity of the landslides. Once the framework of landslide units had been established, it was possible to relate building damage and movement rates to units with known dimensions. From this understanding of ground behaviour it was possible to develop landslide management strategies that reflect variations in stability rather than a blanket approach to the problem.

A search through historical documents, local newspapers from 1855–1989, local authority records and published scientific research, revealed nearly 200 individual incidents of ground movement over the last two centuries. The various forms of movement that have occurred are summarised in Figure 4.24 and include first-time failures off the Chalk Downs, subsidence and joint widening within the Upper Greensand (vent formation; *Photo 7.1*), blocks of material moving en-mass along pre-existing shear surfaces, and the degradation of pre-existing landslide features by a variety of processes such as sliding and falling off scarp faces and mudslides.

Whilst the most serious damage to property has occurred in areas affected by the largest movement rates *(Figure 7.3)*, the situation is not a simple case of extensive damage to property in unstable areas and no damage in more stable areas. Often it is not clear whether some of the reported problems with buildings were a direct result of ground movement or simply due to poor building construction. It is clear, however, that in many areas the type of damage reflects a range of ground movement forms; these include differential vertical and horizontal movement, rotation, torsion, forward tilt and ground heave. There appears to be a strong relationship between the cause of damage (rotation, heave, etc.), the types of landslide movement and a particular geomorphological setting. This indicates that each geomorphological unit identified within the landslide complex *(Figure 8.15)* has its own characteristic range of stress conditions affecting structures and a characteristic type of damage. For example, the types of stresses associated with differential movement of multiple rotational slides includes stress from:

- vertical and horizontal movements
- backtilt, rotation and torsion
- forward tilt
- differential subsidence
- plastic deformation caused by uplift or heave in the toe areas

Figure 8.16 presents a simple example of how contrasting forms of movement and hence building damage can occur within a single landslide system.

This phenomenon can be further illustrated by reference to recorded damage to property within the landslide system in the Ventnor Bay area *(Figure 8.15)* where many buildings have been affected by a range of ground movements. The most seriously affected area was along the seafront where many properties have been tilted forward, probably as a result of heave of the toe of a small landslide system. By contrast, the buildings at the crest of this system have been affected by rotation and settlement. However, towards the middle of the slope, outward movements of

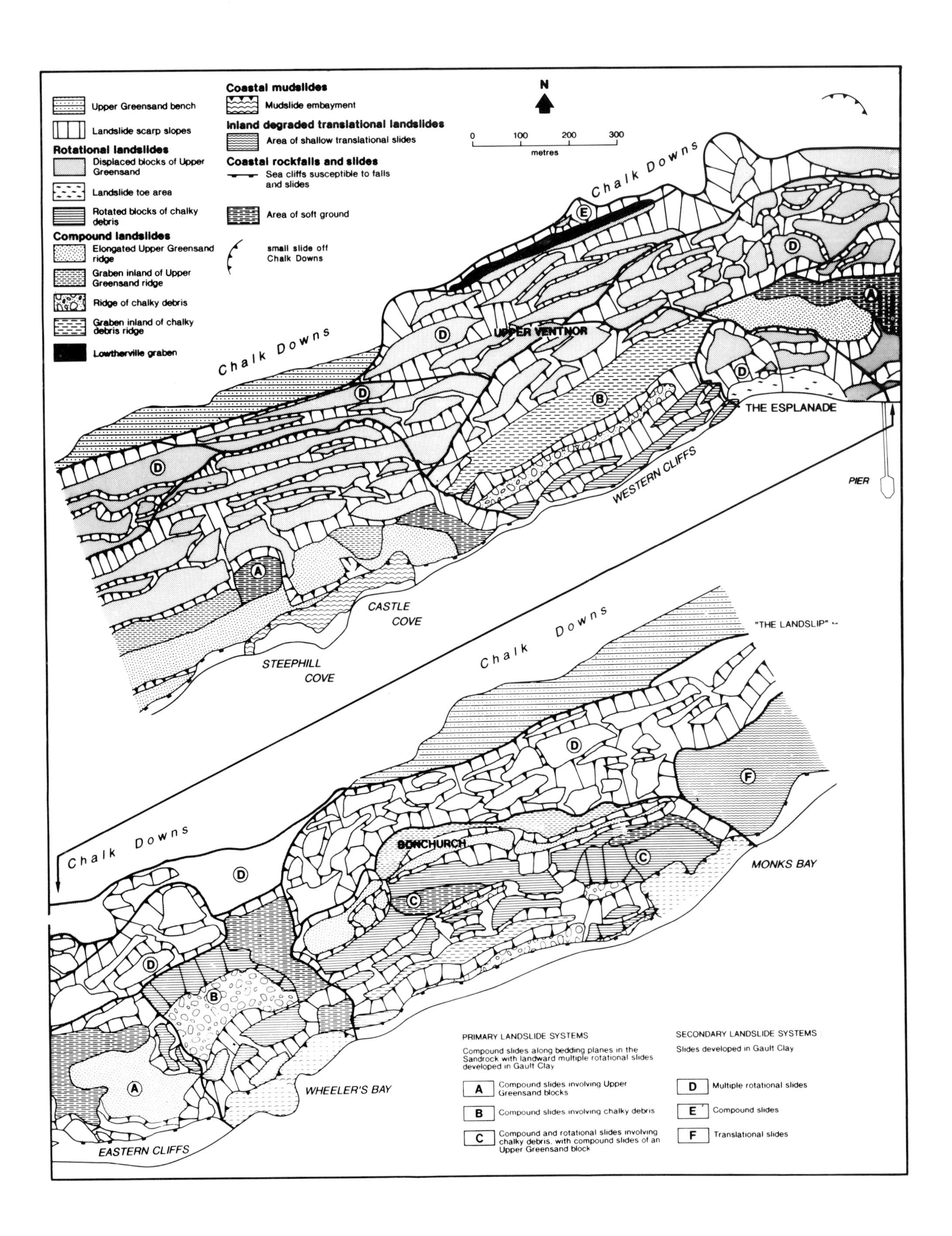

8.15 Simplified geomorphological map of Ventnor, Isle of Wight (after Lee et al., 1991a).

300mm were estimated with only limited evidence of tilting. These three different types of structural damage recorded over a short distance (300m) highlight the range of ground movements that occur within the town and the need to interpret these movements within the framework of the spatial pattern of individual landslide components and in the context of the mechanics of the landslide complex as a whole.

Understanding the geomorphology of the landslide complex at Ventnor has proved to be the key to understanding the nature and pattern of contemporary movements, and was used to form the basis for a 1:2,500 scale **Ground Behaviour** map. This map summarises the nature, magnitude and frequency of **contemporary processes** and their **impact** on the local community, being a synthesis of the following information:

- the nature and extent of individual landslide components which together form the mosaic of landslide features that are the Undercliff at Ventnor (e.g. multiple rotational slides, compound failures and mudslides; *Figure 8.15*);
- the different landslide processes which have operated within the town over the last 200 years *(Figure 4.24)*;
- the location of ground movement events recorded in the last 200 years *(Figure 7.4)*;
- the recorded rates of ground movement, over the last 30–100 years *(Figure 7.3)*;
- the intensity of damage to property caused by ground movement;
- the causes of damage to property as a result of ground movement (e.g. torsion, rotation and heave; *Figure 8.16*);
- the relationship between past landslide events and antecedent rainfall *(Figure 4.25)*.

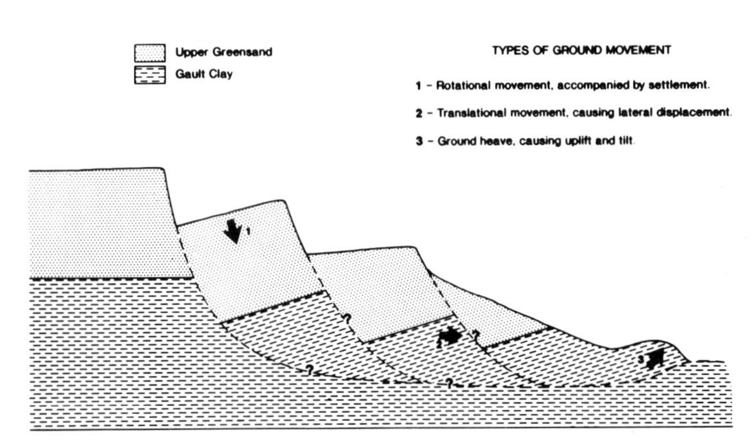

8.16 Schematic section through a multiple rotational slide showing the different types of ground movement and, hence, property damage (after Lee et al., 1991a).

The ground behaviour map clearly shows that the problems resulting from ground movement vary from place to place according to the geomorphological setting. This forms the basis for **landslide management strategies** that can be applied within the context of a zoning framework that reflects the variations in stability rather than a blanket approach to the problem. In the past there appears to have been an ad-hoc response to specific problems in Ventnor, concentrating on emergency action as required, repairing buildings where possible and condemning only those properties damaged beyond repair. Such 'crisis management' responses after the event are a common form of reaction to infrequent problems. However, in areas such as Ventnor, where ground movements are not isolated events but a recurrent problem, there are undoubted benefits to be gained from the formulation of coherent and systematic strategies for managing landslide hazards.

During the study it was recognised that there were considerable opportunities to minimise future damage to new development by incorporating the knowledge of ground behaviour within the existing planning framework. For this purpose a 1:2,500 scale Planning Guidance map was produced which relates categories of ground behaviour to forward planning and development control *(Figure 8.17, Table 8.8)*. The map indicates that different areas of the landslide complex need to be treated in different ways for both policy formulation and the review of planning applications. Areas were recognised which are likely to be suitable for development,

Table 8.8 **Planning guidance categories adopted in Ventnor, Isle of Wight** ***(after Lee et al, 1991a).***

Category	Development Plan	Development Control
A	Areas likely to be suitable for development. Contemporary ground behaviour does not impose significant constraints on Local Plan development proposals.	Results of a desk study and walkover survey should be presented with all planning applications. Detailed site investigations may be needed prior to planning decision if recommended by the preliminary study.
B	Areas likely to be subject to significant constraints on development. Local Plan development proposals should identify and take account of the ground behaviour constraints.	A desk study and walkover survey will normally need to be followed by a site investigation or geotechnical appraisal prior to lodging a planning application.
C	Areas most unsuitable for built development. Local Plan development proposals subject to major constraints.	Should development be considered it will need to be preceded by a detailed site investigation geotechnical appraisal and/or monitoring prior to any planning applications. It is likely that many planning applications in these areas may have to be refused on the basis of ground instability.
D	Areas which may or may not be suitable for development but investigations and monitoring may be required before Local Plan proposals are made.	Areas need to be investigated and monitored to determine ground behaviour. Development should be avoided unless adequate evidence of stability is presented.

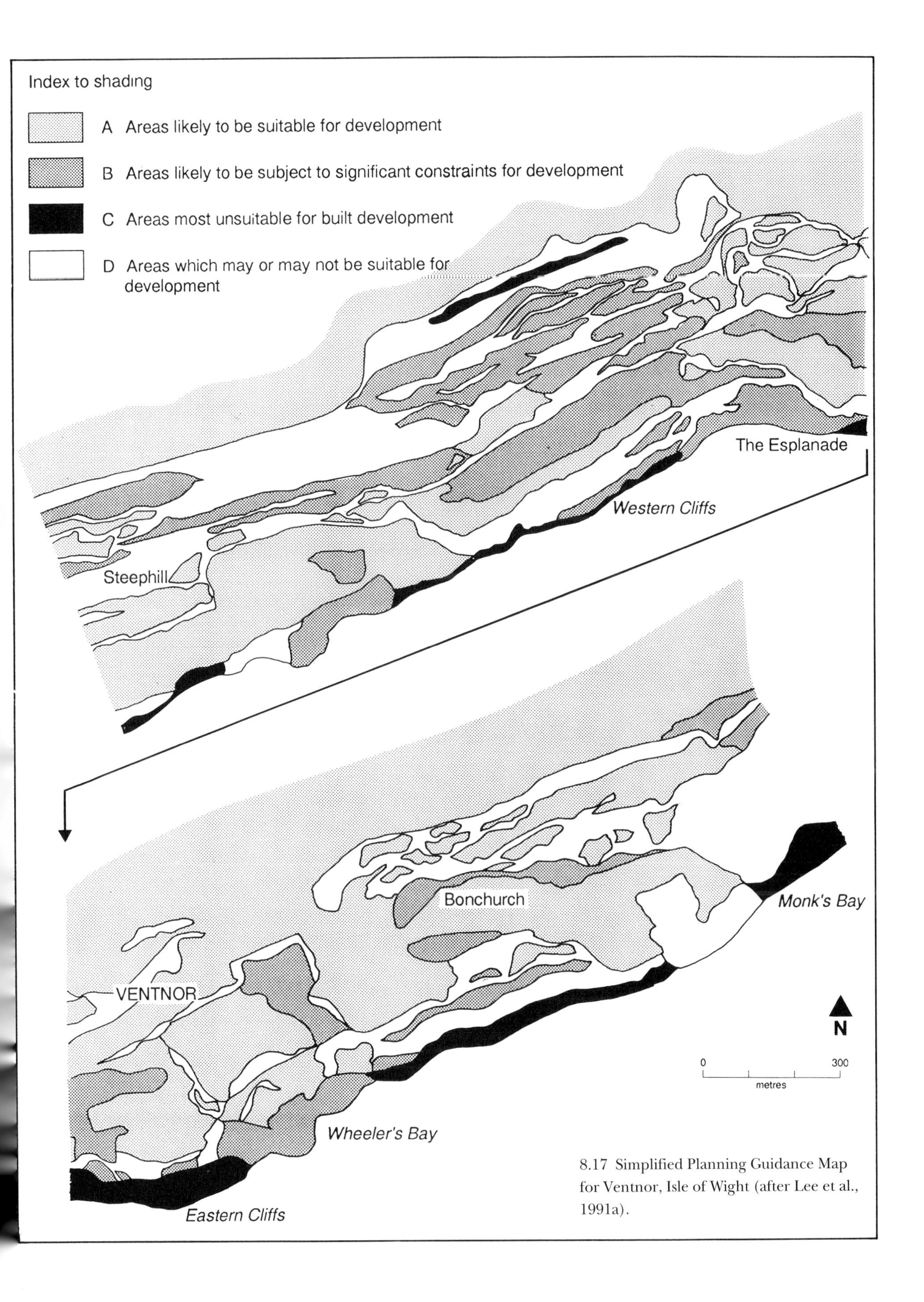

8.17 Simplified Planning Guidance Map for Ventnor, Isle of Wight (after Lee et al., 1991a).

along with areas which are either subject to significant constraints or mostly unsuitable. In some circumstances, it was recognised that the local planning authority may need to withdraw certain permitted development rights which could have an adverse effect on slope stability, by a direction under Article 4 of the GDO. This possibility is currently being considered by South Wight Borough Council and the Joint Island Planning Unit[63].

The mapping at Ventnor also clearly revealed that the problems in the Undercliff are too large-scale and complex to be solved by conventional engineering solutions which would have proved prohibitively expensive. However, available knowledge indicates that many of the problems can be reduced if the local community comes to terms with the situation and learns to live with landslides[74]. Control of construction activity is considered to be especially important in preventing instability with particular emphasis placed on:

- the avoidance of inappropriate cut and fill operations which should be carried out only after due consideration of the geomorphological setting of each site;
- the timing of earth moving operations, which should be restricted during the winter months when slopes appear to be more prone to failure *(Figure 4.25)*;
- the establishment of a code of practice for open trench excavations, such as for the maintenance of gas and water mains.

Because it is known that movements in Ventnor occur during specific rainfall and groundwater conditions, it is vital that a major effort should be directed towards preventing water leakage from service lines and sewers. It was recognised that the flow within the sewerage and water supply network should be monitored to identify areas of leakage which could then be quickly repaired. Soakaways, French drains and other natural percolation methods of disposing of surface water need to be avoided, with storm water outfalls taken down to the shore before being discharged. The importance of preventing water leakage into any landslide system cannot be over-emphasised and at Ventnor it is believed that such an approach is likely to be the most cost-effective way of reducing the future occurrence of damaging ground movements. In addition, good maintenance practice by individual homeowners can be a significant help, as neglect can result in localised instability problems *(Figure 8.18)*.

The importance of getting this message across to the public was recognised by both the DOE and South Wight Borough Council. A free public information leaflet was produced which advised the residents on what the survey revealed and what they should do to control the problems, emphasising the need to prevent water leakage. South Wight Borough Council also funded a 3-month public awareness programme during which time a shop was rented in the High Street (believed to be Britain's first Landslide Shop!) which was manned by two of the team who had carried out the study. Display boards explaining the results of the study were mounted on the walls, together with copies of the 1:2,500 scale Geomorphology, Ground Behaviour and Planning Guidance Maps. Over 2,000 residents visited the shop where they were free to discuss any fears or problems[75].

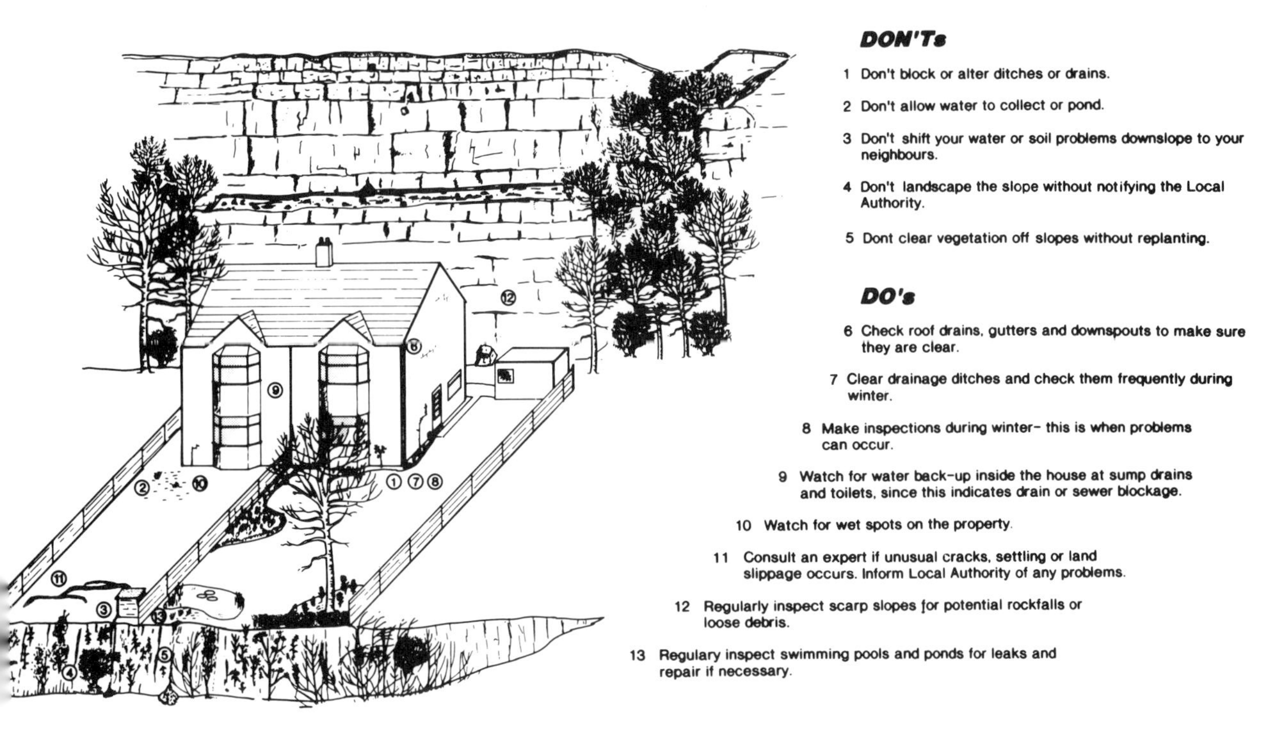

8.18 Suggested good maintenance procedures for homeowners in Ventnor, Isle of Wight (after GSL, 1991; Lee et al., 1991a).

In many cases buildings situated on slow-moving landslides are of unsuitable construction and readily damaged by ground movement. At Ventnor, for example, the most widely used forms of construction – brickwork and strip footings – are inappropriate and particularly vulnerable. Whilst solutions to the problems of unstable ground have long been practised in coal mining areas, there appears to have been no attempt to accommodate movement in the design or construction of property in Ventnor. It is probable, therefore, that improved building standards and an advisory code of practice could be useful in minimising future damage by landslide movement. It is envisaged that such a code of practice could cover the siting of buildings, earthworks, retaining walls, groundwater control, drainage, prohibition of septic tanks and soakaways, flexible service connections, foundation design, building form, and restrictions on height, loading and size of structures[76].

The first-stage proposals developed by the Ventnor study are essentially pragmatic. They are based on the view that landsliding is a natural phenomenon with identifiable causes, mechanisms and patterns of activity. The spatial extent and scale of the problems at Ventnor indicate that total avoidance or abandonment of the site are out of the question, and in any case, wholly unnecessary. Recourse to large-scale, conspicuous engineering structures would be unacceptable in a town dependent on tourism. Instead, co-ordinated measures need to be taken, first to limit the impacts of human activity that exacerbate the probability of ground movement (e.g. slope loading, pipeline leakage, etc.) and then to progressively move to a tighter system of controls based on planning permissions and building regulations that are tailor-made to the variations in stability conditions identified and delimited across the urban area.

The landslide management approaches developed at Ventnor have broader relevance in reducing the impact of landslide problems in urban areas throughout Great Britain. The most effective strategies are likely to involve a combination of planning control, control of construction activity, preventing water leakage, improved building standards and property maintenance. Indeed, the advice given the residents in Ventnor *(Plate 48)* is pertinent to many communities facing landslide problems.

Conclusion

This Chapter has sought to convey the optimistic message that a formidable array of techniques are available to minimise the impact of landslides. Much progress has been made in terms of refining investigation techniques and developing stabilisation measures, although new, more efficient, quicker or cheaper methods could well emerge in the future. Thus it is possible to argue that adequate scientific and technical bases for landslide hazard assessment are already in existence. Indeed, this view is certainly not new, for Varnes[77] was able to write almost a decade ago:

> "... of all natural hazards, slope failures are perhaps the most amenable to measures directed towards avoidance, prevention or correction. Although the areal distribution of places susceptible to slope failure may not be delineated with the same precision as areas subject to high water, they can be located more exactly than areas subject to some other hazards, such as tornadoes and drought. The causes inherent in the terrain are relatively well understood: hence it is feasible to approach landslide hazard mitigation from an areal zonation point of view, with some optimism that hazardous places can be identified" (Varnes, 1984).

Almost ten years later it is possible to be very much more positive, particularly in the light of detailed techniques of landslide hazard assessment that have developed in Hong Kong and elsewhere.

But to understand the basis of a problem is insufficient in itself. What is required is the raised perception of the need to utilise more fully the knowledge and methods that are available in order to achieve greater hazard loss mitigation or, to put it another way, transfer landslide hazard awareness and knowledge from the realms of science and engineering to the realms of the planner and legislator. Such a transfer requires the partial de-mystification of the subject of slope stability so that ground displacements cease to be envisaged as highly technical problems that can only be resolved by engineering solutions. This techno-centric approach has its roots in the past when landsliding frequently was unexpected and often resulted in circumstances where the choice appeared to lie between building structures or abandonment.

Ground stabilisation, however, is often possible by non-structural or so-called **soft engineering** methods, such as drainage or slope regrading that take full account of broader environmental considerations. The very need for stabilisation may be reduced or even removed if unstable ground is identified or avoided. Thus, the increased use of anticipatory measures designed to provide early warning of ground problems at specific sites should result in significant savings. However, this piecemeal, site specific approach is obviously inefficient in areas with widespread slope stability problems where a coherent, well informed policy for land development and construction practice is desirable. Both the Rhondda Landslip Potential Study and the Ventnor Undercliff research, described in this Chapter, indicate different facets of this anticipatory approach which could, with advantage, be applied in adapted form to other instability-prone areas such as along the Scarborough coast *(Plate 48)* and in the Bath area, to name but two. This possibility will be considered further in the next Chapter.

NOTES

1. Skempton et al., 1989
2. Smith, D I, 1989
3. Hawkins, A B, 1977
4. Hutchinson et al., 1980
5. Hutchinson et al., 1973
6. Thompson, 1991
7. Douglas, 1985
8. Allison, J A, et al., 1991
9. Jones, D K C, 1990
10. Alfors, et al., 1973
11. Leighton, 1976
12. Institution of Civil Engineers, 1991
13. National Economic Development Office, 1983
    National Economic Development Office, 1988
14. Johnson, R, 1990
15. National Audit Office, 1989
16. BSI, 1981
17. Geomorphological Service Ltd., 1986-87
18. Doornkamp, 1989
19. Northmore et al., 1978
20. Conway et al., 1980
21. Bentley & Siddle, 1990
22. Conway et al., 1983
    Halcrow, 1985
    Jones, D B, & Siddle, 1988
23. Halcrow 1986
    Halcrow, 1988
24. Head, 1985
    BSI, 1975
25. Smith, G N, 1982
    Bromhead, 1986
    Hoek & Bray, 1977
26. Halcrow, 1981
    Halcrow, 1983
    Halcrow, 1985
27. Siddle, 1986
28. Anon, 1972.
29. Dearman & Fookes, 1974
    Clark, A R & Johnson, 1975
30. Savigear, 1965
    Cooke & Doornkamp, 1990
31. Johnson, R H, 1965b
32. Brunsden & Jones, 1972
33. Brunsden et al., 1975
34. Brunsden et al., 1975
    Demek, & Embleton, 1978
    Doornkamp et al., 1979
    Cooke & Doornkamp, 1990
35. Griffiths & Hearn, 1990
36. Griffiths & Marsh, 1986
37. Lee & Moore, 1989
    Moore et al., 1991a
38. Brunsden et al., 1975
    Jones, D B & Siddle, 1988
39. Geomorphological Services Ltd, 1991
    Lee et al., 1991a
    Lee et al., 1991b
40. Hydraulics Research Ltd, 1991
41. Pethick & Bird, 1993
42. Hutchinson, 1977
    Hutchinson, 1984b
43. Viner-Brady, 1955
44. Early & Skempton, 1972
45. Kelly & Martin, 1986
46. Garrett & Wale, 1985
47. Weeks, 1969
48. Toms, 1953
49. Kelly & Martin, 1986
    Hutchinson & Martin, 1985
50. Robinson B, 1967
51. Hutchinson, 1984a
52. Rogers & Bruce, 1991
53. Ingles & Metclaf, 1972
    Dumbleton, 1962
    Transportation Research Board, 1987
54. Franklin & Denton, 1973
    Franklin, 1977
    Franklin, 1984
55. Franklin, 1984
    Bhandari, 1988
    Krauter, 1988
56. Grainger & Kalaugher, 1991

57. Chandler, J H & Moore, 1989
58. Clark, A R & Guest, 1991
59. Duggleby et al., 1991
60. Town and Country Planning Act, 1990
61. Ministry of Housing and Local Government, 1961
62. Lee & Moore, 1989
    Geomorphological Services Ltd, 1991
63. Jordan, 1991
64. EC Directive 85/337/EEC
65. DOE Circular 15/88
66. Varnes, 1984
67. Hansen, A, 1984
    Geomorphological Services Ltd., 1986-87
68. Siddle et al., 1987
69. Cooke & Doornkamp, 1990
70. Ollier, 1977
71. King, 1987
72. Hansen, A, 1984
    Varnes, 1984
    Brabb, 1984
    Einstein, 1988
73. Siddle et al., 1991
74. Lee et al., 1991c
75. Lee & Moore (in prep)
76. Lee et al., 1991a
    Noton, 1991
77. Varnes, 1984

Chapter 9

Future Prospects

Increasing Landslide Hazards?

Previous chapters in this book have repeatedly emphasised that landsliding has been under-recognised in Great Britain and have argued that this has inevitably resulted in important under-estimations as to the distribution, extent, variety and significance of slope failure. To the majority of the population, landsliding is perceived to be a problem of little importance; a hazard that mainly affects other countries with mountainous zones. As a consequence, ground problems that occur in Great Britain tend to be explained as the result of a wide variety of other factors including erosion, cliff retreat, subsidence, drought, poor foundation design, inadequate site investigation, poor workmanship, settlement, compaction, etc. While this is undoubtedly true in a great number of instances, many movements involve lateral displacement of materials over shear surfaces and therefore could be included under the heading of landsliding. Merely because they do not involve quick movements of large volumes of material is insufficient reason to search for alternative explanations.

Hopefully the three most important points to emerge from the discussions contained in this book are:

(i) that many of the extremely diverse types of slope failure that occur in Great Britain do not conform to popular, stylised conceptions of landsliding largely based on media reports of catastrophic events overseas;

(ii) that a significant proportion of mass movement phenomena actually involve relatively gentle movements which may be extended over a long period of time; and

(iii) that the hazard consists of three distinct components:

   (a) active sites or 'existing landslides' which suffer repeated movement which may be continuous, cyclical or periodic due to the recurring dominance of destabilising forces;

   (b) 'first-time' failures or 'new' landslides created when slope conditions are so altered by natural conditions or human intervention that failure takes place; and

   (c) 'reactivation' of pre-existing 'old' landslides, some of which may be of considerable antiquity, due to changes in environmental conditions caused by natural events or human intervention.

It is the last mentioned, the legacy of past slope instability, that poses the greatest problems because the features are widely distributed but often heavily camouflaged by natural processes of erosion and sedimentation, human practice (e.g. agriculture) and vegetation.

As with any newly identified hazard, or with a hazard that has come to be reassessed as potentially more significant than had previously been thought, the first priority is to obtain better and more reliable information on the scale of the problem, its

9.1 Relationships between forecasting, prediction and adjustments to hazard events.

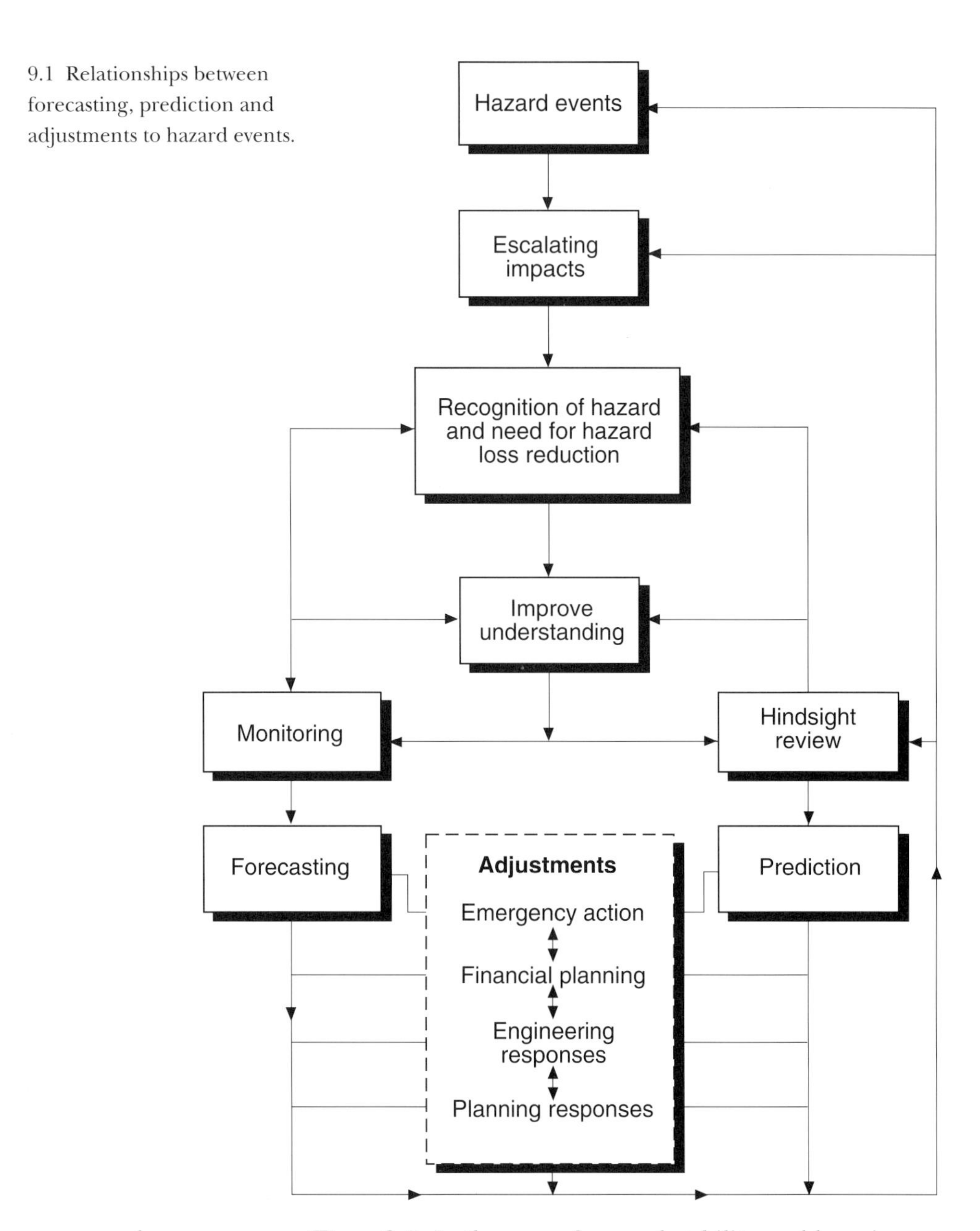

causes and consequences *(Figure 9.1)*. In the case of ground stability problems in Great Britain, information available prior to the DOE initiative in the late 1970s can best be described as superficial and fragmentary. Recognition of landsliding as a threat to property and a cause of recurring costs was largely confined to coastal areas, especially along the rapidly eroding clifflines of Eastern and Southern England stretching from Humberside to Devon. For inland areas, appreciation of instability was generally poor and extremely variable, with some awareness in areas where damage was frequently reported (e.g. the South Wales Coalfield[1], the High Peak in Derbyshire[2], Bath and environs[3]) but generally limited recognition elsewhere. The South Wales Coalfield Landslip Survey[4] and the subsequent , more detailed, Rhondda Valleys Landslip Potential Study[5], have clearly revealed the surprisingly large numbers and density of landslides that can be identified by systematic survey, and the latter study has clearly shown how such information can be used to produce hazard predictions of potential use to planners. Similarly, the

Ventnor Undercliff investigation has revealed the complexity of extensive landslipped ground and shown how detailed geomorphological mapping can produce the ideal framework for the pragmatic management of unstable land, including development control[6].

At a very different level, the nationwide Review of Research on Landsliding in Great Britain[7] has provided the first, coherent synthesis of extant knowledge as to the distribution, character and significance of mass movement. Further, it is important to note that the DOE initiative has not been limited to landsliding, but has included national reviews of the extent, causes and significance to land use planning and development of: mining subsidence; natural underground cavities; foundation conditions; natural contamination; erosion, deposition and flooding; and a preliminary assessment of seismic risk[8] . These, when completed, will provide for the first time the basis for a comprehensive regional assessment of current knowledge regarding ground conditions in Great Britain, essential base-line data for developing strategies designed to minimise future hazard impact/loss.

This note of optimism has to be tempered by the knowledge that prevailing information on landsliding in Britain remains extremely patchy and unsatisfactorily vague, especially as regards the causes and consequences of slope failure. There are, therefore, strong grounds for arguing that many further systematic studies are required before it will be possible to assess the true costs of mass movement to the British economy and how those costs are allocated through society. It may well prove that the present arrangement whereby the costs of landsliding tend to be borne by the contractor, developer or property owner, with a measure of sharing through insurance, contract renegotiation or emergency relief in the case of a disaster, will prove adequate, simple and acceptable for the near future. However, there appear strong grounds for arguing that changing circumstances will inevitably result in growing concerns about landslide damages and lead to greater controls on development in areas of slope instability. Thus, in England and Wales, **PPG 14: Development on Unstable Land, PPG 20: Coastal Planning** and the **1991 Building Regulations** should be seen as the first steps towards a system of greater precaution and tighter development control based on increased levels of scientific knowledge.

There are a number of reasons for this line of argument. First, the increasing dissemination of the results of surveys of landsliding and research into landsliding, together with the recognition of the need to take landsliding into account in the planning process, will inevitably draw attention to the problem of ground instability. As a result of this highlighting, signs of potential problems due to ground movement are likely to be recognised increasingly in other areas, resulting in investigation, publicity and further enhancement of the hazard profile. This snowballing effect is likely to be amplified by the fact that most people have little knowledge of landsliding processes and therefore may become over-concerned about the possible danger, for it is a well known fact that people tend to worry most about phenomena they understand least. This, however, is likely to be a relatively short-term effect, as landsliding in Great Britain is not a major threat to life and will inevitably be displaced from the 'agenda of concern' by new causes of public anxiety.

Second, there are no reasons why the potential for landslide activity should decrease in the future and the most likely scenario is for increased occurrence. There are two main threads to this prediction. The first concerns development pressures. Growing population, greater mobility, wealth generation and increased expectations of quality of life, indicate changing patterns of residence, growth in transport infrastructure and continued pressures for development, all of which involve ground disturbance. Thus human-induced slope instability is certain to increase unless major steps are taken to minimise the problems by means of increased anticipatory planning and better preconstruction investigation.

In addition, it also appears possible that so-called 'naturally induced' landsliding will also increase in the future as a consequence of Global Environmental Change. The debate about the likely outcomes of the 'Greenhouse Effect' or 'Global Heat Trap' in the near future (ie to 2050), has benefited from recent improvements in scientific monitoring and modelling and resulted in clear forecasts as to changing concentrations of greenhouse gasses, generally increasing global temperatures and a sensible range of sea-level change estimations for the next century[9], but less convincing conclusions as to the exact changes in climate that may be anticipated for the British Isles.

The sea-level change scenarios have been based on examination of evidence for historic global sea-level rise (10.5–15.0cm/100 years or 1.05–1.5mm per year with a preference for the range 1.1–1.2mm per year), forecasts of increased concentrations of greenhouse gases, computations of the possible effects on global climates and estimations of the effects on the global hydrological cycle, all of which have been combined to produce forecasts of future sea-level change. The 'best estimate' for global sea-level change, assuming continued accumulation of greenhouse gases with no significant alteration to current practices regarding natural resources management and utilisation - the 'Business-as-Usual' Scenario of the Intergovernmental Panel on Climatic Change (IPCC) - indicates that there will be an acceleration in the rate of global sea-level rise over the next century *(Figure 9.2)* to approaching 80cm/100years (8mm per year) by the year 2100. In the short-term,

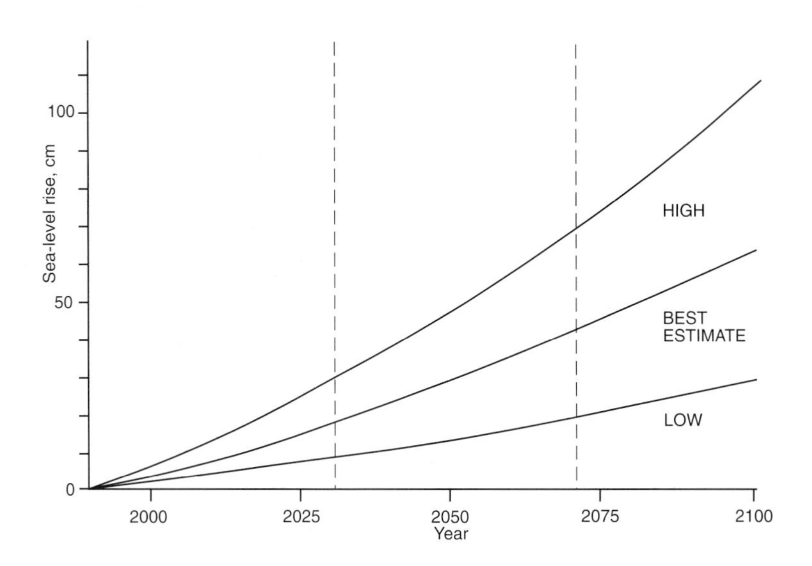

9.2 Global sea level rise, 1990–2100 for the Business-as-usual scenario (IPCC, 1990).

Table 9.1

Factors contributing to sea-level rise in cm, 1985–2030, using Business-as-usual Scenario ***(IPCC, 1990).***

Estimate	Thermal Expansion of Oceans due to warming of sea water	Melting of the Mountain Glaciers	Melting of the Greenland ice-sheet	Melting of the Antarctic ice-sheet	Total
High	14.9	10.3	3.7	0.0	28.9
Best Estimate	10.1	7.0	1.8	–0.6	18.3
Low	6.8	2.3	0.5	–0.8	8.7

however, only three of the four main potential causes of global sea-level rise are considered likely to operate *(Table 9.1)*, although in the longer-term the contribution from the melting of the Antarctic ice-sheet may become significant. Even so, 18cm of sea-level rise over the period 1985–2030 indicates a mean rate of 4mm per year, a significant increase on prevailing conditions.

However, it must be emphasised that these much quoted values relate only to the rising level of the generalised sea-water surface and merely focus on the global warming component of this eustatic effect*. There are also regional scale variations in sea-level rise which IPCC was not able to predict "because of the state of development of computer modelling"[10], as well as local amplification due to changed tidal configurations resulting from dredging and land reclamation. The quoted figures also only have meaning if it is assumed that the coastlines are stable so that eustatic sea-level change equates with the actual observed sea-level change. In reality, the coastlines of Great Britain are for the most part suffering variable rates of vertical displacement due to crustal movements *(Figure 9.3)*. As a consequence of these varied factors, the historically measured rates of sea-level change around the British coast have shown marked spatial variation and tended to depart significantly from the favoured value of 1.1–1.2mm per year. Projecting into the future is therefore problematic at the local level. Interim guidance to authorities for the design of coastal defences in England was issued by the Ministry of Agriculture, Fisheries and Food (MAFF) in November 1991, specifying allowances of 4–6mm per year to 2030 depending on the National Rivers Authority (NRA) region *(Table 9.2)*; figures that were very clearly produced by combining IPCC's estimate of eustatic change with available data on tectonic movements *(Figure 9.3)*. Irrespective of the detail, it becomes clear that accelerating rates of eustatic sea-level rise will be emphasised in Southern and Eastern Britain where the rocks are generally less resistant to erosion than in the north and west. This will inevitably result in progressive coastal erosion along undefended coastlines and shorten the effective life of existing defences[11], especially if the effect of sea-level rise is exacerbated by

* Other components of eustatic change include changing form of ocean basins, sedimentation, volcanic activity, crustal movements and gravitational changes, but their effects are considered minimal over time-scales of 40-100 years.

Table 9.2 **Advised allowances for sea-level rise to 2030** ***(MAFF 1991).***

NRA Region	Allowance
Anglia, Thames, Southern	6mm per year
North West, Northumbria	4mm per year
Remainder	5mm per year

increased storminess. Thus there are strong grounds for arguing that coastal landsliding will continue to pose a management problem for many years to come and may increase in severity in certain locations on the coastline of southern and eastern England.

The impact of global warming on the climate of the British Isles is a matter of greater uncertainty. Best current estimations[12] suggest that the mean temperature of summer months is likely to increase less markedly than for winter months, thus giving rise to a climate with noticeably warmer winters and less noticeably hotter summers. Rainfall predictions indicate that winter precipitation may increase quite markedly, but summer rainfall totals may show little change or actually fall in southern areas. The net result could be a slight shift in climatic character towards that of the Mediterranean area with increasingly marked moisture deficits in the

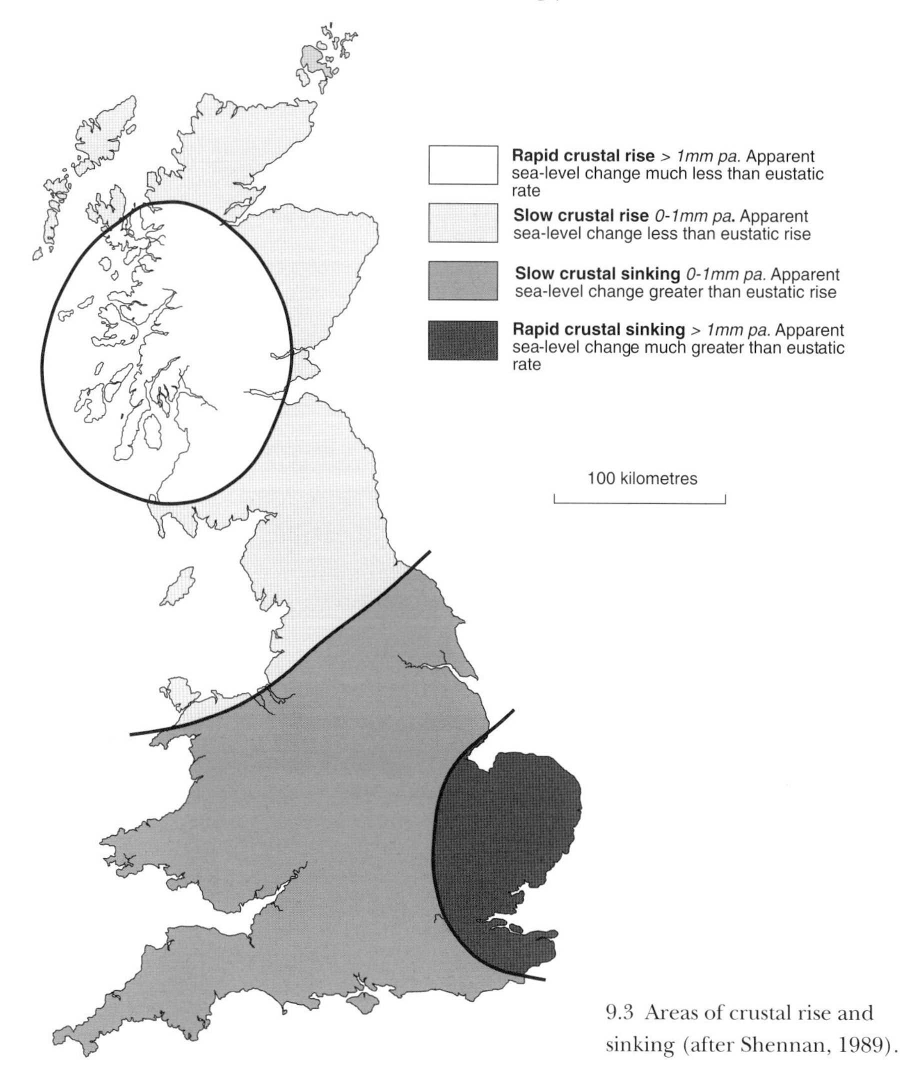

9.3 Areas of crustal rise and sinking (after Shennan, 1989).

summer months *(see Figure 5.22)*. Thus while frost weathering may diminish in importance, dessication may become more important in the southern areas, where the rainfall pattern may also become more dominated by storm events, thereby resulting in marked fluctuations of pore-water pressure. Such changes could well result in increased landsliding on clayey strata in Lowland Britain and exacerbated debris flow activity in upland areas.

Finally, interest in landslide hazard mitigation is also likely to be stimulated by increased vulnerability. The growing costs of infrastructure, the increasing penalties of infrastructure disruption, the greater complexity and inter-connectiveness of economic activity and the ever-growing awareness of insurance and litigation as methods of loss minimisation, point to increasingly careful evaluations of landslide impacts in the future. Thus the extent of vulnerability will be ever more carefully assessed as the value at risk itself grows due to economic development. This should result in the growing appreciation that it is generally more cost-effective to investigate and take account of potential instability early in the design of developments, rather than have to respond to movements during and after construction when the cost penalties may be at a maximum.

The Way Forward

Growing appreciation of risk due to slope instability should inevitably lead to pressures for the adoption of measures designed to reduce hazard loss potential. The range of available responses are outlined in Figure 9.4 and their inter-relationships shown in Figure 9.1.

Clearly the essential preliminary stages of acquiring knowledge about the nature, distribution and significance of the landslide hazard are underway in Great Britain, although much further detail has to be obtained before a clear picture will emerge. It is also apparent that the scientific and technical skills required for avoidance, prevention and correction are also available, although the relative emphasis of the past needs to be changed for the future. In this regard, the need for landsliding to be given greater prominence in the planning process should be examined further.

As was shown in Chapter 8, although landsliding is not mentioned specifically in the Town and Country Planning Act 1990 (as amended), the existing planning system is entirely capable of taking account of slope stability. That it has not always done so to date reflects limited awareness of landsliding in non-coastal situations, the prevailing view that the occurrence of landsliding is a problem for the owner, developer or insurer rather than the planner, and the widespread, but false, view that instability problems are difficult and extremely costly to predict and therefore most easily and cheaply accommodated by maintenance or the employment of remedial measures after damaging events.

These general conclusions were largely borne out by a survey of a very small sample of district and county planners undertaken in 1985 as part of the DOE commissioned **Review of Research in Landsliding in Great Britain**. The purpose of the survey was to ascertain how the problem of landsliding was perceived by planning officers, how the councils reacted to the issues raised and what kinds of decisions they made[13]. A majority of the officers contacted considered that landslides were either not a problem, or did not appear to be a problem in the sense of restricting development options, despite the recorded occurrence of landslide problems in their area and the availability of published information. Of 24 planning officers contacted: 11 considered that, despite some ambiguity in the legislation, the spirit and intent of the Town and Country Planning Acts supported the consideration of landsliding as a relevant planning issue in policy and decisions; 5 officers were unsure of the situation; and 8 considered that landslides were not a legitimate planning issue. The building control and engineering officers contacted all considered that their controls were precise but inadequate in the context of issues relating to landsliding.

The survey also found that although some form of hazard map was in widespread use, especially in mining areas or near industrial complexes, landsliding was rarely included. Indeed, only one district had landslide hazard maps for public inspection (limited to coastal landslides) and two others had landslide hazard maps for part of their areas, but not for public inspection. None of the county councils contacted included hazards as an issue in their structure plans. Out of 24 district councils interviewed, only 4 had policies in Local Plans relating to hazards, only 3 mapped the extent of hazards as a constraint and only one included landslide hazard in local plan policy. The principal development constraint factors were found to be considerations of landscape amenity and conservation, traditional planning constraints that are well understood by the profession and considered more easily defended than a constraint based on hazard.

This last point represents one of the major problems that has to be overcome if landslide hazard assessments are to be incorporated within the planning system and embraced by planning practitioners. The high degree of uncertainty which is associated with landslide forecasting is one impediment. Second, the disparity between the time-scales applicable to geomorphological processes and the time frames of decision-makers often leads to accusations that landslide hazard assessments are "vague and uncertain predictions". Third, the demarcation of zones on maps can cause problems such as adverse effects on property values (blight) and the potential for litigation, if they are not prepared with great care: indeed, some practitioners are reluctant to undertake detailed hazard surveys involving micro-zonation because of these problems. The Ventnor study[6], which defined objectively the nature and scale of contemporary processes operating within different parts of the town and their impact on buildings and structures, is likely to point the way forward in terms of development of hazard maps as landslide management tools. This approach has proved much more understandable and acceptable than hazard assessments in which zones are subjectively described as having "high", "medium" or "low" hazard; such terms have no meaning to the non-scientific community and may actually create a false impression of the real level of hazard in an area. Poorly defined hazard maps can lead to confusion and misunderstanding if they fail to get the message across in a way that can be appreciated by the local community.

Many communities and developments that are currently recognised as vulnerable to landsliding were built before planning control was established in 1947. Examples have been presented throughout this book, from the problems experienced in the South Wales valleys and the Bath area, to the numerous stretches of unstable clifflines along the south and east coasts of England. In recent years, the plight of Fairlight, on the East Sussex coast, and Luccombe, on the Isle of Wight, have both caused considerable distress to local residents. However, the reluctance of some local planning authorities to consider landslides as a planning issue has led to the situation where development has sometimes been permitted, under the planning system, in locations that have subsequently proved vulnerable such as Downderry in Cornwall[14] and Durlston Bay, Dorset[15].

Indeed, it has been recognised that one of the most serious failings of hazard management in Great Britain has been the lack of coordination between land use planning and coastal defence strategy *[16]. In recent years, coast protection or coastal defence policy, has become a very contentious issue, with local property owners interests often coming into direct conflict with recreation, national wildlife and landscape conservation priorities. Vulnerable developments naturally apply pressure for local authorities to provide coast protection measures to safeguard their investments. Such defences are, of course, funded by the public purse through MAFF grants and local authority contributions, with many schemes costing very large sums. Although cost/benefit analyses are undertaken to justify such expenditure, the whole issue of coast defence is becoming increasingly politicised with local councillors and MPs often being drawn into debates over particular schemes as a result of lobbying by those whose house or livelihood is threatened (eg Luccombe, Isle of Wight[17]). In such circumstances, the choice of engineering response has often led to conflict with conservation agencies when the need to conserve internationally important habitats or geological sites has been relegated behind the need to protect property, as in the recent debate over the defence scheme proposed for Chewton Bunny, Christchurch[18].

The Nature Conservancy Council (NCC) have observed[19]:

> "In the event of an objection being raised to coastal protection on geological grounds, the NCC is frequently berated, both privately and in public, by those promoting a particular scheme. Accusations of over-zealous behaviour and bureaucratic insensitivity abound in the one-sided war of words as local residents and their representatives promulgate their case."

The balance between protecting vulnerable properties or communities and the need to preserve the character of the unspoilt coast can also lead to considerable conflict. For example, at Easton Bavents, on the Suffolk coast, the Countryside Commission successfully opposed Waveney District Council's plans to protect a number cliff top homes because of the possible effect of the proposed scheme on an Area of Outstanding Natural Beauty[20]. Such decisions can be justified on the grounds that a

* Land use planning policy is the responsibility of the DOE, and implemented by the local planning authorities. Coastal defence policy is the responsibility of MAFF, who may provide grant aid for the funding of coast protection schemes by maritime district councils (coast protection authorities) and sea defences against flooding by either the NRA, Internal Drainage Boards or local authorities.

coast protection authority is not specifically required to protect all threatened development. However, there is, as yet, no mechanism to provide compensation for affected parties.

The choice of policy option for a particular stretch of coast should depend on the nature of the coast, whether it is developed or undeveloped, land use policies, conservation needs, benefits, costs and resources. It is clear, therefore, that the decision to proceed with an engineering solution needs to be made within the framework of a range of planning policies. However, the unwillingness of some local planning authorities to view landslides as anything other than a problem for developers has led to a tendency to manage the coastline through publicly funded engineering schemes after problems have become apparent, rather than anticipating potential problems in advance by means of effective planning control. Paradoxically, it may have been the very success of engineered defences at certain locations that subsequently led to a lack of control of development in those very same areas at risk. For example, construction of coastal defences often leads to increased pressure for development in what is now perceived to be a safe area. In reality, however, the construction of protection works can only **reduce** the risk of damage. It **cannot eliminate** the risk. Increased investment and density of development behind the defences may only lead to higher losses when, perhaps inevitably, the slopes begin to move again.

These points merely serve to focus attention on two of the major dilemmas currently facing coastal management in Great Britain: poorly coordinated decision making and the difficult management choices arising from the potential effects of sea level rise. The recent report of the House of Commons Environment Committee[10] summarised criticisms of current coastal defence policy as follows:

> "... it is often difficult to distinguish between coast protection and sea defence, that the arrangements can be complex when they are split between MAFF and the local authorities, and how inappropriate it is that coastal defence should be vested in an agricultural Government Ministry when the majority of defences protect urban land and interests" (House of Commons Environment Committee, 1992).

Their conclusion was that all aspects of coastal defence should be brought together under a single body, preferably under the responsibility of the DOE. Such a reorganisation could resolve some, but by no means all, of the current conflicts and could well pave the way for more effective management of coastal instability problems. Such improvements will, however, depend on close co-ordination and co-operation between different interest groups to ensure that acceptable solutions are found to conflicting demands, whilst recognising the significant and diverse impacts that coast protection can have on the coastal zone[16] (*Figure 9.4*).

The potential effects of sea-level rise over the next century are equally problematic. Accelerating rates of transgression to approaching 10mm a year by the end of the next century *(Table 9.1 and Figure 9.2)* indicate greatly increased potential for erosion along undefended soft rock cliffed coastlines. Clearly ubiquitous "hard" defences

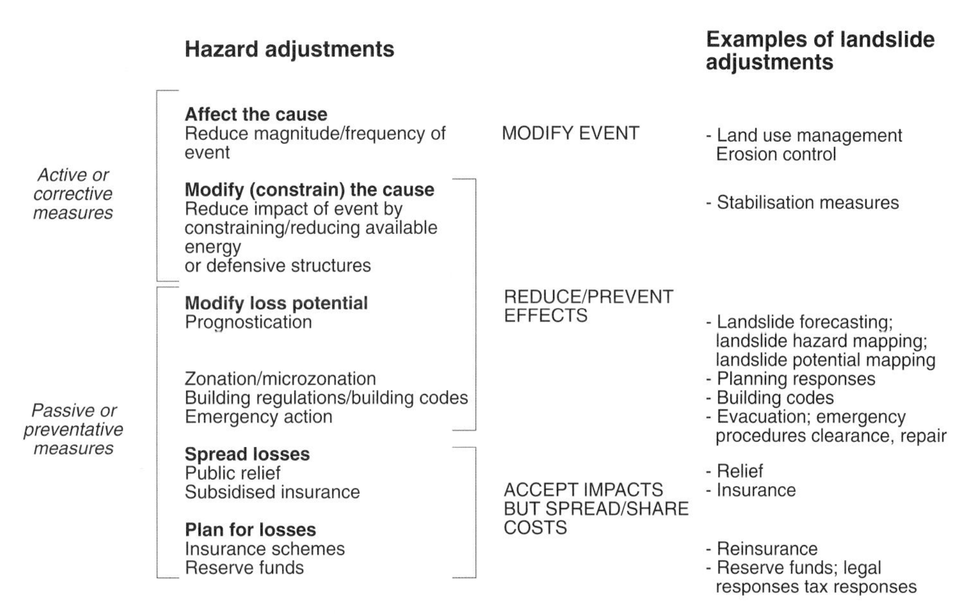

9.4 The range of adjustments to landslide hazard (adapted from Jones, D.K.C., 1991).

would be prohibitively expensive, visually intrusive, aesthetically displeasing and disrupt the supply of sediment to beaches. The available spectrum of options therefore ranges from "defend at all costs" situations where high value property or sites of strategic importance are involved, through "planned retreat" or "managed retreat" options where planned or existing defences will have a finite life before becoming overwhelmed and cliff-top retreat rates rise, to the abandonment option where eroding coastlines are left to evolve naturally because coastal defences would be inappropriate on landscape grounds or too costly in comparison with the benefits to be achieved.

These issues have been highlighted in the MAFF/Welsh Office "Strategy for Flood and Coastal Defences in England and Wales", published in October 1993:

> "Protection against ... erosion can never be absolute. A balance has to be struck between costs and benefits to the nation as a whole. For example, to protect every inch of coastline from change would not only be uneconomic but would work against the dynamic processes which determine the coastline and could have an adverse effect on defences elsewhere and on the natural environment. It is important that .... coastal defence policy and practices contribute to wider social, economic and evnironmental objectives." (MAFF 1993).

Such decisions will prove extremely difficult to make. In many instances, new development may need to be constrained by some form of "development setback policy" involving the establishment of a "setback line" which defines the inland margin of the strip of land within which no development is allowed.

This type of policy has been adopted in Norfolk where new building will not be permitted in areas likely to be affected by erosion during the expected lifetime of the building (Table 9.3). This policy involves establishing the position of a setback line by multiplying the rate of coastal retreat by the expected life of the building.

Table 9.3

Selected local planning authority responses to unstable coastal cliffs in England and Wales (as of Mid-1993).

Area of coastal instability	Counties	Strategic Policies	Comment, including selected district local plan policies
North Yorkshire; Filey to Runswick Bay	North Yorkshire	None	Instability viewed as a local rather than strategic planning matter. **Scarborough BC**; no specific policies, although instability is taken into account when considering individual proposals.
Norfolk; Weybourne to Happisburgh	Norfolk	Yes	Structure plan policy: ● presumpttion against new building in areas likely to be affected by erosion within the expected lifetime of the development. **North Norfolk DC**; proposed local plan policies include: ● presumption against new development or intensification of existing development in areas at risk. ● presumption against development that may increase coastal erosion as a result of changes in surface water run-off.
Suffolk Coast	Suffolk	Yes	Structure plan policy: ● development not acceptable which would be likely to be affected by marine erosion in its lifetime. **Suffolk Coastal DC**; local plan policies: ● stability will be a material consideration. ● the council will expect applicants to provide information relating to erosion rates and the threat to the site or buildings. ● presumpttion against development in Cobbold's Point, Felixstowe until defences are implemented or studies show the threat is not imminent. ● development not permitted on grounds of prematurity at Dunwich and Thorpeness until coastal studies are carried out.

Table 9.3 **Selected local planning authority responses to unstable coastal cliffs in England and Wales (as of Mid-1993).** *(continued)*

Area of coastal instability	Counties	Strategic Policies	Comment, including selected district local plan policies
Thames Estuary	Essex	None	No specific plan policies. **Maldon DC**; local plan policy: ● applications for development in recognised areas of instability should normally be accompanied by a stability report or statement that the site is stable. Any report should indicate how instability will be overcome and show how development will not endanger people or buildings on adjoining land.
	Kent	None	No specific structure plan policies. **Canterbury City Council**; local plan includes restrictions in two areas; East Cliff and Studd Hill: ● the council will safeguard the defined coastal protection zones from development. In unprotected areas the cliff top is in public open space and building would not be permitted, in line with conservation policies. **Swale BC**; Leysdown and Warden Bay local plan includes a development line beyond which no new development will be permitted. The line was established in 1982.
Channel Coast	Kent	None	No specific plan policies. **Shepway DC**; District local plan policy: ● planning permission within defined areas of Sandgate will not be granted until a soil survey clearly demonstrates that the site can be safely developed and that the proposed development will not have an adverse effect on the landslide as a whole.

Table 9.3

Selected local planning authority responses to unstable coastal cliffs in England and Wales (as of Mid-1993). *(continued)*

Area of coastal instability	Counties	Strategic Policies	Comment, including selected district local plan policies
	East Sussex	None	No specific structure plan policies. **Rother DC**; in response to the problems at Fairlight Village the council have formulated a specific policy: ● development not normally allowed in an area south and east of Sea Road, (to be reviewed for incorporating in the 1994 plan). **Hastings BC**; no formal policies, although conditions may be attached to planning permissions requiring adequate site investigation and soil report. Work should not proceed until and unless measures deemed to be necessary by the authority have been incorporated in the development proposals. **Eastbourne BC**; no formal policies.
Christchurch Bay	Hampshire	Yes	Structure plan policy: ● development will be restricted on those parts of the coast where the forces of coastal erosion exist. **New Forest DC**; no formal policies, although there is a commitment to maintaining and improving existing defences or providing new defences in areas of high risk.
	Dorset	None	No specific structure plan policies. **Christchurch BC**; no local plan policies. **Bournemouth BC**; no local plan policies. **Poole BC**; no local plan policies; conditions attached to planning permissions require no soakaways within 400m of cliff top.*

Table 9.3 **Selected local planning authority responses to unstable coastal cliffs in England and Wales (as of Mid-1993).** *(continued)*

Area of coastal instability	Counties	Strategic Policies	Comment, including selected district local plan policies
Glamorgan Cliffs	South Glamorgan	None	No specific structure plan policies. **Vale of Glamorgan BC**; no specific policies but: • cliff tops designated as a coastal conservation zone; development will not be permitted that does not conform with local plan policies.
	Mid Glamorgan	None	No specific structure plan policies. **Ogwr DC**; no specific policies although cliffs protected by conservation policies.

The result is a "variable setback line", depending on the nature of the cliff and the type of building.

Such a policy represents a major departure from past practice and raises important issues regarding compensation claims from affected landowners, the need for reliable information on the rates of coastal retreat to support planning decisions and the best way to explain the local authority's policy to the individuals and communities involved.

Elsewhere there has been a variable response to the planning policy guidance in PPG 14 and PPG 20. Table 9.3 summarises how a selection of different local planning authorities have addressed coastal instability issues within their areas. These range from clear "set-back" strategies for erosion risks in counties such as Norfolk and Suffolk, the need for developers to demonstrate that the site is stable in areas such as Maldon (Essex), Sandgate (Kent) and Hastings (E. Sussex) to no planning policies in many areas of south west England and Wales. It is important to recognise, however, that not all coasts with landslide hazards have been developed to date or experience the same pressures for development. Thus, the absence of planning policies in areas such as South Glamorgan, Dorset and North Yorkshire must not be seen to reflect that potential problems do not exist but rather that they are not priority issues. Processes such as erosion, instability and flooding are natural phenomena; they only become significant hazards when development encroaches into affected areas.

In many cases, undeveloped coastlines have strong protection against unsuitable development through statutory conservation designations such as AONB, SSSI or non-statutory definitions (heritage coasts) and associated planning policies. These conservation related policies can have a similar result to the policies outlined in

Table 9.3 in that they ensure that development does not encroach into vulnerable stretches of coastline.

It seems clear that landslide hazard will be increasingly used in the planning process, with policies addressing landslide issues beginning to be incorporated within development plans *(Table 9.3)*. The study undertaken as part of the **Review of Research into Landsliding in Great Britain**[7] concluded that the Government should issue advice on landsliding in order to increase awareness and reduce ambiguity. It also recommended that requirements specifically related to slope stability problems should be included within building regulations. The first of these recommendations has been fulfilled by the recent issue of **PPG 14** which outlines the effects of instability, why guidance is needed, how instability might be treated in development plans and in considering applications for planning permission, and the important role of expert advice[21].

It appears likely, and is certainly desirable, that the subject matter introduced in PPG 14 will be developed further in a new planning policy guidance specifically on the subject of slope stability. The building regulations have also been amended to include consideration of mass movement in addition to swelling, shrinkage and freezing as causes of ground movement *(see Chapter 8)*. As a consequence of this change, the building regulations will operate in a generally complementary fashion alongside the planning system.

To conclude, therefore, it appears probable that increased awareness of landsliding and better assessments of landslide-induced costs will result in increased levels of adjustment to landslide hazard. There will inevitably be numerous situations where engineering-based solutions will continue to be employed to inhibit slope movements, especially where there is existing development in landslide prone areas. However, the escalating costs of both impact and remedial measures indicate that there will be a shift away from reliance on engineering-based responses and towards anticipatory measures designed to provide better warnings of difficult conditions, better guidance as to required procedures and sufficient data for sensible land management. Hazard mapping, monitoring and planning measures are all likely to increase in profile to minimise what many earth scientists consider to be a readily predictable hazard. Instability problems are not "Acts of God"; unforeseeable, entirely natural events that can, at best, be coped with by avoidance or large-scale engineering works. The role of human activity in initiating or reactivating many slope problems should not be underestimated. Thus the landslide management approaches developed for the Ventnor study have broader relevance, highlighting the need for integrated responses by local authorities, landowners and developers, which seek to minimise the effects of human disturbance on unstable or potentially unstable slopes. The most cost-effective strategies are likely to involve a combination of planning control, preventing water leakage, control of construction activity, improved building standards and better property maintenance. Even in coastal locations there is growing recognition that reducing vulnerability by restricting development in hazardous areas or limiting construction to inland of a safety line "setback" from eroding cliffs is likely to prove preferable to reliance on heavily-engineered, "fortress-like" defences. Thus for landsliding, anticipation is usually preferable to cure.

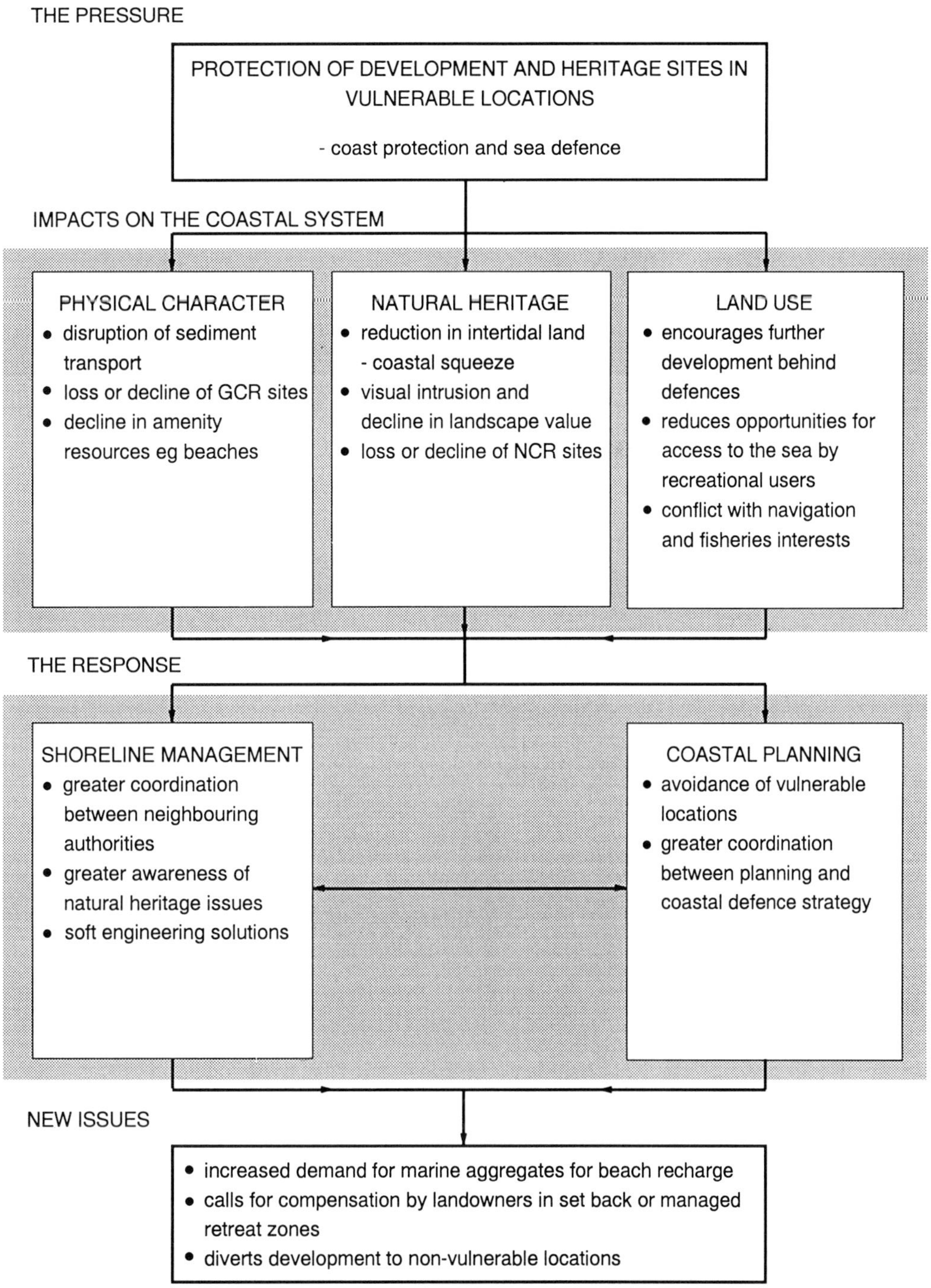

THE PRESSURE

PROTECTION OF DEVELOPMENT AND HERITAGE SITES IN VULNERABLE LOCATIONS

- coast protection and sea defence

IMPACTS ON THE COASTAL SYSTEM

PHYSICAL CHARACTER
- disruption of sediment transport
- loss or decline of GCR sites
- decline in amenity resources eg beaches

NATURAL HERITAGE
- reduction in intertidal land - coastal squeeze
- visual intrusion and decline in landscape value
- loss or decline of NCR sites

LAND USE
- encourages further development behind defences
- reduces opportunities for access to the sea by recreational users
- conflict with navigation and fisheries interests

THE RESPONSE

SHORELINE MANAGEMENT
- greater coordination between neighbouring authorities
- greater awareness of natural heritage issues
- soft engineering solutions

COASTAL PLANNING
- avoidance of vulnerable locations
- greater coordination between planning and coastal defence strategy

NEW ISSUES

- increased demand for marine aggregates for beach recharge
- calls for compensation by landowners in set back or managed retreat zones
- diverts development to non-vulnerable locations

9.5 Coastal hazard management: key issues (after Lee, 1993).

NOTES

1. Bentley & Siddle, 1990
2. Cross, 1988
   Johnson, R H, 1980
   Priestly, undated
   Doornkamp, 1990
3. Hawkins, A B, 1977
4. Conway et al., 1980
5. Halcrow, 1986
   Siddle et al., 1987
   Siddle et al., 1991
6. Geomorphological Services Ltd, 1991
   Lee et al, 1991a
7. Geomorphological Services Ltd, 1986-87
8. Brook, 1992
9. IPCC, 1990
10. House of Commons Environment Committee, 1992
11. Clayton, K M, 1991
12. Warrick & Barrow, 1991
13. Geomorphological Services Ltd, 1986-87. Series D Vol.3: Legislative & Administrative Provisions, Practice in England and Wales.
14. Coard et al, 1987
15. Anon, 1988b
16. Lee, 1993
17. Lee, in press
18. Tyhurt, 1991
19. Anon, 1988a
20. McIlroy, 1990
21. Brook, 1991

Bibliography

A Aarons, A G Weeks and R D Parkes, 1977 Site investigation for the Channel Tunnel British ferry terminal. Ground Engineering, May 1977 pp.43–47.

K J Ackerman and R Cave, 1967 Superficial deposits and structures including landslip in the Stroud district, Gloucestershire. Proceedings of the Geologists Association. Vol.78 pp.567–586.

K Addison, 1987 Debris flow during intense rainfall in Snowdonia, North Wales; a preliminary survey. Earth Surface Processes and Landforms. Vol.12 pp.561–566.

J T Alfors, J L Burnett and T E Gay, 1973. Urban geology masterplan for California. Californian Division of Mines and Geology, Bulletin 198.

J A Allison, J M Mawditt and G T Williams, 1991. The use of bored piles and counterfort drains to stabilise a major landslip: a comparison of theoretical and field performance. In: R J Chandler (ed) Slope stability engineering: developments and applications, Thomas Telford. pp. 369–376.

R J Allison, (ed.) 1990. Landslides of the Dorset coast. British Geomorphological Research Group Field Guide.

S N Al-Saadi, 1982 A method for mapping unstable slopes with reference to the coastline of S.W. Dyfed, Wales. PhD thesis (unpublished), University of Bristol.

F W Anderson and K C Dunham, 1966 The geology of northern Skye. Memoir, Geological Survey of Scotland.

M G Anderson and K S Richards, (eds.) 1987. Slope stability; Geotechnical Engineering and Geomorphology. Wiley.

Anon, 1928a. Fall of cliff in Isle of Wight, 25.7.1928.

Anon, 1928b. Huge Undercliff landslide. Isle of Wight County Press. 28.7.1928.

Anon, 1931. Earthquake felt in Isle of Wight. Isle of Wight County Press, 13 June, 1931.

Anon, 1951. Sussex County Magazine, 25 p.441.

Anon, 1961. The Isle of Wight County Press, 11.2.1961.

Anon, 1972. The preparation of a maps and plans in terms of engineering geology. Quarterly Journal of Engineering Geology. Vol.5 pp.295–382.

Anon, 1988a. The eroding coastline – coastal protection and conservation. Earth Science Conservation. No.24 pp.14–18.

Anon, 1988b. Cliff collapse endangers apartment block. Earth Science Conservation. No.24 pp.35–36.

Anon, 1988c. The Isle of Wight County Press. 15.7.1988.

M A Arber, 1941 The coastal landslips of west Dorset. Proceedings of the Geologists Association. Vol.52 pp.273–283.

M A Arber, 1973 Landslips near Lyme Regis. Proceedings of the Geologists Association. Vol.84 pp.121–133.

R S Arthurton and A J Wadge 1981 Geology of the country around Penrith (Sheet 24). Memoirs of the Geological Survey of Great Britain. London.

E B Bailey and E M Anderson, 1925 The geology of Staffa, Iona and western Mull. Memoir, Geological Survey of Scotland.

P D Baird and W V Lewis, 1957 The Cairngorm floods, 1956: Summer solifluction and distributary formation. Scottish Geographical Magazine Vol.73 pp.91–100.

C K Ballantyne, 1981 Periglacial landforms and environments on mountains in the northern Highlands of Scotland, PhD thesis (unpublished), University of Edinburgh.

C K Ballantyne, 1986 Landslides and slope failures in Scotland: a review. Scottish Geographical Magazine. Vol.102 pp.134-150.

C K Ballantyne, 1987. The present day periglaciation of upland Britain. In: J Boardman (ed.) Periglacial processes and landforms in Britain and Ireland. Cambridge University Press. pp.113–126.

C K Ballantyne, 1990. The late Quaternary glacial history of the Trotternish Escarpment, Isle of Skye, Scotland, and its implications for ice-sheet reconstruction. Proceedings of the Geologists Association. Vol.101 pp.171–186.

C K Ballantyne and M P Kirkbride, 1987 Rockfall activity in Upland Britain during the Loch Lomond Stadial. Geographical Journal. Vol.153 pp.86–92.

M E Barton, 1970 The degradation of the Barton Clay cliffs of Hampshire, England. Proceedings of the 1st International Congress of Engineering Geology, Paris. Vol.1 pp.131–140.

M E Barton, 1973 The degradation of the Barton Clay cliffs of Hampshire. Quarterly Journal of Engineering Geology. Vol.6 pp. 423–440.

M E Barton, 1977 Landsliding along bedding planes. Bulletin of the International Association of Engineering Geology No.16 (Proceedings of the Symposium on Landslides and Other Mass Movements, Prague, Sept. 1977), pp.5–7.

M E Barton, 1984 The preferred path of landslip shear surfaces in over-consolidated clays and soft rocks. In: Proceedings of the IVth International Symposium on Landslides, Toronto. Vol.3 pp.75–79.

M E Barton, 1984. The mechanics of cliff failure at Shanklin, Isle of Wight. In: Proceedings of the IVth International Symposium on Landslides, Toronto. pp.605–609.

M E Barton, 1989 Landsliding of the Upper Greensand scarp slope in East Hampshire. Field Note, Engineering Group of the Geological Society Field Meeting 14–16th April 1989.

M E Barton, 1990. Stability and recession of the Chalk Cliffs at Compton Down, Isle of Wight. In: Chalk. Proceedings of the International Chalk Symposium held at Brighton Polytechnic on 4–7 September 1989, Thomas Telford.

M E Barton and B J Coles, 1983. Rates of movement of soil slopes in southern England using inclinometers and surface peg surveying. Proceedings of the International Symposium on Field Measurements in Geotechnics (Zurich). Balkema, Rotterdam, Vol.1 pp.609–618.

M E Barton and B J Coles, 1984. The characteristics and rates of the various slope degradation processes in the Barton Clay Cliffs of Hampshire. Quarterly Journal of Engineering Geology. Vol.17 No.2 pp.117–136.

S P Bentley and H J Siddle, 1990. The evolution of landslide research in the South Wales Coalfield. Proceedings of the Geologists Association. Vol.101 pp.47–62.

K Beven, A Lawson and A McDonald 1978. A landslip/debris flow in Bilsdale, North York Moors, September 1976. Earth Surface Processes. Vol.3 pp.407–419.

R K Bhandari, 1988. Special lecture: Practical lessons in investigation, instrumentation and monitoring of landslides. In: C Bonnard (ed.) Landslides Proceedings of the Vth International Symposium on Landslides, Laussane. Vol. 3.

R K Bhandari and J N Hutchinson, 1982. Coastal mudflows in the Oligocene clays of Bouldnor Cliff, Isle of Wight. In A Sheko (ed) Landslides and Mudflows. UNESCO-UNEP Seminar, Moscow, pp.176–199.

S J Biczysko, 1981. Relict landslip in West Northamptonshire. Quarterly Journal of Engineering Geology. Vol.14 pp.169–174.

G M Binnie, 1978. The collapse of the Dale Dyke dam in retrospect. Quarterly Journal of Engineering Geology. Vol.11 pp.305–324.

A W Bishop, 1973. The stability of tips and spoil heaps. Quarterly Journal of Engineering Geology Vol.6 pp.335–376.

A W Bishop and L Bjerrum, 1960. The relevance of the triaxial test to the solution of stability problems. ASCE Research Conference on Shear Strength of Cohesive Soils, Boulder Colorado. pp.437–501.

A W Bishop, J N Hutchinson, A D M Penman and H E Evans, 1969. Geotechnical investigation into the causes and circumstances of the disaster of 21st October 1966. Unpublished report for Aberfan Disaster enquiry.

A Blackham, C Davies and J Flenley, 1981. Evidence for Late Devensian landslipping and Late Flandrian forest degeneration at Gormire Lake, North Yorkshire. In: J Neale and J Flenley (eds.) The Quaternary in Britain. Pergamon. Oxford. pp.184–199.

G S Boulton, A S Jones, K M Clayton and M J Kenning, 1977. A British Ice Sheet Model and patterns of glacial erosion and deposition in Britain. In F Shotton (ed) British Quaternary Studies: Recent Advances. Oxford University Press, pp.231–246.

D Q Bowen, J Rose, A M McCabe and D G Sutherland, 1986. Correlation of Quaternary glaciations in England, Ireland, Scotland and Wales. Quaternary Science Reviews, Quaternary Glaciations in the Northern Hemisphere. pp.299–340.

M M Bower, 1959. A summary of available evidence and a further investigation of the causes, methods and results of erosion in blanket peat. MSc thesis, (unpublished) University of London.

D R Bowes, 1960. A bog burst in the Isle of Lewis. Scottish Geographical Magazine. Vol.76 pp.21–23.

W Boyd-Dawkins, 1904. Note on the effect of pressure causing folds at the bottom of valleys (Longsett and Derwent reservoirs). Memoirs of the Proceedings of the Manchester Literature and Philosophical Society. Vol. XLIX, pp.vii–viii.

E E Brabb, 1984. Innovative approaches to landslide hazard and risk mapping. Proceedings of the IVth International Symposium on Landslides, Toronto Vol.1 pp.307–324.

E E Brabb and B L Harrod, 1989. Landslides: extent and economic significance. Balkema, Rotterdam/Brookfields.

G Brannon, 1825 Vectis scenery: being a series of original and select views of the Isle of Wight etc. Wootton Common.

M Bray, 1990. Landslide and littoral zone sediment transport. In: R J Allison (Ed.) Landslides of the Dorset Coast. British Geomorphological Research Group Field Guide, pp.107–119.

M Bray, 1991. Shingle transport budget dynamics on the south-west Dorset coast. PhD thesis, (unpublished). London School of Economics, University of London.

D J Briggs and F M Courtney, 1972. Ridge and trough topography in the North Cotswolds. Proceedings of the Cotteswold Naturalists Field Club. Vol.36 pp.94–103.

British Geological Survey, 1973 (Denbigh) Sheet 107.

British Standards Institution, 1975. BS 1377 Methods of test for soils for civil engineering purposes.

British Standards Institution, 1981. BS 5930 Code of Practice for Site Investigations.

E N Bromhead, 1978a. Large landslides in London Clay and at Herne Bay, Kent. Quarterly Journal of Engineering Geology. Vol.7 pp.291–304.

E N Bromhead, 1978b. Cliff stability at Herne Bay. Civil Engineering and Public Works Review. Vol.67 pp.788–792.

E N Bromhead, 1979. Factors affecting the transition between the various types of mass movement in coastal cliffs consisting largely of overconsolidated clay with special reference to Southern England. Quarterly Journal of Engineering Geology. Vol.12 pp.291–300.

E N Bromhead, 1986. The stability of slopes. Surrey University Press.

E N Bromhead, 1987. Groundwater and Landslides: Principles and Practice. Memoir of the Geological Society of China. Vol.9 pp.147–158.

E N Bromhead, M P Chandler and J N Hutchinson, 1991. The recent history and geotechnics of landslides at Gore Cliff, Isle of Wight. In: R J Chandler (ed) Slope stability engineering: developments and applications, Thomas Telford. pp.189–196.

D Brook, 1991. Planning aspects of slopes in Britain. In: R J Chandler (ed) Slope stability engineering: developments and applications, Thomas Telford. pp.85–94.

D Brook, 1992. Department of the Environment research on geological hazards. Geoscientist. Vol.2, No.1, p44. Abstract of paper presented to the Engineering Group of the Geological Society meeting on Hazards to Development in Great Britain, March 1992.

I J Brown, 1975. Mineral working and land reclamation in the Coalbrookdale Coalfield. PhD thesis (unpublished). Leicester University.

W J Brown, 1966. Contributions to the geomorphology of west Dorset and south Somerset. PhD thesis (unpublished). University of London.

D Brunsden, 1969. Moving cliffs of Black Ven. Geographical Magazine. Vol.41 pp.372–374.

D Brunsden, 1979a. Mass movements. In: C Embleton & J Thornes (eds.) Process in geomorphology. Edward Arnold, London.

D Brunsden, 1979b. Weathering. In: C Embleton & J Thornes (eds.) Process in geomorphology. Edward Arnold, London. pp.73–129.

D Brunsden, 1984. Mudslides. In: D Brunsden & D B Prior (eds.) Slope Instability. Wiley, Chichester, pp.363–418.

D Brunsden and A S Goudie, 1981. Coastal landforms of Dorset. Classic landform guides. British Geomorphological Research Group and Geographical Association.

D Brunsden and D K C Jones, 1972. The morphology of degraded landslide slopes in south-west Dorset. Quarterly Journal of Engineering Geology. Vol.3 pp.205–223.

D Brunsden and D K C Jones, 1976. The evolution of landslide slopes in Dorset. Philosophical Transactions of the Royal Society. London Vol. A283 pp.605–631.

D Brunsden and D K C Jones, 1980. Relative time scales and formative events in coastal landslide systems. Zeitschrift fur Geomorphologie. Supplements NF Vol.34 pp.1–19

D Brunsden and D B Prior, 1984 (eds.) Slope instability. Wiley, Chichester.

D Brunsden, J C Doornkamp, P G Fookes, D K C Jones and J M H Kelly, 1975. Large scale geomorphological mapping and highway engineering design. Quarterly Journal of Engineering Geology. Vol.8 pp.227–254.

D Brunsden, R Gardner, A Goudie and D K C Jones, 1988. Landshapes. David and Charles, Newton Abbot.

J B Burland, T I Longworth and J F A Moore, 1977. A study of ground movement and progressive failure caused by a deep excavation in Oxford Clay. Geotechnique. Vol.27 pp.557–591.

A N Burton, 1969. Air photograph interpretation in site investigation for roads. Road and Road Construction. London March 1969 pp.72–76.

P B Butler, 1983. Landsliding and other large scale mass movements on the escarpment of the Cotswold Hills. BA thesis (unpublished). Hertford College Oxford.

G Cambers, 1973. The retreat of unconsolidated Quaternary Cliffs. PhD thesis (unpublished). University of East Anglia.

W Camden, 1722. Britannia: or a chronological description of Great Britain and Ireland together with the adjacent islands.

P A Carling, 1986a. The Noon Hill flash floods; July 7th 1983. Hydrological and geomorphological aspects of a major formative event in an upland landscape. Transactions of the Institute of British Geographers. Vol.11 pp.105–118.

P A Carling, 1986b. Peat slides in Teesdale and Weardale, Northern Pennines, July 1983: Description and failure mechanisms. Earth Surface Processes and Landforms. Vol.11 pp.193–206.

A M Carson and J Fisher, 1991. Management of landslides within Shropshire. In: R J Chandler (ed) Slope stability engineering: developments and applications, Thomas Telford. pp95–100.

M A Carson and M J Kirkby, 1972. Hillslope form and process. Cambridge.

M A Carson and D J Petley, 1970. The existence of threshold hillslopes in the denudation of the landscape. Transactions of the Institute of British Geographers. Vol.49 pp.71–95.

J H Chandler, 1989. The acquisition of spatial data from archival photographs and the application to geomorphology. PhD thesis (unpublished). City University, London.

J H Chandler and R Moore, 1991. Analytical photogrammetry: a method for monitoring slope instability. Quarterly Journal of Engineering Geology. Vol.22 pp.97–110.

M P Chandler, 1984. The coastal landslides forming the Undercliff of the Isle of Wight. PhD thesis (unpublished). Imperial College, University of London.

M P Chandler and J N Hutchinson, 1984. Assessment of relative slide hazard within a large, pre-existing coastal landslide at Ventnor, Isle of Wight. In: Proceedings of the IVth International Symposium on Landslides, Toronto. Vol.2, pp.517–522.

R J Chandler, 1970a. The degradation of Lias Clay slopes in an area of the East Midlands. Quarterly Journal of Engineering Geology. Vol.2, pp.161–181.

R J Chandler, 1970b. A shallow slab slide in the Lias Clay near Uppingham, Rutland. Geotechnique. Vol.20 pp.253–260.

R J Chandler, 1970c. Solifluction on low-angled slopes in Northamptonshire. Quarterly Journal of Engineering Geology. Vol.3 pp.65–69.

R J Chandler, 1971. Landsliding on the Jurassic Escarpment near Rockingham, Northamptonshire. In: D Brunsden (ed.) Slopes form and process. Institute of British Geographers Special Publication No.3 pp.111–128.

R J Chandler, 1972. Lias Clay: Weathering processes and their effect on shear strength. Geotechnique. Vol.22 pp.403–431.

R J Chandler, 1976. The history and stability of two Lias Clay slopes in the Upper Gwash Valley, Rutland. Philosophical Transactions of the Royal Society, London. Vol. A283, pp.463–491.

R J Chandler, 1977. Back analysis techniques for slope stabilisation works: A case record. Geotechnique. Vol.27 pp.479–495.

R J Chandler, 1984. Recent European experience of landslides in over-consolidated clays and soft rocks. In: Proceedings of the IVth International Symposium on Landslides, Toronto. Vol.1 pp.61–81.

R J Chandler and A W Skempton, 1974. The design of permanent slopes in stiff fissured clay. Geotechnique. Vol.24 pp.457–464.

R J Chandler, G A Kellaway, A W Skempton and R J Watt, 1976. Valley slope sections in Jurassic strata near Bath, Somerset Philosophical Transactions of The Royal Society. London. Vol. A283 pp.527–556.

G P Chattopadhay, 1983. Ploughing blocks on the Drumochter Hills in the Grampian Highlands, Scotland: a quantitative report. Geographical Journal. Vol.149 pp.211–215.

A R Clark and D K Johnson, 1975. Geotechnical mapping as an integral part of site investigation – two case histories. Quarterly Journal of Engineering Geology. Vol.8 pp.211–224.

A R Clark and S Guest, 1991. The Whitby Cliff stabilisation and coast protection scheme. In: R J Chandler (ed) Slope stability engineering: developments and applications, Thomas Telford. pp.283–290.

A R Clark, A B Hawkins and W J Gush, 1979. The Portavadie Dry Dock, West Scotland: A case history of the geotechnical aspects of it's construction. Quarterly Journal of Engineering Geology. Vol.12 pp.301–317.

A R Clark, J S Palmer, T P Firth and G McIntyre, 1990. The management and stabilisation of weak sandstone cliffs at Shanklin, Isle of Wight. In: J C Cripps and C F Moon (eds.) The engineering geology of weak rock. Preprints of the 26th Annual Conference of the Engineering Group of the Geological Society. pp.392–410.

M J Clark, 1980. Property damage by foundation failure. In: J D Doornkamp & K J Gregory (eds.) Atlas of drought in Britain 1975–1976. Institute of British Geographers. pp.63–64.

M J Clark, P J Ricketts and R J Small, 1976. Barton does not rule the waves. Geographical Magazine. Vol.48 pp.580–588.

D J Clayton, 1988. The impact of landslides in Dorset. BSc dissertation (unpublished). Dept. of Geography, University of Nottingham.

K M Clayton, 1980. Coastal protection along the East Anglian coast, UK. Zeitschrift fur Geomorphologie. Supplements NF. Vol.34 pp.165–172.

K M Clayton, 1990. Sea-level rise and coastal defences in the UK. Quarterly Journal of Engineering Geology. Vol.23 pp.279–282.

C T Clough, 1889. The geology of Plashetts and Kielder. Memoirs of the Geological Survey of Great Britain. London.

M A Coard, P C Sims and J L Ternan, 1987. Coastal erosion and slope instability at Downderry, south-east Cornwall – an outline of the problem and its implication for planning. In: M G Culshaw, F G Bell, J C Cripps and M O'Hara (eds). Planning and Engineering Geology. Geological Society, Engineering Geology Special Publication No.4 pp.529–532.

D R Coates, (ed.) 1977. Landslides. Geological Society of America, Reviews in Engineering Geology. Vol.3 pp.1–278.

G W Colenutt, 1928. The cliff-founder and landslide at Gore Cliff, Isle of Wight. Proceedings of the Isle of Wight Natural History and Archaeological Society. Vol.1 pp.561–570.

R Common, 1954. A report on the Lochaber, Appin and Benderloch floods, May 1953. Scottish Geographical Magazine Vol.70 pp.6–20.

B W Conway, 1974. The Black Ven landslip, Charmouth, Dorset. An example of the effect of a secondary reservoir of groundwater in an unstable area. British Geological Survey report No.7413.

B W Conway, 1976. Coastal terrain evaluation and slope stability of the Charmouth – Lyme Regis area of Dorset. British Geological Survey, Geophysical Division, Engineering Geology Unit report EG76/10.

B W Conway, 1977. A regional study of coastal landslips in west Dorset. British Geological Survey, Geophysical Division, Engineering Geology Unit report EG77/10.

B W Conway, 1979. The contribution made to cliff stability by head deposits in the west Dorset coastal area. Quarterly Journal of Engineering Geology. Vol.12 pp.267–275.

B W Conway, A Forster, K J Northmore and W J Barclay, 1980. South Wales Coalfield Landslip Survey. British Geological Survey Special Surveys Division, Report No. EG/80/4.

B W Conway, A Forster and K J Northmore, 1983. A study of the landslipped areas in the South Wales coalfield. British Geological Survey Engineering Geology Unit report EG/83/6.

W D Conybeare, W Buckland and W Dawson, 1840. Ten plates comprising a plan,sections and views representing the changes produced on the coast of east Devon between Axmouth and Lyme Regis by the subsidence of the land ... etc. J Murray, London.

D A Cook, 1973. Investigation of a landslip in the Fuller's Earth clay, Lansdown Bath. Quarterly Journal of Engineering Geology. Vol.6 pp.233–240.

R U Cooke and J C Doornkamp, 1990. Geomorphology in environmental management. 2nd Edition, Oxford University Press.

D Coombs, 1971. Mam Tor: a Bronze Age hillfort? Current Archaeology 3, pp.100–102.

R G Cooper, 1979. Geomorphological studies in the Hambleton Hills, North Yorkshire. PhD thesis (unpublished). Hull University.

R G Cooper, 1980. A sequence of landsliding mechanisms in the Hambleton Hills Northern England, illustrated by features at Peakscar, Hawnby. Geografiska Annaler. Vol.62 pp.149–156.

Countryside Commission, 1992. Heritage coast policies and priorities – policy statement. Countryside Commission.

R E Coxon, 1986. Failure of Carsington Embankment. Report to the Secretary of State for the Environment. Department of the Environment, London.

D T Crisp, M Rawes and D Welch, 1964. A Pennine peat slide. Geographical Journal. Vol.130 pp.519–524.

J E Crofts and D A Berle, 1972. An earth slip at Tiverton, Devon. Geotechnique. Vol.22 pp.345–351.

M Cross, 1988. An Engineering Geomorphological investigation of hillslope stability in the Peak District of Derbyshire. PhD thesis (unpublished). University of Nottingham.

M J Crozier, 1986. Landslides: causes, consequences and environment. Croom Helm, London.

M G Culshaw, 1972. Preliminary report on a landslip survey on the north side of the Ironbridge Gorge. British Geological Survey Engineering Geology Unit Report EG72/3.

M G Culshaw, 1973. Telford landslip investigations. The Lees/Lee Dingle area. A progress report. British Geological Survey Engineering Geology Unit Report EG73/3/1.

P Davies, A T Williams and P Bomboe, 1991. Numerical modelling of lower Lias rock failures in the coastal cliffs of South Wales, UK. In N C Kraus, K J Gingerich and D L Briebel (eds) Coastal Sediments '92. American Society of Civil Engineers, pp.1599–1612.

W R Dearman and N Eyles, 1982. An engineering geological map of the soils and rocks of the UK. Bulletin of the International Association of Engineering Geology, Vol.25 pp.3–18.

W R Dearman and P G Fookes, 1974. Engineering geological mapping for civil engineering practice in the United Kingdom. Quarterly Journal of Engineering Geology. Vol.7 pp.223–256.

M H De Freitas, 1969. The stability of rock cliffs in southwest England. Proceedings of the Geological Society. No.1654 pp.68–70.

M H De Freitas and R J Watters, 1973. Some field examples of toppling failure. Geotechnique. Vol.23 pp.495–514.

F N De Jorge, 1982. Reservoir-induced landslides with special reference to pumped storage schemes. MSc thesis (unpublished). Imperial College, University of London.

F A De Lory, 1956. Long term stability in overconsolidated clays. PhD thesis (unpublished). Imperial College, University of London.

J Demek and C E Embleton, 1978. Guide to medium-scale geomorphological mapping. IGU Stuttgart.

B Denness, 1972. The reservoir principle of mass movement. British Geological Survey Report No.72/7.

B Denness, 1977. The Ironbridge landslide – A case history of instabilities in Carboniferous sediments. Proceedings of the Conference on Rock Engineering, Newcastle University, 4-7th April 1977. British Geotechnical Society.

B Denness, B W Conway, D M McCann and P Grainger, 1975. Investigation of a coastal landslip at Charmouth, Dorset. Quarterly Journal of Engineering Geology. Vol.8 pp.119–140.

Department of the Environment, 1988. Environmental Assessment. DOE Circular 15/88 (Welsh Office Circular 23/88).

Department of the Environment, 1990. Planning Policy Guidance: Development on Unstable Land. PPG14, London, HMSO.

Department of the Environment, 1992. Planning Policy Guidance: Coastal Planning. PPG20, London, HMSO.

N Dixon, 1986. The mechanics of coastal landslides in the London Clay at Warden Point, Isle of Sheppey. PhD thesis (unpublished). Kingston Polytechnic.

N Dixon and E N Bromhead, 1991. The mechanics of first time slides in the London Clay cliff at the Isle of Sheppey, England. In: R J Chandler (ed) Slope stability engineering: developments and applications. Thomas Telford. pp.277–282.

J C Doornkamp, 1988. (ed.) Applied Earth Science Background: Torbay. Department of the Environment.

J C Doornkamp, 1989. New landslide databank for Britain. Civil Engineering Surveyor. July/August pp.13–14.

J C Doornkamp, 1990. Landslides in Derbyshire. East Midland Geographer. Vol.13 pp.33–62.

J C Doornkamp and E M Lee, 1992. Applied geological mapping of the St Helens area, Merseyside. Department of the Environment.

J C Doornkamp, D Brunsden, D K C Jones, R U Cooke and P R Bush, 1979. Rapid geomorphological assessments for engineers. Quarterly Journal of Engineering Geology. Vol.12 pp.189–204.

I Douglas, 1985. Cities and Geomorphology. In: A Pitty (ed.), Themes in geomorphology. Croom Helm, Beckenham pp.233–234

J C Duggleby, P J Argherinos and A J Powderham, 1991. Channel Tunnel: Foundation Engineering at the UK Portal. In: Proceedings of the IVth International Conference on Piling and Deep Foundations, Stresa, Italy.

M J Dumbleton, 1962. Investigations to assess the potentialities of lime for soil stabilisation in the UK. Road Research Laboratory Technical Paper No. 64.

M J Dumbleton and G West, 1976. Preliminary sources of information for site investigations in Britain. Transport and Road Research Laboratory Report 403.

N Duncan, 1984. Slope movements in rock: four case studies. In: Proceedings of the IVth International Symposium on Landslides, Toronto. pp.483–488.

K R Early and A W Skempton, 1972. Investigations of the landslide at Walton's Wood Staffordshire. Quarterly Journal of Engineering Geology. Vol.5 pp.19–41.

K R Early and P G Jordan, 1985. Some landslipping encountered during construction of the A40 near Monmouth. Quarterly Journal of Engineering Geology. Vol.18 pp.207–224.

J R Earp, D Magraw, E G Poole, D H Land and A J Whiteman, 1961. Geology of the country around Clitheroe and Nelson, Sheet 68. Memoirs of the Geological Survey of Great Britain. London.

H H Einstein, 1988. Landslide risk assessment procedure. In C Bonnard (ed.) Proceedings of the Vth International Symposium on Landslides, Lausanne. Vol.2 pp.1075–1090.

H C Englefield, 1816. A description of the principal picturesque beauties, antiquities of the Isle of Wight. Payne and Foss, London.

European Community, 1985. The assessment of the effects of certain public and private projects on the environment. Directive 85/337/EEC.

W A Eyre, 1973. The revetment of rock slopes in the Clevedon Hills for the M5 Motorway. Quarterly Journal of Engineering Geology. Vol.6 pp.223–229.

F Fakhraee, 1979. Stability of slopes in the area of the Blackdown Hills, Somerset. MSc thesis (unpublished). University of Bath.

W Fellinius, 1918. Kaj-och jordrasen: Goteborg. Teknisk Tidsskrift. V.U. Vol.48 pp.17–19.

P G Fookes, W R Dearman and J A Franklon, 1971. Some engineering aspects of rock weathering with field examples from Dartmoor and elsewhere. Quarterly Journal of Engineering Geology. Vol.4 pp.139–185.

P G Fookes, L W Hinch, M A Huxley and N E Simons, 1975. Some soil properties in glacial terrain – the Taff Valley, South Wales. In: Proceedings of the Symposium on the Engineering Behaviour of Glacial Materials. University of Birmingham. Geoabstracts, pp.93–115.

A Forster and K J Northmore, 1986. Landslide distribution in the South Wales Coalfield. In: C S Morgan (ed.) Landslides in the South Wales Coalfield. Polytechnic of Wales Symposium. pp.29–36.

A Forster, P R N Hobbs, R A Monkhouse and R J Wyatt, 1985. Environmental Geology Study; Parts of west Wiltshire and southeast Avon. British Geological Survey Environmental Geological Study.

J A Franklin, 1977. The monitoring of structures in rock. International Journal of Rock Mechanics, Mining Science and Geotechnics. Abstract 14 pp.163–192.

J A Franklin, 1984. Slope monitoring and instrumentation, In: D Brunsden & D B Prior (eds.) Slope instability. Wiley. pp.143–169.

J A Franklin and P A Denton, 1973. The monitoring of rock slopes. Quarterly Journal of Engineering Geology, London. Vol.6 pp.202–211.

C A M Franks, 1986. Mining subsidence and landslips in the South Wales Coalfield. In: C S Morgan (ed.) Landslides of the South Wales Coalfield. Polytechnic of Wales Symposium. pp.225–230.

C A M Franks, S P Bentley and J D Geddes, 1986. Mining histories and their effect on slope stability. In: C S Morgan (ed.) Landslides of the South Wales Coalfield. Polytechnic of Wales Symposium. pp.207–224.

J W Franks and R H Johnson, 1964. Pollen analytical dating of a Derbyshire landslip: the Cowan Edge landslides, Charlesworth. New Phytology. Vol.63 pp.209–216.

D V Frost and J G O Smart, 1979. The geology of the country around Derby. Sheet 125. Memoirs of the Geological Survey of Great Britain. London.

R W Gallois, 1965. The Wealden District. Memoir of the Geological Survey of Great Britain.

C Garrett and J H Wale, 1985. Performance of embankments and cuttings in Gault Clay in Kent. In Failures in Earthworks, Thomas Telford, pp.93–111.

Geoffrey Walton Practice, 1986. Review of current geotechnical practice in British quarries and related interests and requirements of Mineral Planning Authorities and other Statutory Bodies. Department of the Environment.

Geoffrey Walton Practice, 1988. Handbook on the hydrogeology and stability of excavated slopes in quarries for the Department of the Environment. HMSO, London.

Geomorphological Services Ltd, in association with Rendel Palmer and Tritton, 1986–87. Review of Research into Landsliding in Great Britain: Reports to the Department of the Environment.

Series A: Regional review of landsliding – regional atlases of county maps at 1:250,000 scale, regional reports. Landslide database.

Vol. 1. South East England and East Anglia.
Vol. 2. South West England.
Vol. 3. The Midlands.
Vol. 4. Wales.
Vol. 5. Northern England.
Vol. 6. Scotland.

Series B: Causes and mechanisms of landsliding.

Vol. 1. International review.
Vol. 2. Great Britain.

Series C: Landslide investigation techniques and remedial measures – research and practice (2 volumes).

Series D: Landslides and policy.

Vol. 1. Landslide hazard assessment.
Vol. 2. Landslide risk in Britain.
Vol. 3. Legislative and administrative provision: Practice in England and Wales and a review of overseas practice.

Series E: National summary and recommendations.

Geomorphological Services Ltd., 1988. Applied Earth Science Mapping for Planning and Development: Torbay, Devon. Report to the Department of the Environment.

Geomorphological Services Ltd., 1991. Coastal landslip potential assessment: Isle of Wight Undercliff, Ventnor. Department of the Environment.

A J Gerrard and L Morris, 1981. Mass movement forms and processes on Bredon Hill, Worcestershire. Department of Geography, University of Birmingham. Working Paper No.10.

J Gifford, 1953. Landslides on Exmoor caused by the storm of 15th August 1952. Geography. Vol.38 pp.9–17.

T P Gostelow, 1971. A preliminary investigation into the geotechnical properties of the Holderness glacial deposits with particular reference to problems in slope stability. MSc thesis (unpublished). Imperial College, University of London.

T P Gostelow, 1977. A report on valley instability at Blaina, South Wales. British Geological Survey Engineering Geology Unit Report No. 77/11.

Rev. W Gostling, 1756. Letter to Gentleman's Magazine. Vol.XXVI p.160.

A Goudie, 1983. Environmental change. Clarendon Press, Oxford.

P Grainger, 1983. Aspects of the engineering geology of mudrocks with reference to the Crackington Formation of southwest England. PhD thesis (unpublished). Exeter University.

P Grainger and J Harris, 1986. Weathering and slope stability on Upper Carboniferous mudrocks in south-west England. Quarterly Journal of Engineering Geology. Vol.19 pp.155–173.

P Grainger and P G Kalaugher, 1987a. Intermittent surging movements of a coastal landslide. Earth Surface Processes and Landforms. Vol.12 pp.597–603.

P Grainger and P G Kalaugher, 1987b. Cliff-top recession related to the development of coastal landsliding west of Budleigh Salterton, Devon. Proceedings of the Ussher Society. Vol.6 pp.516–522.

P Grainger and P G Kalaugher, 1988. Hazard zonation of Coastal Landslides. In: C Bonnard (ed.) Landslides. Proceedings of the Vth International Symposium on Landsliding, Lausanne. pp. 1169–1174.

P Grainger and P G Kalaugher, 1991. Cliff management: a photogrammetric monitoring system. In: R J Chandler (ed) Slope stability engineering: developments and applications, Thomas Telford. pp.119–124.

A Greenman, 1992. Geotechnical team tackles sludge tip slide. Ground Engineering, March 1992.

K J Gregory, 1963. Contributions to the geomorphology of the North York Moors. PhD thesis (unpublished). London University.

J S Griffiths and G J Hearn, 1991. Engineering geomorphology: a UK perspective. Bulletin of the International Association of Engineering Geology. No.42 pp.39–44.

J S Griffiths and A H Marsh, 1986. BS 5930: The role of geomorphological and geological techniques in a preliminary site investigation. In: A B Hawkins (ed.) Site Investigation Practice: Assessing BS 5930. Geological Society, Engineering Geology Special Publication No.2 pp.261–267.

A T Grove, 1953. Account of a mudflow on Bredon Hill, Worcestershire April 1951. Proceedings of the Geologists Association. Vol.64 pp.10–13.

Sir William Halcrow and Partners, 1981. Preliminary study of possible remedial measures in the landslips at East Pentwyn and Bournville, Gwent.

Sir William Halcrow and Partners, 1983. East Pentwyn landslip: First report on geotechnical investigations.

Sir William Halcrow and Partners, 1985. East Pentwyn and Bournville Landslips Research Project. Department of the Environment and Welsh Office.

Sir William Halcrow and Partners, 1986. Rhondda Landslip Potential Assessment. Department of the Environment and Welsh Office.

Sir William Halcrow and Partners, 1988. Rhondda Landslip Potential Assessment: Inventory. Department of Environment and Welsh Office.

Sir William Halcrow and Partners, 1989. Landslides and Undermining Research Project. Department of the Environment and Welsh Office.

A Hansen, Undated. Unpublished research topic, University of London, Kings College, University of London.

A Hansen, 1984. Landslide hazard analysis. In: D Brunsden & D B Prior (eds.) Slope instability. Wiley, pp.523–602.

M J Hansen, 1984. Strategies for classification of landslides. In: D Brunsden & D B Prior (eds.) Slope instability. Wiley, pp.1–26.

C Harris, 1987. Solifluction and related periglacial deposits in England and Wales. In: J Boardman (ed.) Periglacial processes and landforms in Britain and Ireland, Cambridge University Press. pp.209–223.

C Harrison and J R Petch, 1985. Ground movements in parts of Salford and Bury, Greater Manchester – aspects of urban geomorphology. In: R H Johnson (ed.) The geomorphology of North-West England. University Press, Manchester pp.353–371.

J T Harrison, 1848. West Bay, Portland: Encroachments of the sea. Minutes of the Proceedings of the Institute of Civil Engineers. Vol.7 p.346.

A B Hawkins, 1977. The Hedgemead landslip, Bath. In: J D Geddes (ed.) Large ground movements and structures. Pentech Press, London. pp.472–498.

A B Hawkins, 1988. Stability of inland slopes: some geological considerations. In: C Bonnard (ed.) Landslides. Proceedings of the Vth International Symposium on Landslides, Lausanne. pp.181–186.

A B Hawkins and K D Privett, 1979. Engineering geomorphological mapping as a technique to elucidate areas of superficial structure, with examples from the Bath area of the south Cotswolds. Quarterly Journal of Engineering Geology. Vol.12 pp.221–232.

A B Hawkins and K D Privett, 1981. A building site on cambered ground at Radstock, Avon. Quarterly Journal of Engineering Geology. Vol.14 pp.151–167.

T R W Hawkins, 1985. Influence of geological structure on slope stability in the Maentwrog Formation, Harlech Dome, North Wales. Proceedings of the Geologists Association. Vol.96 pp.289–304.

D Hayward, 1985. Humberside on hunt for cliff slip cure. New Civil Engineer. 15th Aug 1985 pp.24–25.

D Hayward, 1990. Dam double dam. New Civil Engineer, 15th Nov. 1990.

K H Head, 1985. Manual of soil laboratory testing. Pentech Press.

D J Henkel and A W Skempton, 1954. A landslide at Jackfield, Shropshire, in a heavily overconsolidated clay. Proceedings of the European Conference on Stability Slopes Vol.1 pp.90–101.

I E Higginbottom and P G Fookes, 1971. Engineering aspects of periglacial features in Britain. Quarterly Journal of Engineering Geology. Vol.1 pp.85–117.

L W Hinch and P G Fookes, 1989. Taff Vale trunk road stage 4, South Wales: geotechnical design. Proceedings of the Institute of Civil Engineers, Part 1. Vol.86 pp.161–188.

E Hoek and J W Bray, 1977. Rock slope engineering. 2nd Edition. Institute of Mining and Metallurgy, London.

S E Hollingworth, J H Taylor and G A Kellaway, 1944. Large scale superficial structures in the Northamptonshire Ironstone Field. Quarterly Journal of the Geological Society. Vol.100 pp.1–44.

G Holmes, 1984. Rock slope failures in parts of the Scottish Highlands. PhD thesis (unpublished). University of Edinburgh.

G Holmes and J J Jarvis, 1985. Large scale toppling within a sackung type deformation at Ben Attow, Scotland. Quarterly Journal of Engineering Geology. Vol.8 pp.287–289.

P Horswill and A Horton, 1976. Cambering and valley bulging in the Gwash Valley at Empingham, Rutland. Philosophical Transactions of the Royal Society, London Vol.A283 pp.427–462.

House of Commons Environment Committee, 1992. Coastal zone protection and planning. HMSO.

F Hudleston, 1930. The cloudbursts of Stainmore, Westmorland, 18th June 1930. British Rainfall, pp.287–292.

J Hutchins, 1803. The history and antiquities of the County of Dorset. Vol.2 pp.354–371. J Nichols and Son.

J N Hutchinson, 1962. Building Research Station Note C890. Report on visit to landslide at Lyme Regis, Dorset.

J N Hutchinson, 1965a. The stability of cliffs composed of soft rocks, with particular reference to the coasts of South-East England. PhD thesis (unpublished). University of Cambridge.

J N Hutchinson, 1965b. A reconnaissance of coastal landslides in the Isle of Wight. Building Research Station, Note No. EN 11/65.

J N Hutchinson, 1965c. A survey of coastal landslides in Kent. Building Research Station Note No. 35/65.

J N Hutchinson, 1967. The free degradation of London Clay cliffs. Proceedings of the Geotechnical Conference of Oslo, Vol.1 pp.113–118.

J N Hutchinson, 1968. Mass Movement. In: R W Fairbridge (ed.) Encyclopaedia of Geomorphology. Reinhold Publications, USA, pp.688–696.

J N Hutchinson, 1969. A reconsideration of the coastal landslides at Folkestone Warren, Kent. Geotechnique. Vol.19 pp.6–38.

J N Hutchinson, 1970. A coastal mudflow on the London Clay cliffs at Beltinge, North Kent. Geotechnique. Vol.20 pp.412–438.

J N Hutchinson, 1972. Field laboratory studies of a fall in Upper Chalk Cliffs at Joss Bay, Isle of Thanet. Stress Strain Behaviour of Soils. In: Proceedings of the Roscoe Memorial Symposium, Cambridge Session. Vol.6 pp.692–706.

J N Hutchinson, 1973. The response of London Clay cliffs to differing rates of toe erosion. Geologia Applicata e Idrogeologia. Vol.8 pp.221–239.

J N Hutchinson, 1974. Periglacial solifluction: an approximate mechanism for clayey soils. Geotechnique. Vol.24 pp.438–443.

J N Hutchinson, 1976. Coastal landslides in cliffs of Pleistocene deposits between Cromer and Overstrand, Norfolk, England. In: N Janbu, F Jorstad and B Kjaernsli (eds). Laurits Bjerrum Memorial Volume. Contributions to Soil Mechanics, Norwegian Geotechnical Institute Oslo. pp.155–182.

J N Hutchinson, 1977. The assessment of the effectiveness of corrective measures in relation to geological conditions and types of slope movement. Bulletin of the International Association Engineering Geology. Vol.16 pp.131–155.

J N Hutchinson, 1980. Various forms of cliff instability arising from coast erosion in the UK. Fjellsprengnings – teknikk – Bergmekanikk – Geoteknikk, 1979, 19.1–19.32. Trundheim: Tapir, for Norsk Jord-og Fjellteknikk Forbund tilknyttet N.I.F.

J N Hutchinson, 1982. Slope failures produced by seepage erosion in sands. In: A Sheko (ed.) Landslides and mudflows. UNESCO-UNEP Seminar, Moscow, pp.250–268.

J N Hutchinson, 1983a. Engineering in a landscape. Inaugural lecture, 9 October 1979. Imperial College, University of London.

J N Hutchinson, 1983b. A pattern in the incidence of major coastal mudslides. Earth Surface Processes and Landforms Vol.8 pp.391–397.

J N Hutchinson, 1984a. Landslides in Britain and their countermeasures. Journal of the Japan Landslide Society. Vol.21 pp.1–21.

J N Hutchinson, 1984b. An influence line approach to the stabilisation of slopes by cuts and fills. Canadian Geotechnical Journal. Vol.21 pp.363–370.

J N Hutchinson, 1986. Cliffs and shores in cohesive materials; geotechnical and engineering geological aspects. In: M G Skafel (ed.) Proceedings of the Symposium on Cohesive Shores, May 5th-7th 1986 Burlington, Ontario. pp.1–44.

J N Hutchinson, 1987a. Some coastal landslides of the southern coast of the Isle of Wight. In: K E Barker (ed.) Wessex and the Isle of Wight. Quaternary Research Association, Cambridge. pp.123–1335.

J N Hutchinson, 1987b. Mechanisms producing large displacements in landslides on pre-existing shears. Memoir of the Geological Society of China. Vol.9 pp.175–200.

J N Hutchinson, 1988. General report: Morphological and geotechnical parameters of landslides in relation to geology and hydrogeology. In: C Bonnard (ed.) Landslides. Proceedings of the Vth International Symposium on Landslides, Lausanne. pp.3–35.

J N Hutchinson, 1991. The landslides forming the South Wight Undercliff. In: R J Chandler (ed) Slope stability engineering: developments and applications, Thomas Telford. pp.157–168.

J N Hutchinson and M J Hughes, 1968. The application of micropalaeontology to the location of a deep-seated slip surface in the London Clay. Geotechnique. Vol.18 pp.508–510.

J N Hutchinson and R K Bhandari, 1971. Undrained loading, a fundamental mechanism of mudflows and other mass movements. Geotechnique. Vol.21 pp.353–358.

J N Hutchinson and T P Gostelow, 1976. The development of an abandoned cliff in London Clay at Hadleigh, Essex. Philosophical Transactions of the Royal Society, London. Vol.A283 pp.557–604.

J N Hutchinson and A Thomas-Betts, 1990. Technical note: extent of permafrost in southern Britain in relation to geothermal flux. Quarterly Journal of Engineering Geology. Vol.23 pp.387–390.

J N Hutchinson, S H Somerville and D J Petley, 1973. A landslide in periglacially disturbed Etruria Marl at Bury Hill, Staffordshire. Quarterly Journal of Engineering Geology. Vol.6 pp.377–404.

J N Hutchinson, E N Bromhead and J F Lupini, 1980. Additional observations on the Folkestone Warren landslides. Quarterly Journal of Engineering Geology. Vol.13 pp.1–31.

J N Hutchinson, M P Chandler and E N Bromhead, 1981a. Cliff recession on the Isle of Wight, S W coast. Tenth International Conference on Soil Mechanics and Foundation Engineering, Rotterdam. pp.429–434.

J N Hutchinson, E N Bromhead and M P Chandler, 1981b. Report on the coastal landslides at Bonchurch, Isle of Wight. Unpublished report to Lewis and Duvivier, Consulting Engineers.

J N Hutchinson, M P Chandler and E N Bromhead, 1985a. A review of current research on the coastal landslides forming the Undercliff of the Isle of Wight, with some practical implications. Paper presented to the conference on problems associated with the coastline, Newport, Isle of Wight.

J N Hutchinson, C Poole, N Lambert and E N Bromhead, 1985b. Combined archaeological and geotechnical investigations of the Roman fort at Lympne, Kent. Brittania XVI pp.209–236.

J N Hutchinson, E M Lee and D J Petley, 1988. Report on a preliminary appraisal of the stability of the reservoir complex on Churchdown Hill, Gloucestershire. Confidential report by the Dept. of Civil Engineering, Imperial College, London.

J N Hutchinson, E N Bromhead and M P Chandler, 1991. Investigations of landslides at St Catherine's Point, Isle of Wight. In: R J Chandler (ed) Slope stability engineering: developments and applications, Thomas Telford. pp.151–161.

Hydraulics Research Ltd, 1991. A guide to the selection of appropriate coast protection works for geological SSSIs. Nature Conservancy Council.

O G Ingles and J B Metcalf, 1972. Soil stabilisation, principles and practice, Butterworths.

J L Innes, 1982. Debris flow activity in the Scottish Highlands. PhD thesis (unpublished). Cambridge University.

J L Innes, 1983. Lichenometric dating of debris flow deposits in the Scottish Highlands. Earth Surface Processes and Landforms. Vol.8 pp.579–588.

Institute of Civil Engineers, 1991. Inadequate site investigation. Report by the Ground Board of ICE, Thomas Telford.

Intergovernmental Panel on Climatic Change, 1990. Climatic change: an IPCC assessment. Cambridge University Press.

J F Jackson, 1928. Pressure ridges in the Gore Cliff landslide. Proceedings of the Isle of Wight Natural History and Archaeological Society. Vol.1 pp.611–612.

I Jackson and D J D Lawrence, 1990. Geology and land-use planning: Morpeth-Bedlington-Ashington. British Geology Survey Technical Report WA/90/14.

C Jaeger, 1972. Rock mechanics and engineering. Cambridge University Press, London.

A Jenkins, P J Ashworth, R I Ferguson, I C Grieve, P Rowling and T A Stott, 1988. Slope failure in the Ochil Hills, Scotland, November 1984. Earth Surface Processes and Landforms. Vol.13 pp.69–76.

M A Johnson and J R Rodine, 1984. Debris flow. In: D Brunsden & D B Prior (eds.) Slope instability. Wiley. pp.257–361.

R Johnson, 1990. Symposium report on quality management in geotechnical engineering. Ground Engineering. Vol.23 p.23.

R H Johnson, 1965a. The glacial geomorphology of the west Pennine slopes from Cliviger to Congleton. In: J B Wilton and P D Wood (eds). Essays in Geography for Austin Miller. University of Reading. pp.58–82.

R H Johnson, 1965b. A study of the Charlesworth landslips near Glossop, North Derbyshire. Transactions of the Institute of British Geographers. Vol.37 pp.111–126.

R H Johnson, 1980. Hillslope stability and landslide hazard – a case study from Longendale, North Derbyshire, England. Proceedings of the Geologists Association. Vol.91 pp.315–325.

R H Johnson, 1987. Dating of ancient, deep-seated landslides in temperate regions. In: M G Anderson & K S Richards (eds.) Slope Stability. Wiley, London. pp.561–600.

R H Johnson and R D Vaughan, 1983. The Alport Castles, Derbyshire : A South Pennine slope and its geomorphic history. East Midland Geographer. Vol.8 pp.79–88.

R H Johnson and S Walthall, 1979. The Longdendale landslides. Geological Journal. Vol.14 pp.135–158.

D B Jones and H J Siddle, 1988. Geotechnical parameters for stabilization measures to a landslide. In: C Bonnard (ed.) Landslides. Proceedings of the Vth International Symposium on Landslides, Lausanne. pp.193–198.

D B Jones, D J Reddish, H J Siddle and B N Whittaker, 1991. Landslides and undermining: slope stability interaction with mining subsidence behaviour. 7th ISRM Congress, Aachen.

D K C Jones, 1981 Geomorphology of the British Isles – South East and Southern England. Methuen.

D K C Jones, 1985. Subsidence. In: P G Fookes and P R Vaughan (eds.) A handbook of engineering geomorphology. Surrey University Press. pp.284–297.

D K C Jones, 1990. Grounds for concern: landsliding in Great Britain. Structural Survey. Vol.9 pp.226–236.

D K C Jones, 1991. Environmental hazards. In R J Bennett and R C Estall (eds) Global change and challenge. pp 27–56.

D K C Jones, 1992. Landslide hazard assessment in the context of development. In G J H McCall, D J C Laming & S C Scott (eds) Geohazards, Chapman and Hall, pp.117–141.

D K C Jones, E M Lee, G J Hearn and S Genc, 1988. The Catak landslide disaster, Trabzon Province, Turkey. Terra Nova. Vol.1 pp.84–90.

M E Jones, R J Allison and J Gilligan, 1983. On the relationships between geology and coastal landforms in central southern England. Proceedings of the Dorset Natural History and Archaeological Society. Vol.105 pp.107–118.

M S Jordan, 1991. The implications of the Ventnor study for the planning authority. In Coastal Instability and Development Planning: Papers and Proceedings of the SCOPAC Conference, Southsea, pp.68–73.

M D Joyce, 1969. A geological study of the boulder clay cliffs of Holderness, East Yorkshire. MSc thesis (unpublished). University of Leeds.

P G Kalaugher and P Grainger, 1981. A coastal landslide at Westdown Beacon, Budleigh Salterton, Devon. Proceedings of the Ussher Society. Vol.5 pp.217–221.

P G Kalaugher, P Grainger and R L P Hodgson, 1987. Cliff stability evaluation using geomorphological maps based on oblique aerial photographs. In: M G Culshaw, F G Bell, J C Cripps & M O'Hara (eds.) Planning and engineering geology. Geological Society, Engineering Geology Special Publication No.4 pp.157–167.

G A Kellaway and J H Taylor, 1952. Early stages in the physiographic evolution of a portion of the East Midlands. Quarterly Journal of the Geological Society. Vol.108 pp.343–367.

G A Kellaway and J H Taylor, 1968. The influence of landslipping on the development of the City of Bath, England. Report of the International Geological Congress, 23rd session, Czechoslovakia. Vol.12 pp.65–76.

J M H Kelly and P L Martin, 1986. Construction works on or near landslides. In: C S Morgan (ed.) Landslides in the South Wales Coalfield. Polytechnic of Wales Symposium. pp.85–103.

T C Kenney, 1984. Properties and behaviour of soils relevant to slope instability. In: D Brunsden & D B Prior (eds.) Slope instability. Wiley. pp.27–65.

C Kidson, 1953. The Exmoor storm and the Lynmouth floods. Geography. Vol.38 pp.1–9.

R B King, 1987. Review of geomorphic description and classification in land resource surveys. In: V Gardner (ed.) International Geomorphology 1986 Part II. Wiley pp.384–403.

M J Kirkby, 1973. Landslides and weathering rates. Geologia Applicata e Idrogeologia, Bari, Vol 8, pp.171–183.

A Knowles, 1973. Observations on ancient landslides and related phenomena in the Chew and Greenfield valleys, the West Riding of Yorkshire. MA thesis (unpublished). University of Manchester.

G Knox, 1927. Landslides in the South Wales Valleys. Transactions of the South Wales Institute of Engineers Vol.43 pp.161–247, 257–291.

E Krauter, 1988. Special lecture: Applicability and usefulness of field measurements on unstable slopes. In: C Bonnard (ed.) Landslides. Proceedings of the Vth International Symposium on Landslides, Lausanne. Vol.2 pp.367–373.

G N Lacey, 1972. Observations on Aberfan. Journal of Psychosomatic Research. Vol.16 pp.257–260.

T W Lambe and R V Whitman, 1979. Soil Mechanics. Wiley.

H Lapworth, 1911. The geology of dam trenches. Transactions of the Institute of Water Engineers. Vol.16 pp.25–66.

C J Lawrence, 1972. Terrain evaluation in West Malaysia: Part I: Terrain classification and survey methods. Report LR 506 (Transport and Road Research Laboratory, Crawthorne).

R Leafe, 1991. The English Nature view. In Coastal Instability and Development Planning: Papers and Proceedings of the SCOPAC Conference, Southsea, pp.91–106.

E M Lee, 1993. Coastal planning and management: a review. Report by Rendel Geotechnics to the Department of the Environment. HMSO.

E M Lee, in press. Landsliding and insurance: the problems of Luccombe, Isle of Wight. Paper presented to the Engineering Group of the Geological Society, 1991.

E M Lee and J S Griffiths, 1989. Landsliding in the Vale of Fernhurst. Field Note. Engineering Group of the Geological Society Field Meeting 14–16th April 1989.

E M Lee and R Moore, 1989. Report on the study of landsliding in and around Luccombe Village. Department of the Environment. HMSO.

E M Lee and R Moore, (In prep.) Getting the message across: ground movement and public perception.

E M Lee, J C Doornkamp, J S Griffiths and D Traghiem, 1988. Environmental geology mapping for land use planning purposes in the Torbay area. Proceedings of the Ussher Society. Vol.7 pp.18–25.

E M Lee, J C Doornkamp, D Brunsden and N H Noton, 1991a. Ground movement in Ventnor, Isle of Wight. Department of the Environment.

E M Lee, R Moore, D Brunsden and H J Siddle, 1991b. The assessment of ground behaviour at Ventnor, Isle of Wight. In: R J Chandler (ed) Slope stability engineering: developments and applications, Thomas Telford. pp.207–212.

E M Lee, R Moore, N Burt and D Brunsden, 1991c. Strategies for managing the landslide complex at Ventnor, Isle of Wight. In: R J Chandler (ed) Slope stability engineering: developments and applications, Thomas Telford. pp.219–225.

E M Lee, J C Doornkamp and R Moore, 1991d. The wider implications of the Ventnor study. In Coastal Instability and Development Planning: Papers and Proceedings of the SCOPAC Conference, Southsea, pp.14–19.

F B Leighton, 1976. Urban landslides: targets for land-use planning in California. Geological Society of America Special Paper 173. pp.89–96.

J J Lowe and M J C Walker, 1985. Reconstructing Quaternary environments. Longman, London.

H MacKinder, 1902. Britain and the British seas. Oxford.

Malvern Gazette and Ledbury Reporter. £¼m reservoir plan abandoned after 'chance in a million' landslip. April 23, 1981.

V J May, 1964. A study of recent coastal changes in southern England. MSc thesis (unpublished), University of Southampton.

V J May, 1971. The retreat of chalk cliffs. Geographical Journal. Vol.137 pp.203–206.

A McGown, A Salvidor-Dali and A M Radwan, 1974. Fissure patterns and slope failures in till at Hurlford. Quarterly Journal of Engineering Geology, Vol.7 pp.57–67.

A J McIlroy, 1990. Clifftop families admit defeat in battle with advancing sea. Daily Telegraph, 8.10.90.

G McIntyre and R G McInnes, 1991. A review of instablity on the southern coasts of the Isle of Wight and the role of the local authority. In R J Chandler (ed) Slope Stability Engineering: developments and applications. Thomas Telford. pp.237–244.

J N McLaren, 1983. The instability of slopes in the North-east of the Isle of Portland, Dorset. MSc thesis (unpublished), University of Surrey.

S Middleboe, 1986. Progressive failure led to Carsington collapse. New Civil Engineer. 24 July 1986.

S Middleboe, 1988. Designer pays £3.25M over Carsington dam. New Civil Engineer. 12 May 1988.

J Miller, 1974. Aberfan – a disaster and its aftermath. Constable, London.

Ministry of Agriculture, Fisheries and Food/Welsh Office, 1993. Strategy for flood and coastal defence in England and Wales. MAFF Publications.

Ministry of Housing and Local Government, 1961. Town and County Planning Acts 1947–1959: Surface development in Coal Mining Areas. Circular 44/61, HMSO.

W A Mitchell, 1991. Quaternary features in Upper Wensleydale. PhD thesis (unpublished). Luton College of Higher Education.

M S Money, 1979. Coastal and inland landslides of North East Yorkshire. Guide to Excursion "A". International Association of Engineering Geology / Engineering Group of the Geological Society. Symposium on Engineering Geological Mapping. 2nd–6th September 1979.

R Moore, 1986. The Fairlight landslips: the location, form and behaviour of coastal landslides with respect to toe erosion. Kings College, London. Occasional paper No.27 43pp.

R Moore, 1988. The clay mineralogy, weathering and mudslide behaviour of coastal cliffs. PhD thesis (unpublished). Kings College, University of London.

R Moore, 1991. The chemical and mineralogical controls upon the residual strength of pure and natural clays. Geotechnique. Vol.41 pp.35–47.

R Moore, E M Lee and F Longman, 1991a. The impact, causes and management of landsliding at Luccombe, Isle of Wight. In: R J Chandler (ed) Slope stability engineering: developments and applications, Thomas Telford. pp.225–230.

R Moore, E M Lee and N H Noton, 1991b. The distribution, frequency and magnitude of ground movements at Ventnor, Isle of Wight. In: R J Chandler (ed) Slope stability engineering: developments and applications, Thomas Telford. pp.231–236.

C S Morgan, 1986 (ed.) Landslides in the South Wales Coalfield. Polytechnic of Wales Symposium.

L Morris, 1974.The geomorphology of Bredon Hill. In Worcester and its Region: Field studies in the former county of Worcestershire. The Worcester Branch of the Geographical Association, pp.174–183.

R N Mortimore, 1986. Stratigraphy of the Upper White Chalk of Sussex. Proceedings of the Geologists Association. Vol.97 pp.97–139.

R N Mortimore, 1990. Chalk or chalk? In: R N Mortimore (ed.) Chalk. Thomas Telford. pp.15–45.

R Muller, 1979. Investigating the age of a Pennine landslip. Mercian Geologist. Vol.7 pp.211–218.

D Nash, 1987. A comparative review of limit equilibrium methods of stability analysis. In: M G Anderson and K S Richards (eds.) Slope stability. Wiley. pp.11–75.

National Audit Office, 1989. Quality control of road and bridge construction. HMSO, London.

National Economic Development Office, 1983. Faster building for industry. NEDO, London.

National Economic Development Office, 1988. Faster building for commerce. NEDO, London.

New Civil Engineer, 1990a. China clay spoil slip alarms experts. 15 Feb 1990.

New Civil Engineer, 1990b. Second china clay spoil heap fails. 22 Feb 1990.

New Civil Engineer, 1990c. Hidden menace. 20 Sept 1990.

J Newberry and A B Baker, 1981. The stability of cuts on the M4 north of Cardiff. Quarterly Journal of Engineering Geology. Vol.14 pp.195–205.

M D Newson, 1975. The Plynlimon Floods of August 5/6th 1973. Natural Environment Research Council Institute of Hydrology, Wallingford. Report No.26.

E F P Nickless, 1982. Environmental geology of the Glenrothes district, Fife Region. Institute of Geological Sciences Report No. 82/15.

K J Northmore and A Forster, 1986. West Glamorgan – a case study. The Glynrhigos Farm landslip. In: C S Morgan (ed.) Landslides in the South Wales Coalfield. Polytechnic of Wales Symposium. pp.139–151.

K J Northmore, S V Duncan and T P Gostelow, 1978. Rhondda and North-East survey areas interim report 1978, Landslip Survey of South Wales Coalfield. British Geological Survey Report No. 78/17.

N H Noton, 1991. Living with landslip: Ventnor. In: R J Chandler (ed) Slope stability engineering: developments and applications, Thomas Telford. pp.244–250.

C D Ollier, 1977. Terrain classification principles and applications. In J R Hails (ed.) Applied Geomorphology, Elsevier. pp.277–316.

C D Ollier, 1984. Weathering. 2nd Edition. Longman, London.

C W Osman, 1917. The landslips of Folkestone Warren and thickness of the lower chalk and Gault near Dover. Proceedings of the Geologists Association. Vol.28 pp.59–84.

M J Palmer, 1991. Ground movements of the Encombe landslip at Sandgate, Kent. In: R J Chandler (ed) Slope stability engineering: developments and applications, Thomas Telford. pp.291–296.

C D Parks, 1991. A review of the possible mechanisms of cambering and valley bulging. In: A Forster, M G Culshaw, J C Cripps, J A Little & C Moon (eds.) Quaternary Engineering Geology 25th Annual Conference of the Engineering Group of the Geological Society, Edinburgh. pp.373–380.

A Penck, 1894. Morphologie der erdoberflache (2 vols.).

A D M Penman, 1986. On the embankment dam. Geotechnique. Vol.36 pp.303–348.

S Penn, C J Royce and C J Evans, 1983. The periglacial modification of the Lincoln scarp. Quarterly Journal of Engineering Geology. Vol.16 pp.309–318.

A H Perry, 1981. Environmental hazards in the British Isles. George Allen & Unwin.

J Pethick and F Burd, 1993. Coastal defence and the environment: a guide to good practice. MAFF Publications.

D J Petley, 1984. Ground investigation, sampling and testing for studies of slope stability. In: D Brunsden & D B Prior (eds.) Slope instability. Wiley. pp.67–101.

J Pitts, 1974. The Bindon Landslip of 1839. Proceedings of the Dorset Natural History and Archaeological Society. Vol.95 pp.18–29.

J Pitts, 1979. Discussion of 'The contribution made to cliff instability by head deposits in the west Dorset coastal area'. Quarterly Journal of Engineering Geology. Vol.12 pp.277–279.

J Pitts, 1981. Landslides of the Axmouth-Lyme Regis Undercliff National Nature Reserve, Devon. PhD thesis (unpublished). Kings College, University of London.

J Pitts, 1983. The temporal and spatial development of landslides in the Axmouth – Lyme Regis undercliffs, National Nature Reserve, Devon. Earth Surface Processes and Landforms. Vol.8 pp.589–603.

J Pitts and D Brunsden, 1987. A reconsideration of the Bindon landslide of 1839. Proceedings of the Geologists Association. Vol.98 pp.1–18.

D M Potts, G T Dounias and P R Vaughan, 1990. Finite element analysis of progressive failure of Carsington embankment. Geotechnique. Vol.40 pp.79–101.

R C Preece, 1980 The biostratigraphy and dating of a Postglacial slope deposit at Gore Cliff, near Blackgang, Isle of Wight, southern England. Journal of Archaeological Science. Vol 7 pp.255–265.

R J Price, 1983. Scotland's environment during the last 30,000 years. Scottish Academic Press.

M Priestly, Undated. The impact of mass movement; a case study of the future of the A625, North Derbyshire. The Frontier. Birmingham University.

R S Pugh, A G Weeks and D E Hutchinson, 1991. Landslip and remedial works in Wadhurst Clay. In: R J Chandler (ed) Slope stability engineering: developments and applications, Thomas Telford. pp.377–382.

R H Pullan, 1986. Gwent – a case study. Landslips in the Blaina area. In: C S Morgan (ed.) Landslides in the South Wales Coalfield. Polytechnic of Wales Symposium. pp.171–177.

Rendel Geotechnics, 1992. Coastal planning and policy: A review for the Department of the Environment. HMSO, London.

Rhondda Leader, 1909. The Pentre disaster. 13 Feb 1909.

A Richards, 1971. The evolution of marine cliffs and related landforms in the Inner Hebrides. PhD thesis (unpublished). University of Wales Aberystwyth.

K S Richards and N R Lorriman, 1987. Basal erosion and mass movement. In: M G Anderson & K S Richards (eds.) Slope stability. Wiley. pp.331–357.

B Robinson, 1967. Landslip stabilisation by horizontally bored drains. Highways and Public Works. Vol.35 pp.3–37.

D A Robinson and R B G Williams, 1984. Classic landforms of the Weald. The Geographical Association – Landform Guides No.4.

M Robinson, 1977. Glacial limits, sea-level changes and vegetational development in part of Wester Ross. PhD thesis (unpublished). University of Edinburgh.

C D F Rogers and C J Bruce, 1991. Slope stabilisation using lime. In: R J Chandler (ed) Slope stability engineering: developments and applications, Thomas Telford. pp.395–402.

W C Rouse, 1969. An investigation of the stability and frequency distribution of slopes in selected areas of West Glamorgan. PhD thesis (unpublished), University of Wales.

W C Rouse, 1984. Flowslides. In: D Brunsden & D B Prior (eds.) Slope instability. Wiley. pp.491–522.

W C Rouse and E M Bridges, 1986. Landslide susceptibility in the South Wales coalfield. In C S Morgan (ed.) Landslides in the South Wales coalfield. Polytechnic of Wales Symposium, pp.189–200.

I T Rozier and M J Reeves, 1979. Ground movement at Runswick Bay, North Yorkshire. Earth Surface Processes Vol.4 pp.275–280

H Russell, 1992. Race to clear sludge landslide. New Civil Engineer, 20 February 1992, p.5.

R A G Savigear, 1952. Some observations on slope development in South Wales. Transactions of the Institute of British Geographers. Vol.18 pp.31–51.

R A G Savigear, 1965. A technique of morphological mapping. Annals of the Association of American Geographics. Vol.53 pp.514–438.

M J Selby, 1982. Hillslope materials and processes. Oxford University Press.

J Sgambatti, 1979. A geotechnical study of an area around Robins Hood Bay, north-east Yorkshire. MSc thesis (unpublished). Leeds University.

C F S Sharpe, 1938. Landslides and related phenomena. Pageant, New Jersey.

I Shennan, 1989. Holocene crustal movements and sea-level change. Journal of Quaternary Science. Vol.4 pp.77–89.

F W Sherrel, 1971. The Nag's Head landslips, Cullompton by-pass, Devon. Quarterly Journal of Engineering Geology. Vol.4 pp.37–74.

H J Siddle, 1986. Groundwater tracing in a South Wales landslide. In: A L H Gameson (ed.) Tracers for the water industry. Water Research Centre Environment. pp.133–138.

H J Siddle, H J Payne and M J Flynn, 1987. Planning and development control in an area susceptible to landslides. In: M G Culshaw, F G Bell, J C Cripps & M O'Hara (eds.) Planning and Engineering Geology. Geological Society Engineering Geology Special Publication No.4, pp.247–253.

H J Siddle, M D Turner and S P Bentley, 1989. Computer aided landslide potential mapping and its application to planning and development control. International Conference on Computers in Urban Planning, Hong Kong. pp.1–11.

H J Siddle, D B Jones and H R Payne, 1991. Development of a methodology for landslip potential mapping in the Rhondda Valley. In: R J Chandler (ed) Slope stability engineering: developments and applications, Thomas Telford. pp.137–142.

R C Sidle, A J Pearce and C L O'Loughlin, 1985. Hillslope stability and land use. American Geophysical Union, Washington D.C.

P Sims and L Ternan, 1988. Coastal erosion: protection and planning in relation to public policies – a case study from Downderry, South-east Cornwall. In J M Hooke (ed) Geomorphology in Environmental Planning. pp.231–244. J Wiley and Sons.

I G Simmons and P R Cundill, 1974. The late Quaternary vegetational history of the North York Moors. II. Pollen analysis of landslip bogs. Journal of Biogeography Vol.1 pp.253–261.

A W Skempton, 1948. The rate of softening in stiff fissured clays, with special reference to London Clay. Proceedings of the IInd International Conference on Soil Mechanics and Foundation Engineering, Rotterdam, Vol.2, Paper No.IVC6, pp.50–53.

A W Skempton, 1953. Soil mechanics in relation to geology. Proceedings of the Yorkshire Geological Society. Vol.29 pp.33–62.

A W Skempton, 1964. Long term stability of clay slopes. Geotechnique. Vol.14 pp.77–101.

A W Skempton, 1977. Slope stability of cuttings in Brown London Clay. Proceedings of the 9th International Conference on Soil Mechanics, Tokyo. Vol.3 pp.261–270.

A W Skempton, 1985. Geotechnical aspects of the Carsington Dam failure. Proceedings of the 11th International Conference on Soil Mechanics and Foundation Engineering, San Francisco. Vol.5 pp. 2581–2591.

A W Skempton and J D Brown, 1961. A landslip in boulder clay at Selset, Yorkshire. Geotechnique. Vol.11 pp.280–293.

A W Skempton and P La Rochelle, 1965. The Bradwell Slip: A short term failure in London Clay. Geotechnique. Vol.15 pp.221–242.

A W Skempton and D J Petley, 1967. The strength along structural discontinuities in stiff clays. Proceedings of the Geotechnical Conference, Oslo. Vol.2 pp.29–46.

A W Skempton and J N Hutchinson, 1969. Stability of natural slopes and embankment foundations. State of the art report. Proceedings of the 7th International Conference on Soil Mechanics and Foundation Engineering, Mexico. pp.291–340.

A W Skempton and A G Weeks, 1976. The Quaternary history of the Lower Greensand escarpment and Weald Clay vale near Sevenoaks, Kent. Philosophical Transactions of the Royal Society. London. Vol.A283 pp.493–526.

A W Skempton, A D Leadbeater and R J Chandler, 1989. The Mam Tor landslide, North Derbyshire. Philosophical Transactions of the Royal Society. London. Vol.A329 pp.503–547.

D I Smith, 1984. The landslips of the Scottish Highlands in relation to major engineering projects. British Geological Survey Project 09/LS. Unpublished report for the Department of the Environment.

G N Smith, 1982. Elements of soil mechanics for civil and mining engineers. CLS.

C L So, 1966. Some coastal changes between Whitstable and Reculver, Kent. Proceedings of the Geologists Association. Vol.77 pp.475–490.

Soil Survey of England and Wales, 1984. Soils and their use in England and Wales. Regional Bulletins 10–15.

R G Soutar, 1989. Afforestation and sediment yields in British fresh waters. Soil Use and Management. Vol.5 pp.82–86.

South Wales Echo, 1961. Streets of fear. 6 Jan 1961.

L Starkel, 1966. The palaeogeography of mid and eastern Europe during the last cold stage and West European comparisons. Philosophical Transactions of the Royal Society, London. Vol.B280 pp.351–372.

I Statham, 1976. Debris flows on vegetated screes in the Black Mountain, Carmarthenshire. Earth Surface Processes Vol.1 pp.173–180.

C Stevens, 1992. The open coastline. In: M G Barrett (ed) Coastal zone planning and management. Thomas Telford. pp.91–99.

H E Steward and J C Cripps, 1983. Some engineering implications of chemical weathering of pyritic shale. Quarterly Journal of Engineering Geology. Vol.16 pp.281–289.

G J Strachan, 1976. Debris flows in Wester Ross. MA dissertation (unpublished).

A S Subramaniam and R W Carr, 1983. A55 Pwll Melyn slip and remedial works: a case history. Quarterly Journal of Engineering Geology. Vol.16 pp.53–63.

J H Tallis and R H Johnson, 1980. The dating of landslides in Longendale, North Derbyshire using pollen analysis techniques. In: R Cullingford, D Davidson and J Lewis (eds.) Timescales in Geomorphology. Wiley. pp.189–205.

W L Tamblyn, 1978. Cambering and valley deformation processes of the rocks of Lower and Middle Jurassic age. MSc thesis (unpublished). Imperial College, University of London.

J D Tankard, 1973. Reports of H.M. Inspectors of Mines and Quarries for 1972. Southern District. Department of Trade and Industry.

R K Taylor and D A Spears, 1970. The breakdown of British Coal Measure rocks. International Journal of Rock Mechanics and Mining Science. Vol.7 pp.481–501.

R K Taylor and J C Cripps, 1987. Weathering effects: slopes in mudrocks and over-consolidated clays. In M G Anderson & K S Richards (eds.) Slope stability. Wiley. pp.405–445.

K Terzaghi, 1950. Mechanisms of landslides. Geological Society of America. Engineering Geology (Berkley Vol.) pp.83–123.

K Terzaghi and R B Peck, 1948. Soil mechanics in engineering practice. Wiley and Sons, New York.

R P Thompson, 1991. Stabilisation of a landslide on Etruria Marl. In: R J Chandler (ed) Slope stability engineering: developments and applications, Thomas Telford. pp.403–408.

A H Toms, 1948. The present scope and possible future development of soil mechanics in British Railway civil engineering construction and maintenance. Proceedings of the 2nd International Conference on Soil Mechanics and Foundation Engineering, Rotterdam. Vol.4 pp.226–237.

A H Toms, 1953. Recent research into the coastal landslides of Folkestone Warren, Kent. Proceedings of the 3rd International Conference on Soil Mechanics and Foundation Engineering, Zurich. Vol.2 pp.288–293.

M J Tooley, 1978. Sea level changes, North-West England during the Flandrian stage. Clarendon, Oxford.

W Topley, 1893. The landslip at Sandgate. Proceedings of the Geologists Association. Vol.13 pp.40–47.

Transportation Research Board, 1987. Lime stabilisation state-of-the-art report: reactions, properties, design and construction. National Research Council, Washington D C USA.

L Tufnell, 1972. Ploughing blocks with special reference to north-west England. Biuletyn Periglacjalny Societas Scientiorum Lodziensis. Vol.21 pp.237–270.

M F Tyehurst, 1991. Planning aspects of the Chewton Bunny inquiry. In Coastal Instability and Development Planning: Papers and Proceedings of the SCOPAC Conference, Southsea, pp.78–90.

H Valentin, 1971. Land loss at Holderness. In: J A Spears (ed.) Applied coastal geomorphology. London. Macmillan. pp.116–137.

D J Varnes, 1958. Landslide types and processes. In: E B Eckel (ed.) Landslides in engineering practice. US National Academy of Sciences, Highway Research Board, Special Report 29, pp.20–47.

D J Varnes, 1978. Slope movement, types and processes. In: R L Schuster & R J Krizek (eds.) Landslides, analysis and control. US Academy of Sciences, Transportation Research Board Special Report 176, pp.11–33.

D J Varnes, 1984. Landslide hazard zonation: a review of the principles and practice. International Association of Engineering Geology Commission on Landslides and other Mass Movements on Slopes. Natural Hazards, United Nations Economic, Scientific and Cultural Organisation.

P R Vaughan, 1965. Field measurements in earth dams. PhD thesis (unpublished). Imperial College, University of London.

P R Vaughan, 1976. Appendix : The deformations of the Empingham Valley slope. Philosophical Transactions of the Royal Society, London Vol.A283, pp.451–462.

P R Vaughan, 1991. Stability analysis of deep slides in brittle soil: lessons from Carsington. In: R J Chandler (ed) Slope stability engineering: developments and applications, Thomas Telford. pp.1–13.

P R Vaughan and H J Walbancke, 1975. The stability of cut and fill slopes in Boulder Clay. In: Proceedings of the Symposium on the Engineering Behaviour of Glacial Materials, University of Birmingham. Geoabstracts. pp.185–195.

N E V Viner-Brady, 1955. Folkestone Warren landslips: Remedial measures 1948-1954. Proceedings of the Institution of Civil Engineers, Railway Paper. Vol.57 pp.429–441.

K L Wallwork, 1960. Some problems of subsidence and land use in the mid-Cheshire industrial area. Geographical Journal. Vol.126 pp.191–199.

W H Ward, 1948. A slip in a flood defence bank constructed on a peat bog. Proceedings of the IInd International Conference on Soil Mechanics and Foundation Engineering, Rotterdam. Vol.II pp.19–23.

R A Warrick and E M Barrow, 1991. Climatic change scenarios for the UK. Transactions of the Institute of British Geographers, Vol.16 (4) pp.387–399.

G T Warwick, 1964. Relief and structure. In J W Watson and J B Sissons (eds) The British Isles: a systematic geography, pp.91–109, Nelson.

W D A Waters, 1982. Coast Protection Act 1949: Report of Survey (Wales).

R J Watters, 1972. Slope stability in the metamorphic rocks of the Scottish Highlands. PhD thesis (unpublished). Imperial College, University of London.

W Watts, 1905. Geological notes on sinking Langsett and Underbank concrete trenches in the Little Don Valley. Transactions of the Institute of Mining Engineering. No.31 pp.668–680.

A G Weeks, 1969. The stability of natural slopes in south-east England as affected by periglacial activity. Quarterly Journal of Engineering Geology. Vol.2 pp.49–61.

W B Whalley, 1984. Rockfalls. In: D Brunsden & D B Prior (eds.) Slope instability. Wiley. pp.217–256.

G White, 1788. The Natural History and Antiquities of Selborne. Letter XLV.

J L Whitehead, 1911. The Undercliff of the Isle of Wight: Past and Present. Ventnor, Winchester and London.

A Whittaker, 1972. Geology of Bredon Hill, Worcestershire. Bulletin of the Geological Survey of Great Britain. Vol.42 pp.1–49.

A Whittaker and G W Green, 1983. Geology of the country around Weston-super-Mare Sheet 279 (new series) and parts of 263 & 295. Memoirs of the Geological Survey of Great Britain. London.

J Whittow, 1977. Geology and scenary in Scotland. Penguin.

J Whittow, 1980. Disasters; London. Penguin.

A T Williams and P Davies, 1984. Cliff failure along the Glamorgan Heritage Coast, Wales, UK. In: Communications du Colloque; Movements de Terrains. Caen, Mars 1984. Association Francaise Geographie Physique, Serie Document due BRGM No.83, pp.109–119.

A T Williams and P Davies, 1987. Rates and mechanisms of coastal cliff erosion in Lower Lias rocks. In N C Kraus (ed) Coastal Sediments '87. American Society of Civil Engineers, pp.1855–1870.

A T Williams, N R Morgan and P Davies, 1991. Recession of the littoral zone cliffs of the Bristol Channel, UK. In: O T Magoon (ed) Coastal Zone '91. American Society of Civil Engineers, pp.2394–2408.

W W Williams, 1956. An east coast survey – some recent changes in the coast of East Anglia. Geographical Journal Vol.122 pp.317–334.

V Wilson, F B A Welch, J A Robbie and G W Green, 1958. Geology of the country around Bridport and Yeovil Sheets 327 & 312. Memoirs of the Geological Survey of Great Britain. London.

A W Woodland, 1986. The South Wales Coalfield: an introduction to its geology and landslips. In C S Morgan (ed.) Landslides of the South Wales Coalfield. Polytechnic of Wales Symposium. pp.9–17.

S W Wooldridge, 1950. Some features in the structure and geomorphology of the country around Fernhurst, Sussex. Proceedings of the Geologists Association. Vol.61 pp.165–187.

M J Wright, 1982. A stability analysis of the cliffs along the coastline at Robin Hood's Bay, north-east Yorkshire. MSc thesis (unpublished). Leeds University.

M Youdale, 1989. Wall of death. Yorkshire Post. Feb 1989.

Rev G Young and J Bird, 1822. A geological Survey of the Yorkshire coast, Whitby.

Q Zaruba and V Mencl, 1976. Landslides and their control. Elsevier/Academia, Prague.

Appendix A

A.1. Principal areas of reported coastal landslide activity

A.2. Major concentrations of reported inland landsliding

A.3. Phases of landslide activity in Great Britain

See following pages.

Table A.1: **Principal areas of reported coastal landslide activity**

Location	Geology	General description	Sources (not exclusive)
NORFOLK; between Weyburne and Happsburgh	Pleistocene tills, laminated clays, sands and gravels overlying Chalk.	A long history of cliff retreat along this exposed coastline has been described by Hutchinson (1976), which includes the loss of villages (Shipden) and buildings. The average rate of retreat is around 1.1 m/year, mainly through deep-seated rotational and compound landslides.	Cambers, 1973 Hutchinson, 1976
SUFFOLK; Dunwich to Covehithe	Pleistocene tills; soft sands and gravels.	Rapid erosion of 10m high sandy cliffs, around 1m/yr. Dunwich now comprises 30–40 houses, having been a small city around 1000 years ago.	Williams, 1956
THAMES ESTUARY; Suffolk-North Kent	London Clay.	Variable rates of marine erosion of the 25–45m high London Clay cliffs has resulted in contrasting landslide systems comprising deep-seated rotational slides (erosion exceeds weathering) mudslides (erosion in balance with weathering) and shallow part-successive rotational, part translational slides (free degradation with zero erosion).	Hutchinson, 1965a, 1967, 1968, 1970, 1973, 1986 Hutchinson & Bhandari, 1971 Hutchinson & Gostelow, 1976 Bromhead, 1978a, 1987b, 1979 Dixon, 1986 Dixon & Bromhead, 1991
CHANNEL COAST; Sandwich to Seaford	Varied Cretaceous sedimentary rocks.	Folkestone Warren; large multiple rotational landslide complex developed in Gault Clay and the overlying Chalk.	Hutchinson, 1969 Hutchinson et al., 1980
		Hythe-Sandgate; large slides developed in the Lower Greensand Sandgate Beds.	Topley, 1893 Hutchinson, 1965a Palmer, 1991
		Fairlight Glen; large landslide complex developed in Lower Cretaceous Ashdown Sands and Fairlight Clay, comprising mudslides and deep-seated rotational failures. Cliff recession has been around 1m/year over the last century, with the loss of the famous Lovers' Seat.	Robinson & Williams, 1984 Moore, 1986
		Beachy Head to Brighton; rockfalls off the Chalk cliffs	May, 1964 May, 1971
ISLE OF WIGHT UNDERCLIFF; Blackgang to Luccombe	Chalk, Upper Greensand, Gault Clay and Lower Greensand.	Large landslide complex comprising multiple rotational slides with shear surfaces in the Gault Clay, compound slides with basal shear surfaces in the Lower Greensand, large rockfalls off the Upper Gransand and rear scarp and mudslides. This landslide has caused considerable disruption to development and infrastructure in the area, particularly in the town of Ventnor, at Luccombe and Blackgang.	Hutchinson, 1965b, 1983b, 1991 Chandler, M.P., 1984 Hutchinson et al., 1981a, 1985a Chandler & Hutchinson, 1984 Lee & Moore, 1989 Lee et al., 1991a, b, c Moore et al., 1991a, b Bromhead et al., 1991 Chandler, M.P. et al., 1991

ISLE OF WIGHT, Shanklin to Sandown	Cretaceous Lower Greensand.	Rockfalls controlled by stress relief, frost action and seepage erosion threatening downslope developments. Currently in process of stabilisation.	Barton, 1984 Clark et al., 1990
CHRISTCHURCH BAY – BOURNEMOUTH BAY	Tertiary Barton Beds.	Rapidly retreating Barton Clay cliffs (1.9m/year). Lithological variations have been accentuated by landsliding, producing a benched profile. Degradation processes include mudslides, compound failures, rockfalls and debris slides.	Barton, 1970, 1973 Barton & Coles, 1983, 1984
WEYMOUTH BAY	Variable Cretaceous and Upper Jurassic sedimentary rocks.	A wide variety of landslides have been reported along this section of coastline, including mudslides in Warbarrow Bay, developed in the Wealden Clay, rockfalls off the Chalk cliffs, debris slides off the Purbeck and Portland Beds, and compound slides involving the Corallian at Red Cliff.	Brunsden & Goudie, 1981 Jones, M.E. et al., 1983 Moore, 1988 Bromhead, 1986
ISLE OF PORTLAND	Upper Jurassic Purbeck Beds, Portland Beds and underlying Kimmeridge Clays.	Large complex landslides, occasionally involving topping failures as at Great Southwell.	Bromhead, 1986 Clayton, D.J., 1988 Harrison, 1848
LYME BAY	Upper Greensand and Gault overlying Lias clays and Triassic mudstones	A series of massive landslide complexes have developed along this stretch of coast, from Bridport to Sidmouth. These include Fairy Dell, Black Ven and the Landslip Nature Reserve west of Lyme Regis. A wide variety of landslide processes have been recorded, including mudsliding, multiple rotational sliding and block sliding. The characteristic form along this coast is a series of undercliffs which highlight lithological variations. Cliff retreat rates of between 0.4–0.5m/year were recorded by Brunsden & Jones (1980) at Fairy Dell. In West Bay landsliding has caused considerable problems for coastal developments, with rates of cliff retreat of up to 2.8m/year recorded.	Brunsden & Jones, 1976, 1980 Conway, 1974, 1976, 1977, 1979 Arber, 1941, 1973 Pitts, 1979, 1983 Allison, 1990 Chandler, J.H., 1989
BUDLEIGH SALTERTON	Budleigh Salterton Pebble Beds and Littleham Mudstone Formation.	A variety of landslide forms occur on the rapidly eroding cliffs west of Budleigh Salterton, including mudslides and rockfalls.	Grainger & Kalaugher, 1987a, b Kalaugher & Grainger, 1981 Kalaugher et al., 1987
TORBAY	Varied sedimentary rocks; Permian breccias, Devonian limestones and mudrocks.	Numeroius rockfalls, debris slides and mudslides along the Torbay coastline. There is a clear relationship between landslide form and geology (lithology and structure).	Doornkamp, 1988 Lee et al., 1988
CORNWALL; Downderry	Devonian slates overlain by solifluction deposits.	Rotational slides, wedge failures and debris slides on the 10m high cliffs. Recession around 0.1m/yr.	Coard et al, 1987 Sims and Ternan, 1988

Table A.1: **Principal areas of reported coastal landslide activity – continued**

Location	Geology	General description	Sources (not exclusive)
BUDE BAY & BIDEFORD BAY	Interbedded Carboniferous sandstones and shales of the Culm Measures.	Rockfalls and toppling failures off the hard rock cliffs	De Freitas, 1969 De Freitas & Watters, 1972
WATCHET BAY, SOMERSET	Lower Lias Clays.	Debris slides and rockfalls off the coastal cliffs.	Whittaker & Green, 1983
GLAMORGAN; Nash Point to Barry	Lower Lias limestones and shales.	Rockfalls, topples and translational slides; erosion rates of 0.3–0.7m/yr.	Williams & Davies, 1984 Williams & Davies, 1987 Williams et al, 1991 Davies et al, 1991
ST. BRIDES BAY, DYFED	Varied Silurian, Devonian and Carboniferous rocks	Rockfalls, debris slides and complex landslides developed on the cliffs within St. Brides Bay.	Al Saadi, 1982
THE GWYNEDD COAST	Pleistocene superficial deposits.	Complex landslides and debris slides off the coastal cliffs.	Waters, 1982
HOLDERNESS, HUMBERSIDE	Pleistocene tills, with layers of sand, silt and gravels.	Rapidly retreating cliffline, with rate of retreat of up to 12m/year. Over the last 300 years at least 30 villages have been lost to the North Sea. Retreat occurs through a sequence of deep-seated rotational slides, with occasional debris and mudslides.	Joyce, 1969 Hutchinson, 1986 Richards & Lorriman, 1987
NORTH YORKSHIRE; FILEY TO RUNSWICK BAY	Variable sequence of superficial deposits (glacial tills and proglacial lake deposits), Middle Jurassic sandstones and Lias clays.	This stretch of coastline has a long history of landslide problems, especially in Runswick Bay where the old village slid into the sea in 1689. Nearby Kettleness was lost in 1829. Landslide problems are also significant at Whitby, Scarborough, Cayton Bay and Filey. The main types of reported falure include complex slides, mudslides and debris slides.	Sgambatti, 1979 Wright, 1982 Rozier & Reeves, 1979 Clark & Guest, 1991
INNER HEBRIDES	Variable sequence of Tertiary igneous rocks and Permian to Jurassic sedimentary rocks.	**Skye**; large multiple rotational slides developed in Tertiary basalts overlying Jurassic sedimentary rocks. Examples such as the Storr and Quirang at Trotternish are amongst the largest landslides in Great Britain **Mull**; rockfalls and complex slides of the basalt cliffs of west and south Mull, as in the Wilderness area. **Arran**; rockfalls off the New Red (Permian) sandstone cliffs of south Arran.	Richards, 1971 Anderson & Dunham, 1966 Bailey & Anderson, 1925

Table A.2: **Major Concentrations of Reported Inland Landsliding**

Location	Geology	General description	Sources (not exclusive)
LONDON BASIN	London Clay	Many shallow compound landslides, slab slides and successive rotational failures on slopes above 8°. It is likely that such slides are the result of reduction of shear strength through weathering and high pore water pressures. A significant number of cut slope failures have also occurred, often following a considerable delay after excavation, as at Wembley Hill in 1918 (14 year delay) and Kensal Green (29 year delay).	Hutchingson, 1967 Skempton, 1948 Skempton, 1964 De Lory, 1956 Allison, J.A. et at., 1991
THE WEALD	Lithologically varied Cretaceous strata	A wide variety of landslide forms ranging from cambered strata, multiple rotational failures to shallow debris slides have been reported. The main concentrations of landslides are: *Lower Chalk escarpment* **Folkestone**; a number of large landslides have been identified below the escarpment, probably with the basal shear surface in the Gault Clay. These landslides have had design implications for the Channel Tunnel terminals. *Upper Greensand escarpment*; landsliding is ubiquitous along the Upper Greensand escarpment in the western Weald, and involves large rotational and compound failures with shear surfaces in the Gault Clay. *Lower Greensand escarpment;* this escarpment has experienced widespread major landsliding, particularly involving cambering and multiple rotational failure. In general the failures have developed in the Weald and Atherfield Clays and the overlying Hythe Beds. Examples of this type of landsliding can be seen at Sevenoaks, Kent, along the abandoned cliff between Lympne and Hythe and in the Vale of Fernhurst. *Central Weald;* numerous shallow rotational and translational slides have been reported on the Weald Clay, Wadhurst Clay, Ashdown Beds and Purbeck Beds.	Aarons et al., 1977 Barton, 1989 Hutchinson et al., 1985b Robinson & Williams, 1984 Skempton & Weeks, 1976 Wooldridge, 1950 Hansen (undated) Pugh et al., 1991
EAST DEVON PLATEAU	Upper Greensand and occasionally Chalk, overlying Liassic and Triassic clays and marls	Large number of rotational and translational slides have been identified on valley-side slopes of the dissected east Devon plateau and adjacent areas of west Dorest and south Somerset	Wilson et al., 1958 Brown, 1966 Fakhraee, 1979 Griffiths & Marsh, 1986
CENTRAL DEVON	Upper Carboniferous Crackington Formation.	Numerous active and degraded ancient landslides have been recorded by Grainger (1983) in a 25km^2 study area west of Exeter. Most of these failures are shallow debris slides, often involving the movement of periglacially remoulded clay soils in response to groundwater conditions. Similar failures have caused problems during building development and highway construction elsewhere in central Devon.	Grainger, 1983 Grainger & Harris, 1986

Table A.2: **Major Concentrations of Reported Inland Landsliding – continued**

Location	Geology	General description	Sources (not exclusive)
EXMOOR	Middle Devonian shales and slates	Shallow debris slides involving failure of the weathered layers of bedrock are common on valley slopes over 20°. Many such slides were associated with the Exmoor strorm and Lynmouth floods of 15th August 1952.	Gifford, 1953 Kidson, 1953 Carson & Petley, 1970
COTSWOLDS	Lias clays overlain by oolitic limestones	Landslides are common throughout the Cotswolds, occurring both on the scarp face and on the flanks of the dip slope valleys. The most frequent forms of failure include cambered strata, multiple rotational slides and shallow debris slides. The main areas of reported landslides are: *South Cotswolds;* in the Avon Valley and its tributaries, around Bath. Considerable river downcutting during the Pleistocene has initiated widespread cambering and subsequent deep-seated rotational failures. Landslides have caused considerable problems for development in this area with many failures apparently triggered by man's activities, such as the Hedgemead landslide of the 1870's. *Gloucestershire;* cambering, large scale rotational failures and mudslides are common along the Cotswolds escarpment, particularly north of Stroud. Similar forms of mass movement also occur in the deeply incised valleys which dissect the dipslope, such as the Frome Valley. The distribution of landslides in this area appears to be largely controlled by the presence of either Upper Lias clays or Fullers Earth clays. *Bredon Hill and other outliers;* the flanks of the isolated hills in front of the escarpment are heavily disturbed by cambering, rotational failures and mudslides. *Oxfordshire & Warwickshire;* cambering and deep-seated rotational failures are common along the escarpment, as at Edge Hill and in valleys such as the Evenlode.	Ackerman & Cave, 1967 Briggs & Courtney, 1972 Butler, 1983 Chandler, R.J. et al., 1976 Forster et al., 1985 Gerrard & Morris, 1981 Hawkins, A.B. & Privett, 1979
SOUTH WALES COALFIELD	Varied sequences of sandstones, shales, silty mudstones, seat earths and coals of the Carboniferous Coal Measures	Around 1,000 discrete landslides have been identified on the steep slopes of the South Wales valleys. The majority of these are essentially shallow translational failures occuring in the superficial 'head' or drift deposits which mantle the slopes. In general slopes over 17° are close to limiting equilibrium for this type of failure. However, significant number of deep-seated rotational or compound fialures have been identified, such as at East Pentwyn and Bournville near Blaina, Gwent. It is considered probable that subsidence of shallow mine workings has contributed to the recent instability, as well as a possible increase in rainfall intensity. In addition to these natural slope failures there have been many problems related to the failure of colliery waste tips, such as occurred at Aberfan in 1966. The South Wales Coalfield has been extensively mapped to delineate	Knox, 1927 Conway et al., 1980 Bishop et al., 1969 Sir W Halcrow & Ptns, 1988 Jones, D B & Siddle, 1988 Siddle et al., 1987, 1991 Jones D B et al., 1991

Table A.2: **Major Concentrations of Reported Inland Landsliding – continued**

Location	Geology	General description	Sources (not exclusive)
		the nature and extent of landsliding, by the British Geological Survey. Recently landslide hazard maps have been produced for planning purposes in the Rhondda Valley by Sir William Halcrow & Partners.	
NORTH WALES [VALE OF CLWYD & DENBIGHSHIRE MOORS]	Glacial tills overlying Silurian Ludlow Beds.	Large numbers of slope failures reported by British Geological Survey during the course of its mapping programme. Only limited information is available, but presumably they involve either shallow rotational or translational failures off the flanks of drumlins.	British Geological Survey, 1973
IRONBRIDGE GORGE	Lithologically varied Carboniferous Coal Measures (Middle and Upper) comprising sandstones, shales, marls coal seams and seat earths	The valley-side slopes of Ironbridge Gorge are characterised by numerous rotational and translational failures e.g. the Jackfield landslide and the failure at Jockey Bank. Although the principal factor affecting the stability of the slopes has been the geologically recent diversion of the River Severn through the gorge, there is evidence which suggests that mining and building activities have contributed to contemporary landsliding.	Brown, 1975 Culshaw, 1972, 1973 Denness, 1977 Henkel & Skempton, 1954 Carson & Fisher, 1991
EAST MIDLANDS PLATEAU	Middle and Upper Lias clays overlain by Inferior Oolite sands and limestones.	The Northamptonshire Sand Ironstone is widely affected by cambering, with numerous records of valley bulges, gulls, dip and fault structures and cambered blocks. In addition the Lias Clay slopes are mantled by numerous shallow and deep-seated rotational failures, slab slides and periglacial mudslides. Degraded slope failures are especially common on the flanks of valleys which cut through the Inferior Oolite escarpement, such as the Gwash in Leicestershire.	Chandler R J, 1970a, b, c 1971, 1976, 1977 Hollingworth et al., 1944 Horswill & Horton, 1976 Kellaway & Taylor, 1952 Tamblyn, 1978
THE PENNINES	Variable Carboniferous sedimentary rocks with occasional Silurian rocks, Devonian lavas and granites	Landsliding is widespread throughout the Pennines, especially where the uplands have been dissected by glacial and fluvial processes to form steep sided valleys mantled with glacial deposits. Major concentrations of landslides occur in the following locations: *High Peak;* large numbers of landslides have been reported in the High Peak, where competent massive strata alternate with thick sequences of weaker shales and mudstones. A wide variety of landslide forms occur including cambering and valley bulging, rotational failure (e.g. The Binns), compound slides (e.g. Alport Castles, Laddow Rocks), debris slides and mudslides. *Central Pennines;* a range of rotational slides, translational slides and complex failures have been identified in the incised valleys of the central Pennines, especially the Chew and Greenfield valleys near Oldham, Airedale and on the flanks of Ilkley Moor. *The Forest of Bowland;* in the deeply incised Ribble and Calder valleys there are many landslides which have been recognised as part of the British Geological Survey mapping programme.	Cross, 1988 Frost & Smart, 1979 Johnson, 1965b Johnson, 1980 Johnson & Vaughan, 1983 Johnson & Walthall, 1979 Earp et al., 1961 Knowles, 1973 Mitchell, 1991 Arthurton & Wadge, 1981 Clough, 1889 Doornkamp, 1990

Continued overleaf

Table A.2: **Major Concentrations of Reported Inland Landsliding – continued**

Location	Geology	General description	Sources (not exclusive)
THE PENNINES – *continued*		*Northern Pennines;* large numbers of landslides, mainly translational or rotational in form have been recognised in the valleys of the northern Pennines, such as Upper Teesdale, Upper Wensleydale and the Vale of Eden. *Northumbrian Fells;* the steep slopes in the North Tyne and Rede valleys are the location of numerous landslides, recognised by the British Geological Survey. Although information about these slides is limited, it is considered likely that they involve either translational or rotational mechanisms.	
NORTH YORK MOORS	Jurassic sandstones and limestones overlying Lias Clays.	Landsliding is widespread along the steep western scarp of the Hambleton Hills, where large multiple rotational slides, cambering and toppling failures have been recorded. One of the most recent large failures was reported by John Wesley in his Journal of 1755 as having occurred at Whitestone Cliff. Large numbers of landslides have also been reported in the Cleveland Hills, particularly in Eskdale, including debris flows and rotational failures generally associated with the outcrop of the Alum Shales.	Gregory, 1963 Cooper, 1979, 1980 Blackham et al., 1981
SCOTTISH HIGHLANDS	Moinean and Dalradian metamorphic rocks	Rock slides and debris flows are common throughout the Scottish Highlands where the metamorphic rocks have been deeply incised by glacial erosion to create steep, unstable valley-side slopes. Rock slope failures are apparently more frequent on schists and within the limits of the Loch Lomond Advance glaciers. Debris flow activity is widespread, with failures usually triggered by intense rainstorms, as occurred in the Cairngorm Floods of 1956. However, it is believed that debris flow activity has increased over the last 250 years probably as a result of land management practices on the moorlands, especially burning.	Common, 1954 Baird & Lewis, 1957 Holmes, 1984 Innes, 1982 Innes, 1983 Smith, 1984 Watters, 1972 Robinson, 1977

Table A.3: **Phases of landslide activity in Great Britain**

Phase		Date	Conditions	Landslide activity
I	WOLSTONIAN GLACIATION	125,00+B.P.	Cold periglacial environment with ice-sheets developed over northern Britain. Southern Britain affected by widespread permafrost.	Presumed to be a major period of development of cambered strata. Examples include:- Southern Cotswolds: Chandler, R.J. et al., 1976; Hawkins & Privett, 1979 Jurassic escarpment of Northamptonshire and Leicestershire; Horswill & Horton, 1976; Chandler, R.J., 1976 Hythe Beds escarpment, Sevenoaks, Kent: Skempton & Weeks, 1976 Periglacial solifluction is also believed to have been an important slope process at this time (e.g. Skempton & Weeks, 1976).
II	EARLY DEVENSIAN	11,000–60,000 B.P.	Long period of cold sub-arctic and cold periglacial environments, separated by warmer interstadial phases (Chelford & Wretton Interstadials).	The Chelford and Wretton interstadials are presumed to be a period of accelerated landslide activity, but no dated evidence. Solifluction described by Chandler, R.J. (1976) and Chandler, R.J. et al (1976).
III	MIDDLE DEVENSIAN	58,000–43,000 B.P.	Period of cold periglacial conditions, with associated permafrost, with the warmer Upton Warren Interstadial around 43,000 B.P.	Presumed to be a period of accelerated landslide activity, but no dated evidence.
IV	LATE MIDDLE DEVENSIAN	40,000–25,000 B.P.	Cold arctic conditions, with polar desert conditions in parts of Britain. Mean annual temperatures between –8°C and –12°C, and precipitation of 200–300mm per year.	Presumed to be a period of landslide activity, but no dated evidence. However, Hutchinson & Gostelow (1976) ascribe the formation of the first bench at the foot of the Hadleigh Castle slope to this time (c.27,000 B.P.).
V	LATE DEVENSIAN	18,000–11,000 B.P.	Decay of the Devensian ice sheets accompanied by warmer interstadial conditions (the Windermere Interstadial). Periglacial conditions close to retreating ice margins.	Considered to be one of the most important phases of landslide activity on Great Britain, although precise dates are rare. Examples include: Southern Cotswolds: Hawkins & Privett, 1979; Chandler, R.J. et al., 1976 South Wales: Conway et al., 1980. Whitestone Cliffs, North Yorkshire: Blackham et al., 1981. Wensleydale, N. Yorkshire: Mitchell, 1991. Hythe Beds escarpment, Sevenoaks, Kent: Skempton & Weeks, 1976. Gwash Valley, Leicestershire: Chandler, R.J., 1976; Horswill & Horton, 1976. Scotland: Talus slopes outside the limits of the Loch Lomond Stadial (Ballantyne & Kirkbride, 1987).

Table A.3: **Phases of landslide activity in Great Britain — continued**

Phase	Date	Conditions	Landslide activity
VI LOCH LOMOND STADIAL	11,000–10,000 B.P.	Cold periglacial conditions, discontinuous permafrost and small glaciers in highland areas of Scotland, Northern England and North Wales.	This is a period of widespread slope instability throughout Great Britain. Ballantyne & Kirkbride (1987) demonstrated that rockfall activity in upland Britain was two orders of magnitude higher than present day, with rockwall retreat rates of 1.5–4.0mm/year. Other examples include: The Bury Hill landslide: Hutchinson et al., 1973 The Waltons Wood landslide: Early & Skempton, 1972 The Rockingham escarpment: Chandler, R.J., 1971 Hadleigh Castle: Hutchinson & Gostelow, 1976.
VII LATE BOREAL-EARLY ATLANTIC	7,500–6,000 B.P.	Climate slightly warmer and wetter than present. Rising sea levels.	This period coincides with one of Starkel's (1966) phases of frequent landsliding in Europe. Examples include: Hadleigh Castle: Hutchinson & Gostelow, 1976 High Peak, Derbyshire: e.g. Alport Castles (8300–7600 BP; Tallis, 1983) Mam Nick (8000 BP; Skempton et al., 1989) Coomes Rock (7190 BP; Frank & Johnson, 1980) Bradwell Sitch (7100 BP; Tallis & Johnson, 1980) Millstone Rock (6500 BP; Tallis & Johnson, 1980) South Pennites; Buckstones Moss, c.7,500 BP (Muller, 1979) North York Moors; the Blakey Landslip c.7,500 BP (Simmons & Cundill, 1974) This period is also associated with the initiation of major coastal landslide complexes in southern England e.g. the Isle of Wight Undercliff (Hutchinson, 1987a).
VIII SUB-BOREAL	5,500–3,000 B.P.	Sea levels approaching present day position. Climate was relatively warm and dry.	Initiation of major coastal landslide complexes e.g. Folkestone Warren: Hutchinson, 1969; Hutchinson et al., 1980 Lympne: Hutchinson et al., 1985b Fairy Dell, Dorset: Brunsden & Jones, 1976; 1980 Probably a period of relative stability for inland slopes, although Simmons & Cundill (1974) record a date of c.5,500 BP for the initiation of the St Helens landslide in the North York Moors. Movements have also been recorded in the High Peak, e.g. Mam Tor (3600 BP; Skempton et al., 1989), Ladlow Racks (4000–5000 BP; Tallis & Johnson, 1980).
IX EARLY SUB-ATLANTIC	2,500–2,000 B.P.	The Sub-Atlantic period marked a sharp deterioration in climate with increased rainfall and cooler temperatures. Widespread forest clearance associated with the introduction of agriculture.	Examples of landsliding during this period include: Hadleigh Castle: Hutchinson & Gostelow, 1976 The Rockingham escarpment: Chandler, R.J., 1971 Upper Bradwell Sitch, High Peak c. 1970 BP: Tallis & Johnson, 1980 Mam Tor, High Peak c. 2,000 BP: Johnson, 1987. During this period the second major phase of landslide activity occurred along the Isle of Wight Undercliff, with debris apron formation arund 2,500–2,000 BP and deep-seated failures around 1,800 BP e.g. at St Catherine's Point (Hutchinson, 1987a).

Table A.3: **Phases of landslide activity in Great Britain — continued**

Phase	Date	Conditions	Landslide activity
X LITTLE ICE AGE	A.D. 1550–A.D. 1850	Britain suffered an appreciable climatic deterioration, characterised by cold winters and wet summers, and was probably a time of abnormally high groundwater levels.	A large number of major landslides have been reported as occurring during this period including: 1571 – the Wonder at Marcle, Woolhope Dome 1639 – Folkestone Warren (Hutchinson, 1969) 1755 – Whitestone Cliff, North York Moors (John Wesley, 1755) 1773 – The Birches, Ironbridge Gorge (Rev. John Fletcher of Madeley) 1774 – The Hawkley slip near Selbourne, Hampshire (Gilbert White, 1778; Barton, 1989) 1790 – Beacon Hill, Bath (Kellaway & Taylor, 1968) 1799 – Rocken End, Isle of Wight Undercliff (Webster in Englefield, 1816) 1810/1818 – The Landslip, Isle of Wight Undercliff (Webster in Englefield, 1816; Brannon, 1825) 1839 – (The following large slides all took place on Dec 24/25, 1839): Bindon, East Devon (Conybeare & Dawson, 1840; Pitts & Brunsden, 1987) Folkestone Warren (Hutchinson, 1965c) Gore Cliff, Isle of Wight (Conybeare & Dawson, 1840) This period was also marked by a dramatic increase in debris flow activity in Scotland (Innes, 1982; 1983).
XI MODERN	1870–PRESENT	Rapid industrialisation and urban development of much of Great Britain.	Increase in landslide activity, often as a direct consequence of man's actions, including: **Coal mining –** Bournville, South Wales 1882 (Gostelow, 1977) East Pentwyn, South Wales 1954 (Jones, D.B. & Siddle, 1988) Aberfan, South Wales 1966 (Bishop et al., 1969) **Road construction –** A625 at Mam Tor Derbyshire, 1912–1979 when closed (Priestly undated; Skempton et al., 1989). M5 Motorway at Waltons Wood, 1962 (Early & Skempton, 1972) A21 Sevenoaks Bypass, 1965 (Skempton & Weeks, 1976) M4 Motorway, Burderop and Hodson, 1969 (Hutchinson, 1987b) **Housing development –** Hedgemead landslide, Bath (1870's; Hawkins, 1977) Bury Hill, West Midlands 1960 (Hutchinson et al., 1973) Ewood Bridge (Douglas, 1985) Brierley Hill, West Midlands (Thompson 1991) Gypsy Hill, London (Allison, J.A. et al., 1991) **Land use change –** Debris flows in Scotland (Innes, 1982; 1983).

NOTE: Coastal landsliding has been a continuous, albeit intermittent process since initiation of the clifflines and landslide complexes between 8,000-3,000 B.P. [Only larger coastal events are listed.]

Printed in the United Kingdom for HMSO.
Dd.0297773, 12/94, C8, 3396/4, 5673, 282015.